Ultrafast Dynamics in Molecules, Nanostructures and Interfaces

Selected Lectures Presented at Symposium on
Ultrafast Dynamics of the 7th International Conference
on Materials for Advanced Technologies

Series in Optics and Photonics

Series Editor: S L Chin *(Laval University, Canada)*

Published

Vol. 1   Fundamentals of Laser Optoelectronics
by S. L. Chin

Vol. 2   Photonic Networks, Components and Applications
edited by J. Chrostowski and J. Terry

Vol. 3   Intense Laser Phenomena and Related Subjects
edited by I. Yu. Kiyan and M. Yu. Ivanov

Vol. 4   Radiation of Atoms in a Resonant Environment
by V. P. Bykov

Vol. 5   Optical Fiber Theory:
A Supplement to Applied Electromagnetism
by Pierre-A. Bélanger

Vol. 6   Multiphoton Processes
edtied by D. K. Evans and S. L. Chin

Vol. 7   An Introduction to Optoelectronic Sensors
edited by G. C. Righini, A. Tajani and A. Cutolo

Vol. 8   Ultrafast Dynamics in Molecules, Nanostructures and Interfaces
edited by G. G. Gurzadyan, G. Lanzani, C. Soci and T. C. Sum

Series in Optics and Photonics – Vol. 8

Ultrafast Dynamics in Molecules, Nanostructures and Interfaces

Selected Lectures Presented at Symposium on Ultrafast Dynamics of the 7th International Conference on Materials for Advanced Technologies

Singapore   30 June–5 July 2013

Editors

G. G. Gurzadyan
Nanyang Technological University, Singapore

G. Lanzani
Istituto Italiano di Tecnologia, Italy

C. Soci
Nanyang Technological University, Singapore

T. C. Sum
Nanyang Technological University, Singapore

World Scientific

NEW JERSEY · LONDON · SINGAPORE · BEIJING · SHANGHAI · HONG KONG · TAIPEI · CHENNAI

Published by

World Scientific Publishing Co. Pte. Ltd.

5 Toh Tuck Link, Singapore 596224

USA office: 27 Warren Street, Suite 401-402, Hackensack, NJ 07601

UK office: 57 Shelton Street, Covent Garden, London WC2H 9HE

British Library Cataloguing-in-Publication Data
A catalogue record for this book is available from the British Library.

Series in Optics and Photonics — Vol. 8
ULTRAFAST DYNAMICS IN MOLECULES, NANOSTRUCTURES AND INTERFACES
Selected Lectures Presented at Symposium on Ultrafast Dynamics of the 7th International
Conference on Materials for Advanced Technologies

Copyright © 2014 by World Scientific Publishing Co. Pte. Ltd.

ISBN 978-981-4556-91-0

Printed in Singapore by B & Jo Enterprise Pte Ltd

v

Contents

Preface

Many natural processes can be described as a sequence of events, the earliest of which takes place on very short time scale. Laser pulses are the only tool that can be adopted to study such phenomena in real time. This has motivated a huge effort in generation, characterization, detection and application of ultrashort laser pulses. The ultrafast science community has grown in size and impact since 1960, when the first ps optical pulses were produced. The set of techniques and experiments now available is very broad. Laser sources are becoming more and more user friendly, and experiments are well characterized. Yet a distinct specialization is still required and this justifies the gathering of a dedicated community. Far from being a close enclave however ultrafast science is at the frontier of many research areas and provide a reliable contribution to modelling and understanding of phenomena.

The recent 7th International Conference on Materials for Advanced Technologies (ICMAT) in Singapore has hosted a Symposium on Ultrafast Dynamics that has been particularly successful in terms of attendance and quality of the talks. This is once again evidence of a lively community able to bring innovation and results.

This prompted us to collect part of the contributions in a monograph with tutorial introduction that can help the newcomers to approach the field as well us the expert to be up dated.

The present volume comprises articles that contain detailed description of the ultrafast methods in spectroscopy, i.e. polarization resolved pump-probe spectroscopy, fluorescence up-conversion, real-time vibrational spectroscopy, multidimensional 2D and 3D optical spectroscopy, ultrafast optical microscopy, vibrational spectroscopy via second harmonic and sum-frequency generation on surfaces, transient absorption data analysis. Fundamentals of the light matter interaction are discussed and various applications, e.g. in organic photovoltaics, vision, photosynthesis, singlet fission, interfaces, semiconductor nanowires, plasmonics are elaborated.

Video-recording of all invited talks and oral presentation can be found online: http://tinyurl.com/m22ejru.

Sponsorship of Materials Research Society of Singapore (MRS-S), EINST Technology, Pte Ltd (Singapore) and Analytical Technologies, Pte Ltd (Singapore) in publication of the present book is gratefully acknowledged. We would like to thank Miss Lin Ma for her generous contribution in the preparation of this volume.

Gagik G. Gurzadyan,
Nanyang Technological University, Singapore
gurzadyan@ntu.edu.sg

Guglielmo Lanzani,
Istituto Italiano di Tecnologia, Italy
Guglielmo.Lanzani@iit.it

Cesare Soci,
Nanyang Technological University, Singapore
csoci@ntu.edu.sg

Tze-Chien Sum,
Nanyang Technological University, Singapore
tzechien@ntu.edu.sg

1 November 2013

Publication of the present volume is sponsored by

EINST Technology Pte Ltd

Analytical Technologies Pte Ltd

MRS
Materials Research Society
SINGAPORE
ICMAT 2013
7TH INTERNATIONAL CONFERENCE ON MATERIALS FO
30 JUNE - 5 JULY, SUNTEC SINGAPORE | www.mrs.o
INFORMATION
PREPAID REGISTRATION BY LAST/FAMILY NAME
K IN REGISTRATION/
MRS
Materials Research Society

1. Jingzhi Shang (Singapore). 2. Liang Cheng (Singapore). 3. Lin Ma (Singapore). 4. Shuo Dong (Singapore). 5. Nguyen Thi Hong (Singapore). 6. Paola Lova (Singapore). 7. Guozhong Xing (Australia). 8. Takayoshi Kobayashi (Japan). 9. Majid Panahandeh-Fard (Singapore). 10. Elbert E. M. Chia (Singapore). 11. Zhaogang Nie (Singapore). 12. Zhi-Heng Loh (Singapore). 13. Cesare Soci (Singapore). 14. Leonas Valkunas (Lithuania). 15. Maxim Pshenichnikov (Netherlands). 16. Alexander Van Rhijn (USA). 17. Vidmantas Gulbinas (Lithuania). 18. Thomas Elsaesser (Germany). 19. Daniele Brida (Germany). 20. Shaul Mukamel (USA). 21. Marius Franckevicius (Lithuania). 22. Giulio Cerullo (Italy). 23. Gagik Gurzadyan (Singapore). 24. Guglielmo Lanzani (Italy). 25. Larry Luer (Spain). 26. Ilka Kriegel (Germany). 27. Antoinette Taylor (USA). 28. Richard Averitt (USA). 29. Ajay Ram Srimath Kandada (Italy). 30. Prashanth Upadhya (India). 31. Tze Chien Sum (Singapore). 32. Marcelo Alcocer (Italy). 33. Rohit Prasankumar (USA). 34. Raavi Sai Santosh Kumar (USA). 35. Paolo Maioli (France). 36. Xinfeng Liu (Singapore). 37. Chee Yong Neo (Singapore). 38. Jing Ngei Yip (Singapore). 39. Guichuan Xing (Singapore). 40. Wee Kiang Chong (Singapore). 41. Teck Wee Goh (Singapore). 42. Ho Fai Leung (Singapore).

FEMTOSECOND REAL-TIME VIBRATIONAL SPECTROSCOPY USING ULTRAFAST LASER PULSES

TAKAYOSHI KOBAYASHI[†]

*Advanced Ultrafast Laser Research Center, University of Electro-Communications,
1-5-1, Chofugaoka, Chofu, Tokyo 182-8585, Japan
kobayashi@ils.uec.ac.jp*

JUAN DU

*State Key Laboratory of High Field Laser Physics, Shanghai Institute of Optics and
Fine Mechanics, Chinese Academy of Sciences, Jiading, Shanghai 201-800, China
dujuan@mail.siom.ac.cn*

Ultrashort visible pulsed laser with constant-phase broad spectrum and broadband detector with ability of simultaneous detection of full spectral region of the pulse make time-resolved vibration spectroscopy a powerful method by detecting the vibrational amplitude in real time. It enables observation of ultrafast change in the molecular structure and lattice configuration to be used in photo-physics, photo-chemistry, and photo-biology research. We have generated stable visible to near-infrared sub-5-fs laser pulses using a noncollinear optical parametric amplifier (NOPA) and used these ultrashort laser pulses to study a series of ultrafast electronic relaxation and vibrational dynamics in polymers. Our findings can be extended to other condensed matter such as molecules and biopolymers.

1. Introduction

The multidisciplinary field of ultrafast optical science has rapidly evolved since the generation and development of ultrashort laser pulses. The use of ultrashort laser pulses to excite matters with femtosecond time scales and probe the subsequent ultrafast processes, i.e. ultrafast spectroscopy, has opened up completely new fields of research in physics, chemistry, and biology. For example, it can determine the molecular structural changes that occur during a chemical reaction, including the structures of transition states (TSs), an ability that has long been sought by chemists. The information provides synthetic chemists with a strategy for the synthesis of new molecules important in medical

---

[†] This work was partially supported by the joint research project of the Institute of Laser Engineering, Osaka University under Contract No. A3-01.

and pharmaceutical sciences. It is very difficult to obtain a clear guiding principle in organic synthesis using conventional methods that are usually based on laborious trial and error.

Compared with electron and/or X-ray diffraction methods, ultrafast laser spectroscopy offers much higher time resolutions [1]-[3] and can be used to measure various materials including amorphous materials and liquids, which are typically difficult to analyze by X-ray or electron diffraction. As we reported in our previous paper [4], measurements of time-resolved absorption spectral changes clarified ultrafast changes in the frequencies of the in- and out-of-plane bending modes associated with structural changes during photoisomerization, such that even the structure in the transition state could be determined.

The generation of ultrashort laser pulses is always the initial step in realizing direct observations by ultrafast spectroscopy. We have stably generated visible to near-infrared sub-5-fs laser pulses using a noncollinear optical parametric amplifier (NOPA) [5]-[7] and used them to study ultrafast processes that proceed after ultrashort excitation, primary relaxation, and long-term subsequent dynamics.

This review is organized as follows. First, we review the development of ultrashort pulse light sources, focusing on the NOPA. Then, we describe the principles of real-time vibrational spectroscopy and compare how it differs from conventional vibrational spectroscopic techniques (e.g., time-resolved IR absorption and Raman spectroscopies). The next section explains how real-time vibrational spectroscopy probes the modulation of electronic transition probabilities, which reflect wave-packet motion. Examples of applications of real-time spectroscopy are provided and, finally, future prospects are summarized and discussed.

2. Femtosecond Pump Probe Experimental Setup Based on NOPA

2.1. *Development of Ultrashort Laser Pulse*

The design of our home-made NOPA system is as follows. The pump source of this NOPA system is a commercially supplied regenerative amplifier (Spectra Physics, Spitfire) with central wavelength, pulse duration, repetition rate, and average output power of 800 nm, 50 fs, 5 kHz, and 750 mW, respectively. The input light is separated into two beams. The first beam propagates through a 2-mm-thick sapphire plate to generate a white-light continuum as a seed of signal. A cut-off filter that can block the wavelength longer than 750 nm is established after the sapphire to prevent an intense spike around 800 nm. The second beam propagates through a 0.4-mm-thick BBO crystal (29.2°z-cut) to yield second harmonic light for the NOPA pump, following the type-I (o+o→e) phase-match condition. To achieve effective and stable optical parametric amplification, a quartz block is used to stretch the pump pulse to 200 fs before introducing the first beam to the sapphire plate, making the duration comparable to that of the white-light continuum, which is broadened due to group-velocity inherent to the broadened spectrum.

A prism with a 45° apex angle is arranged after a quartz block plate so that the pre-tilted pulse front of the pump matches the signal in the crystal. Tilting the signal pulse front to compress the signal pulse into a sub-5-fs pulse was first considered by Kobayashi and Shirakawa *et al.* [8][9], who proposed using a pump beam with a tilted wave front to prevent the signal pulse from tilting in space (which causes the angular dispersion of the amplified pulse). This configuration (known as pulse-front matching) was implemented by sending the pump beam through a prism and adjusting the pulse-front tilt using a telescope consisting of two convex lenses [10]. Another function of the prism is for the angular dispersed pump pulse excitation to extend the gain band width by distributed incident angle of pump pulse resulting in the distributed phase matching condition [6][7].

The pre-tilted pulse is then focused on a 1-mm-thick BBO crystal (31.5°z-cut) to overlap spatially with a white-light continuum beam. The delay length introduced by an optical delay line (OD1) in the femtosecond white-light continuum beam is adjustable for optimum temporal overlap. To achieve higher signal energy, the remaining pump and signal are reflected after the first OPA processing and are focused again on a BBO crystal at a lower position. The temporal overlapping of pump and signal can be regulated by the optical delay line (OD2) in the pump beam path.

The amplified signal just after the BBO crystal is about 0.3 μJ. A pair of ultra-broadband chirped mirrors was designed to compensate for group delay (GD), group delay dispersion (GDD), and third-order dispersion (TOD) in the NOPA system, followed by a pair of prisms. One of the chirped mirrors exhibits a GD property with several reflections to compensate for the chirp of the NOPA output by combining the prism pair, the air, and the beam splitter with the spectral region from 480 to 760 nm. The other chirped mirror was designed to have a GDD and TOD with -45 fs^2 and 20 fs^3, respectively, covering the spectral region from 500 to 780 nm. These chirped-mirror and prism pairs compose the compressor system. The chirped mirrors have high reflectivity (R>99%) and the prism pair is the main source of loss in the present system. The spectrum of the NOPA output can be adjusted by slightly changing the noncollinear angle and the prism for the pump angular chirping. During the experimental procedure, the best noncollinear angle was judged by adjusting the fluorescence ring width to the thinnest case. Theoretically, the pulse width could be as short as 5 fs calculated for the spectrum of 520-720 nm of the output. However, in some experimental cases, the spectral range of the NOPA output was adjusted to the longer wavelength, extending from 556 to 753 nm to achieve higher absorbance of some samples by spectral matching. The pulse width in this case was about 7 fs, which was slightly broader than the optimum case. In this way the spectral range can be slightly adjusted by small angle tuning at the expense of a reduction in spectral width that results in pulse durations of small increments.

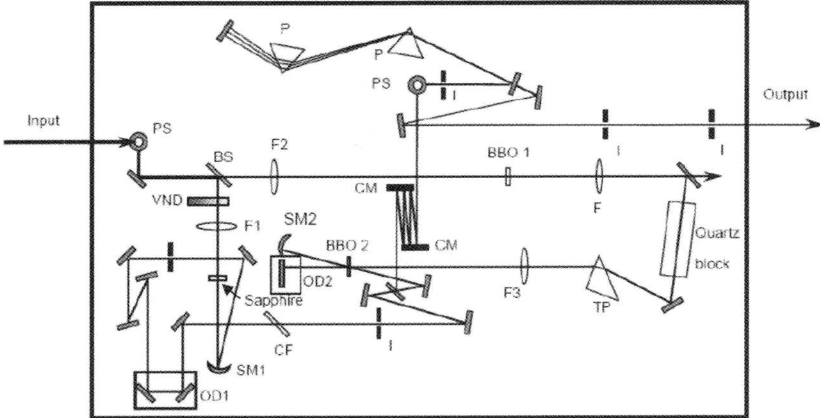

Figure 1. Diagram of NOPA system. PS, periscope; BS, beam splitter; VND, variable neutral-density filter; F, focusing lens (F1: r = 100 mm; F2: r = 400 mm; F3: r = 250 mm); SM, spherical mirrors (SM1: r = 120 mm; SM2: r = 100 mm); I, iris; Sapphire: plate of sapphire to obtain broad band signal; BBO 1, non-linear crystal of β-barium borate to generate the second harmonic light for pumping the noncollinear parametric amplifier crystal; BBO 2, non-linear crystal of β-barium borate used for noncollinear parametric amplification; Quartz block, Block of fused quartz for pulse stretching; CF, cut-off filter (cut to 750 nm); OD1, optical delay line adjustable for first path OPA; OD2, adjustable optical delay line for second path OPA; TP, 45-degree apex angle prism for pulse-front tilting and angularly dispersed pump pulse; CM, chirped mirror; P, 60-degree prism; The elements of thin rectangular shapes without a label are plane mirrors.

2.2. *Development of a Broadband Detector*

A pump-probe experimental setup is shown in Figure 1. For the pump-probe experiment, the output from the NOPA is separated into pump and probe beams by a beam splitter at a branching ratio of 5:1. A frequency divider is synchronized with the NOPA output pulse and controls the chopper with 2.5 kHz, which is half the repetition frequency of the NOPA. Both pump and probe beams are focused on the sample surface over areas slightly larger and smaller than about 50 μm in diameter, respectively, using a 127 mm (0.5 in) parabolic mirror. The spatial overlap of the pump and probe beams on the sample is checked and guaranteed by a pinhole with a 0.1 mm diameter. The transmitted probe light beam is guided through the fiber to the detection system.

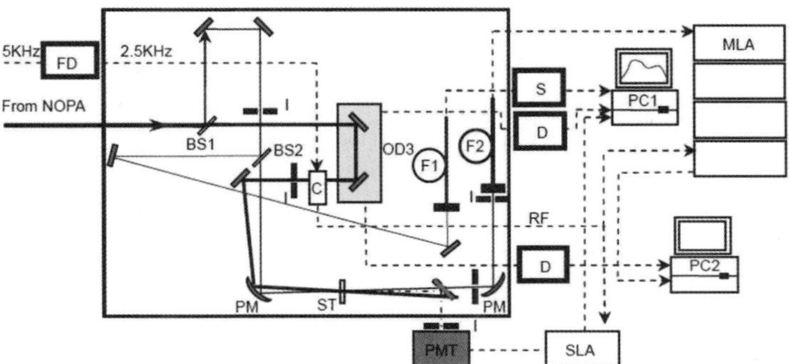

Figure 2. Diagram of pump probe setup. Solid lines represent beam pathways and dashed lines denote electric connection among equipment. BS1 and BS2, beam splitters; OD3, adjustable optical delay line for pump; C, chopper; I, iris; PM, parabolic mirror; F1 and F2, optical fibers for signal detecting; PC1 and PC2, personal computers; PMT, photo multiplier tube; D, optical delay controller; S, spectrometer; SLA, single lock-in amplifier; MLA, multi lock-in amplifier. ST, Sample for pump-probe measurement and a BBO crystal for the measurement of auto (cross) -correlation of pump and probe pulses. Thick solid lines after BS1 pump pulse beam line; thin solid lines after BS1, probe pulse beam line; Thin boxes of rectangular shapes without a label, plane mirrors; Dashed lines, electric connections between the elements of the elements used for pump-probe experiment and autocorrelation measurement.

Two computers (PC1 and PC2) are used for different purposes. PC1 is used to measure pulse width from the NOPA before the pump-probe experiment, and to measure the NOPA's real-time monitoring stability during the experiment. Pump and probe beams are spatially overlapped in a BBO crystal on the sample stage to generate the sum frequency. By motor driving the optical delay line (OD3 in Figure 2) in the pump beam path, the auto (cross) -correlation trace of pump and probe can be measured and thus the pulse width from the NOPA can be checked before and after the pump-probe experiment.

The second computer (PC2 in Figure 2) is used to record the DC (laser and transmitted spectrum) and AC (ΔT) from the multi-lock-in amplifier. The signal is spectrally dispersed using a polychromator (JASCO, M25-TP) over 128 channels with a spectral resolution of 1.5 nm, and each channel is detected by an avalanche photodiode in conjunction with a lock-in amplifier locked onto the 2.5-kHz reference frequency set by an optical chopper that modulated the pump beam. The detection spectral range is tunable by a relatively small amount through changes in the angle of the dispersed grating of the polychromator, according to the NOPA spectrum. PC2 is also connected to control OD3 in changing the delay time and measuring the data of normalized transmittances ($\Delta T(\mathrm{t})/T$). From the value the "apparent time-resolved absorbance change" is calculated with $\Delta A = (1/2.303) \log [T(\Delta/T+T)]$. The polarizations of both pump and probe are horizontal. Perpendicular configuration of pump-probe polarizations can be easily constructed using a set of periscope. The meaning of apparent is that the absorbance change observed at one specific wavelength does not necessary be due to the change in the absorption transition probability but it may be due to spectral shift or even to the change of probe spectral shape by phase modulation as discussed in terms of molecular phase modulation. In this case the effect is probe spectrum dependent and hence it is not intrinsic effect of the sample being studied.

2.3. *Advantages of Femtosecond Real-time Vibrational Spectroscopy*

The femtosecond pump-probe real-time spectroscopy technique used in our study can be applied to investigate both the vibrational and electronic dynamics in molecules and polymers using the same laser and detection systems under the same condition. Compared with conventional vibrational spectroscopies such as infrared absorption and Raman scattering techniques, this method has the following advantages.

(1) Spontaneous Raman signals even in the condition of resonance Raman are very frequently overwhelmed by the fluorescence signal, especially in the case of highly fluorescent molecules. In contrast, in the case of real-time vibrational spectroscopy, the effect of spontaneous fluorescence can be almost totally eliminated due to a much more intense and spatially coherent incident probe beam directed to the detection system, while the overall spontaneous fluorescence emitted are emitted to broad angular directions.

(2) The low frequency modes can easily be studied by pump-probe as long as a few quanta of the modes can be covered within the width of the laser spectrum with a nearly constant phase. However, detection by Raman scattering method is difficult due to the intense Rayleigh scattering of the excitation beam. For example, modes with a frequency lower than 200 cm^{-1} are difficult to be detected using conventional Raman spectroscopy, but easy to measure using real-time spectroscopy.

(3) As a pure time domain technique, the pump-probe method enables the direct observation of vibronic dynamics, including time-dependent instantaneous

frequencies. In cases of conventional time-resolved vibrational spectroscopy, time-dependent frequencies can be detected with a much longer time step than the pulse duration. Therefore, the change to be followed is in the time step of the sub-picosecond regime. Hence, it is difficult to detect change in this vibrational frequency within the oscillation period. However, in the case of real-time spectroscopy, using a time step of 0.1 or 0.2 fs (as used in our group) results in a very small detectable change in frequency shifting that is nearly continuous in the real-time domain. In the case of time-resolved Raman spectroscopy, Raman spectra at different delay times can be measured but the difference in the Raman spectrum among the various delay times is not easily controlled in the sub-1-fs time range. Therefore, molecular structural change information, such as in the transition state, cannot be detected. The group of Mathies recently developed stimulated Raman spectroscopy using two pump pulses and one probe pulse[11]. One of the two pump pulses is used for electronic excitation with ~30 fs and the other is a narrow band ($<$10 cm^{-1}) with a ~2 ps duration for the Raman pump. The probe pulse has a broad band (~800cm^{-1}) with femtosecond duration (~20 fs). By combining the sharp spectrum of the Raman pump and the short pulse duration of the probe pulse this spectroscopic method claims to have a time resolution of about 20-30 fs and a spectral resolution of 10 cm^{-1}[11]. Even with this method, however, it is not possible to get better resolution than 20-30 fs. Hence, the gradual change in instantaneous frequency, which may appear when the system is moving on the potential surface during chemical reaction, cannot be observed.

(4) The real-time spectroscopy can provide the information on the vibrational initial phase, which cannot be obtained with conventional Raman or infrared spectroscopy because these methods are intensity detection schemes measuring Raman scattering intensity or infrared absorption intensity. Although the Raman scattered signal light and infrared (IR) absorbed probe light contain the information of the real and imaginary parts of the field, the detectors of Raman scattered light and absorbed IR light can only measure their intensities. Here, using the case of IR absorption, a more detailed explanation is provided in Figure 3. The macroscopic polarization induced by IR optical field E is given by $P(\omega) = \chi(\omega)E(\omega)$ as the response that IR absorptive material has to the field. Then, this polarization drives the Maxwell equation, which leads to the IR radiation field $E(\omega, d)$ after the incident beam field $E(\omega, 0)$ is absorbed by uniform IR active material with a thickness of d. In this process, the absorptivity defined by $\alpha(\omega) = (2/d)\ln(|E(\omega, 0)/E(\omega, z)|)$ is given by $\alpha(\omega) = k\chi''(\omega)/n^2$, where k is the wave number, n is the refractive index, and $\chi''(\omega)$ is the imaginary part of the susceptibility $\chi(\omega) = \chi'(\omega) + i\chi''(\omega)$ and is related to the

real part $\chi'(\omega)$ through the Kramers-Kronig relations. The relations indicate that the same amount of information on the optical (in this case IR) properties of materials is contained in $\chi'(\omega)$ and $\chi''(\omega)$. Given that only the intensity reduction $|E(\omega,0)|^2/|E(\omega,z)|^2 = \exp(\alpha(\omega)d)$ can be measured by an ordinary photo detector such as a photomultiplier and a photodiode, the phase shift due to the refractive index introduced by the interaction with the material cannot be detected. This means that in the measurement procedure, these methods lose as much as 50% of the information on vibrational behavior through the projection of the wave packet onto the axis of the phase space for the amplitude radius with an expense of angle as shown in Figure 3. In this figure, the optical field of probe light (both before and after the absorption process) is represented by the dot at the crossing point of the circle and radial arrow. This point shows the phase and amplitude of the field. If it is possible, by measuring the complex field, information of the complex susceptibility $\chi(\omega) = \chi'(\omega) + i\chi''(\omega)$ of the material can be obtained by comparing the measured complex fields before and after the sample.

However, using the intensity detector as being done in ordinary case after observation of both before and after the sample, the information on the phase is lost. At first sight, this may look like a collapse of wave function in quantum mechanics, but it is in the opposite direction. In the case of quantum mechanics, the distribution of the wave packet has finite distribution in the phase space and after measurement it collapses to a single point. In "classical" photo detection scheme, before measurement the light has a well-defined phase and amplitude, but the former becomes ambiguous after measurement, even using photon quantum counters; it means that a half of the information is lost.

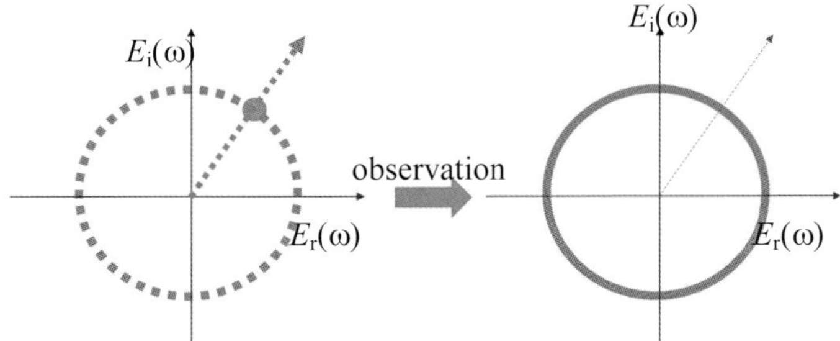

Figure 3. Conceptual description of the observation process using photo detector of optical field to obtain the optical properties described by the susceptibility; $\chi(\omega) = \chi'(\omega) + i\chi''(\omega)$.

(5) Using real-time spectroscopy, we can investigate the dynamics of vibrational modes coupled with the electronic transition. The vibronic coupling can be studied in relation to the decay mechanism of the electronic excited states. Strong vibronic coupling to some specific modes is considered to be an implementation of the electronic relaxation channel. Therefore, simultaneous electronic decay and vibrational dynamics have outstanding advantages over other time-resolved vibrational spectroscopy. The real-time vibrational spectroscopy provides the two dynamic processes simultaneously using the same measuring system under exactly the same experimental conditions, such as laser power, temperature, and probing sensitivity.

3. Principle of Real-time Vibration Spectroscopy: Modulation of Electronic Transition Probability by Wave-Packet Motion

There are two methods for studying the vibrational dynamics of excited electronic (or excitonic) states in condensed matters. One is an analysis in the frequency-domain and the other is an analysis in the time-domain. The former method is the most frequently used through practices such as time-resolved Raman scattering spectroscopy [12] and time-resolved infrared absorption spectroscopy [13]-[15]. The latter is real-time vibration spectroscopy used to gather information on the real-time amplitudes of vibration through the modulation of the electronic transition probability, including the vibrational initial phase after impulsive excitation. They can also provide information on the vibronic coupling strength and dynamics of the vibrational modes in both ground and excited electronic states. They have been used in the real-time vibrational spectroscopy of many molecular and polymer systems [16]-[19].

In most molecular systems, the wave function of a molecular electronic state can be factored into the electronic part with nuclear coordinates as parameters and the nuclear component using the Born-Oppenheimer approximation, except when the electronic state exhibits degeneracy:

$$\Psi(q,Q) = \Phi(Q,q)\chi(Q). \tag{1}$$

Here, q and Q represent the electron and nuclear coordinates, respectively. The transition dipole between two electronic states under the Born-Oppenheimer approximation is given by the following in a case of Condon approximation, which assumes the independence of the transition dipole moment from the nuclear coordinates:

$$<\Psi_1(q,Q)|eq|\Psi_2(q,Q)>_{Q,q} = <\chi_1(Q)<\Phi_1(Q,q)|eq|\Phi_2(Q,q)>\chi_2(Q)> \cong <\Phi_1(Q,q)|eq|\Phi_2(Q,q)><\chi_1(Q)|\chi_2(Q)> \tag{2}$$

The electronic wave function is mixed with a third electronic state, which causes the electronic transition probabilities to be exchanged between the two relevant electronic states. Following the wave-packet motion, the molecular structure is deformed and the corresponding electronic wave functions and energies of these states are modified, resulting in the modulation of the transition probability.

The origins of the intensity modulation can be classified into two cases: when the Condon approximation is satisfied and when it is not satisfied. The former case is due to the time-dependent Franck-Condon (FC) overlap that results from wave-packet formation. The wave function of the wave packet is a linear combination of the wave functions of the vibrational levels and the distributed vibrational quantum numbers of the relevant mode. The latter case occurs when the electronic wave function is mixed with a third electronic state, which causes electronic transition probabilities to be exchanged between the two relevant electronic states. Following the wave-packet motion, the molecular structure is deformed and the corresponding electronic wave functions and energies of these states are modified, resulting in the modulation of the transition probability.

3.1. *The Case of Condon Approximation Satisfied*

This subsection discusses an electronic transition in a two-level system when the Condon approximation is satisfied. In this case, the integrated intensity of the absorbance change over the relevant Franck-Condon distribution with different vibrational quantum numbers remains constant during wave-packet motion. The two-level system has two harmonic potential curves for four typical cases in the ground and lowest excited states, as shown in Figure 4.

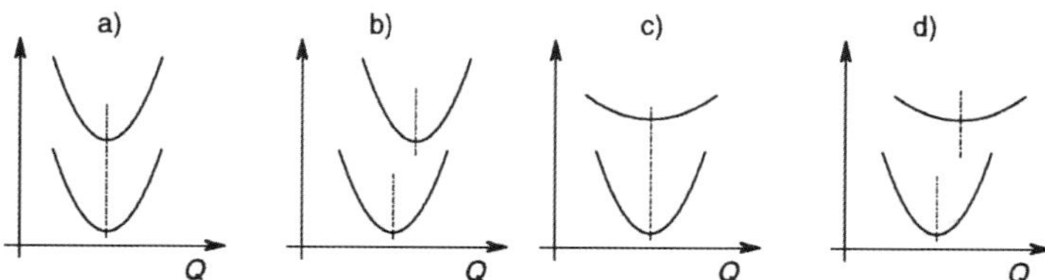

Figure 4. Four typical cases for the minima and curvatures of the potential curves: a) no change in the potential minimum position or curvature upon photoexcitation; b) a change in potential minimum position with no change in curvature; c) a change in curvature with no change in the potential minimum position; and d) with a change in both the potential minimum position and the curvature. The dash-dotted lines indicate the minimums of the potential surfaces.

Here, we only discuss the wave packet generated in the ground state but a similar discussion applies when the wave packet is generated in the excited state. There are four typical cases for the minima and curvatures of the potential curves: a) neither the potential minimum position nor the curvature change upon photoexcitation; b) the potential minimum position changes, but the curvature remains the same; c) the curvature changes, but the potential minimum position remains the same; and d) both the potential minimum position and the curvature change.

There are four different cases relevant to the abovementioned characteristics of the potential curves: a′) generation of a vibrational eigenstate with no wave-packet motion due to the selection rule for allowed transitions; b′) generation of a wave packet in the ground state, which starts to move along the potential curve; c′) generation of a wave packet in the ground state, whose size starts to breathe after photoexcitation; and d′) generation of a wave packet in the ground state, which starts to move along the potential curve and breathe at the same time.

Intensity modulation can be introduced by the time-dependent overlap of the molecular wave functions between the initial and final electronic states associated with the transition. In such a case, the modulation mechanism for the transition probability is the time-dependent FC overlap factor. When the molecular vibration is Fourier analyzed to separate into many modes, it can be described in terms of a one-dimensional harmonic potential curve. The FC factor is then associated with the motion of a wave packet moving between two equal-energy points on the potential curve along one of the normal coordinates on which the wave packet is located [31]. The time-dependent FC factor, $F(t)$, related to the initial and final states (i and f, respectively) coupled with the transition, is given by Equation (3):

$$F(t) = \sum_l \sum_m c_l^* c_m \left\langle \tilde{\chi}_l^i(\hat{Q}) \mid \tilde{\chi}_m^f(\hat{Q}) \right\rangle = \sum_l \sum_m c_l^* c_m \left\langle \chi_l^i(\hat{Q}) \mid \chi_m^f(\hat{Q}) \right\rangle e^{i(l-m)\omega_v t} \qquad (3)$$

The nuclear wave function is given by

$$\left| \tilde{\chi}_\alpha(\hat{Q}) \right\rangle = \left| \chi_\alpha(\hat{Q}) \right\rangle e^{i\alpha\omega_v t} \qquad (4)$$

The nuclear wave function has a vibrational quantum number of $\alpha = (l, m)$, and ω_v is the frequency of the molecular vibration relevant to the vibronic coupling. The equation describing the spectral change owing to the motion of a wave packet induced by impulsive excitation depends on the FC factor, which is the coefficient of the sinusoidal function in Equation (3). The coefficients $c_l^* c_m$ are determined by the pump laser spectrum and the cross section of the ground state absorption. If i and f are in the ground and excited states, respectively, there are two possible ways for the wave packet to be generated by the pump: either in i state by coherent vibronic excitation or in f state by the stimulated Raman process. Wave-packet motion then starts along the potential curve of the corresponding state.

We discuss the case in which the potential minimum is displaced along the potential surface of the i state with respect to that of the f state (Figure 4b). This displaced potential surface case is referred to as the D case in the discussion below about the transition probability modulation mechanism. If the shift owing to wave-packet motion is small, then the spectral change will be small and it can be obtained from the first derivative of the absorption spectrum. The probe wavelength dependence of vibrational amplitude shows a π phase jump at the peak wavelength of the absorption band [21,22].

When there is no displacement between the ground and excited states (Figure 4c; referred to as the ND case below), the wave-packet motion does not involve oscillations between two turning points of equal energy on the potential curve of either the excited state or the ground state. In such a case, the principal wave-packet motion arises from the second-order difference between the initial and final states. The second-order difference is then given by the breathing of the wave packet, in which the width of the wave packet oscillates in time at the vibrational frequency. This mechanism has higher vibronic coupling, with respect to the spectral change, than the FC case as it involves higher-order wave-packet motion upon photoexcitation. The initial and final states are expected to have similar potential curves, except when very large geometrical relaxation occurs after excitation. The difference between the potential energies can then be expanded as a Taylor series in terms of the normal coordinate (Q), the nth term of which is proportional to the nth derivative ($\partial^n A(\omega)/\partial\omega^n$) of the absorption spectrum descried as $A(\omega)$ [22]. The components of the power series are classified in terms of the vibronic coupling strength. In most cases, the first-order term has the highest coupling strength.

As mentioned above, in the case of breathing, the spectral change is approximately determined by the second derivative of the relevant transition spectrum. When the wave-packet motion is in the ground state, the relevant spectrum is the transition spectrum from the ground state to the relevant excited state. If the wave-packet motion is in the excited state, then the vibrational amplitude depends on the second derivative of the stimulated emission spectrum and/or the bleaching spectrum.

The above discussion is expressed in Equation (5), which shows the modulation of the absorbance change ($\Delta A(\omega)$) as a function of the probe frequency ω:

$$\delta \Delta A(t;\Omega,\omega) \cong \Delta A(t;\Omega,\omega) - \Delta A_0(\omega) = \left(\delta\omega \frac{d\Delta A_0(\omega)}{d\omega} + \delta\Delta\omega \frac{d^2\Delta A_0(\omega)}{d\omega^2} \right) \cos(\Omega t + \phi). \quad (5)$$

Here, Ω is the frequency of the relevant molecular vibrational mode and $\Delta A_0(\omega)$ is the difference absorption spectrum without molecular vibrations. This spectrum can be obtained by smoothing the real-time vibrational trace over several vibrational periods at each probe wavelength. It may be a composite of the gain (due to the stimulated emission) and bleaching (owing to ground-state depletion) spectra. $\delta\Delta A_0(\Omega;\omega)$ is the amplitude of the Fourier transformation (FT) of the absorbance change as a function of the molecular vibrational frequency Ω. $\delta\Delta\omega$ is the change in the bandwidth ($\Delta\omega$) of the absorption, gain, or bleaching spectrum.

Thus far, we have considered a case in which the Condon approximation is satisfied. This is referred to as the C mechanism, and it can be sub-classified into D and ND cases. For the D case, the probe frequency dependence of $\delta\Delta A(\Omega;\omega)$ is mainly given by the first derivative of $\Delta A_0(\omega)$, whereas in the ND case, the main contribution is given by the second derivative of $\Delta A_0(\omega)$. The configuration shown in Figure 3d is a mixture of D and ND cases, thus it is expected to have a mixed probe wavelength dependence on the first and second derivatives.

3.2. *Cases in which the Condon Approximation is Unsatisfied*

In this subsection, deviation from the Condon approximation is introduced by considering another electronic state (a third state) that is radiatively coupled to the other two states, between which the transition intensity is being monitored. Here, we discuss a three-level system that includes the third state. The system comprises three electronic states, S_0, S_1, and S_2 (listed in order of increasing energy). The vibronic coupling is assumed to be predominantly between the S_1 and S_2 states. The following three wave functions are considered:

$$\left| S_0(\hat{q},\hat{Q}) \right\rangle = \left| \psi_0(\hat{q},\hat{Q}) \right\rangle \left| \chi_n^0(\hat{Q}) \right\rangle, \quad (6)$$

$$\left| S_1(\hat{q},\hat{Q}) \right\rangle = \frac{1}{\sqrt{1 + (H_{vib}/\Delta E_{21})}} \left(\left| \psi_1(\hat{q},\hat{Q}) \right\rangle - \frac{H_{vib}}{\Delta E_{21}} \hat{Q} \left| \psi_2(\hat{q},\hat{Q}) \right\rangle \right) \left| \chi_m^l(\hat{Q}) \right\rangle, \quad (7)$$

$$\left| S_2(\hat{q},\hat{Q}) \right\rangle = \frac{1}{\sqrt{1 + (H_{vib}/\Delta E_{21})}} \left(\left| \psi_2(\hat{q},\hat{Q}) \right\rangle - \frac{H_{vib}}{\Delta E_{21}} \hat{Q} \left| \psi_1(\hat{q},\hat{Q}) \right\rangle \right) \left| \chi_m^2(\hat{Q}) \right\rangle. \quad (8)$$

Here, $\left|\psi_X(\hat{q},\hat{Q})\right\rangle$ and $\left|\chi_k^X(\hat{Q})\right\rangle$ respectively represent the electron and nuclei wave functions for state X (= 0, 1, and 2 for S_0, S_1, and S_2, respectively) as a function of the electron and nuclear coordinate operators, $\hat{q}$ and $\hat{Q}$, respectively. In the nuclear wave function $\left|\chi_k^X(\hat{Q})\right\rangle$, the vibrational quantum number is denoted by the suffix k (= n and m). ΔE_{21} is the energy difference between states S_2 and S_1 and H_{vib} is the interaction Hamiltonian. The states S_1 and S_2 can be any two electronic excited states in general (i.e., the electronic state S_2 can be a higher excited state than the S_1 state). In this way, the spectral change integrated over the relevant spectral range does not have constant oscillator strength, but rather it varies with the vibrational motion of the mode, whose interaction H_{vib} contributes to the mixing of the two electronic states. The mixing is controlled by the symmetries of the vibrational mode and the electronic states.

Using the above equations, the transition dipole moment between $\left|S_0(\hat{q},\hat{Q})\right\rangle$ and $\left|S_1(\hat{q},\hat{Q})\right\rangle$ can be written as:

$$\left\langle S_0\left|\hat{\mu}\right|S_1\right\rangle = \left\langle\varphi_0\left|\left\langle\chi\left|\hat{\mu}(\left|\varphi_1\right\rangle - \beta\hat{Q}\left|\varphi_2\right\rangle)\right|\chi_m^l\right\rangle\right.$$

$$= \left\langle\varphi_0\left|\hat{\mu}\right|\varphi_1\right\rangle\left\langle\chi_n^0\left|\chi_m^l\right\rangle - \beta\left\langle\varphi_0\left|\hat{\mu}\right|\varphi_2\right\rangle\left\langle\chi_m^0\left|\hat{Q}\right|\chi_m^l\right\rangle\right.,$$

$$\beta = H_{vib}/\Delta E_{21}.$$

The transition probability is given by Equation (9):

$$\left|\left\langle S_0\left|\hat{\mu}\right|S_1\right\rangle\right|^2 = \left|\left\langle\varphi_0\left|\hat{\mu}\right|\varphi_1\right\rangle\right|^2 \left|\left\langle\chi_m^0\left|\chi_m^l\right\rangle\right|^2 - \beta^*\left\langle\varphi_0\left|\hat{\mu}\right|\varphi_1\right\rangle\left\langle\varphi_0\left|\hat{\mu}\right|\varphi_2\right\rangle^*\left\langle\chi_m^0\left|\chi_m^l\right\rangle\right.$$

$$\left\langle\chi_m^0\left|\hat{Q}\right|\chi_m^l\right\rangle^* - \beta\left\langle\varphi_0\left|\hat{\mu}\right|\varphi_1\right\rangle^*\left\langle\varphi_0\left|\hat{\mu}\right|\varphi_2\right\rangle\left\langle\chi_m^0\left|\chi_m^l\right\rangle^*\left\langle\chi_m^0\left|\hat{Q}\right|\chi_m^l\right\rangle + h.o.$$

$$\tag{9}$$

The discussion becomes complicated when both S0 → S1 and S1 → S2 are allowed transitions. Therefore, we use the modified phenomenological equation (Equation (7)) as Equation (10):

$$\delta\Delta A(t;\Omega,\omega) \cong \left(\frac{\delta(\mu^2)}{\mu^2}\Delta A_0(\omega) + \delta\omega\frac{d\Delta A_0(\omega)}{d\omega} + \delta\Delta\omega\frac{d^2\Delta A_0(\omega)}{d\omega^2}\right)\cos(\Omega t + \phi). \tag{10}$$

The first term represents the modulation of the transition dipole moment operator $\hat{\mu}$ of the relevant electronic transition. The non-conserved oscillator strength will make the Condon approximation invalid, which implies that the integrated intensity over the electronic transition is not constant during the vibrational period (this is referred to as the non-Condon (NC) mechanism below). This mechanism can also be sub-classified into D and ND cases.

The dependence of $\delta\Delta A(\Omega; \omega)$ on the probe photon frequency for the NC mechanism is given as follows. In the D case, $\delta\Delta A(\Omega; \omega)$ depends on the zeroth and first derivatives of $A_0(\omega)$, whereas it depends on the zeroth and second derivatives of $A_0(\omega)$ for the ND case. For both the D and ND cases for the NC mechanism, the zeroth-order dependence of the probe wavelength was added to fulfill the non-conserved integrated transition probability in the NC mechanism. The zeroth-order derivative can also be understood as the time-dependent contribution of the third electronic state, which varies periodically with the vibrational frequency. These phenomena have been described as alternating intensity borrowing and returning [23]-[25]. The process is also described as dynamic intensity borrowing (and returning).

4. Examples of Real-Time Spectroscopy Applications

4.1. *Ultrafast Electronic Relaxation and Vibrational Dynamics in a Polyacetylene Derivative*

The combination of the large third-order nonlinearities generated by the electronic correlation and the ultrafast response is quite attractive for basic techniques such as optoelectro-switching and optical information processing. Many experimental and theoretical studies have been conducted to clarify the mechanisms responsible for the macroscopic nonlinear properties characteristic of this class of materials [26]. The optical nonlinearities of the one-dimensional conjugated polymers are closely related to geometrically relaxed excitations such as a pair of solitons, and polarons, and a self-trapped exciton (STE). An STE is equivalent to an exciton polaron and a neutral bipolaron. This nonlinearity is due to strong coupling between electronic excitations and lattice vibrations [27]. A free exciton formed in such a one-dimensional system spontaneously relaxes within ~ 100 fs due to the absence of a barrier between the free exciton and STE minimums of the potential curves [10], changing the optical properties of the conjugated polymers and inducing the absorption coefficients and refractive indices [28,29]. The relaxation dynamics of photoexcitations in polydiacetylenes and polythiophenes are thus related to the ultrafast nonlinear response dynamics, which are considered useful for ultrafast optical switching. Their formation and relaxation processes are, therefore, quite essential and one of the most fundamental research subjects in the field of dynamics in polymer systems. The change is induced not only by the localized electronic excitation, but also by the vibrational excitation coupled to the electronic excitation through vibronic coupling, which starts to modulate the molecular structure [23,24][30,31]. The structural modulation changes the energy-level scheme and the transition probability. Then the former changes the electronic spectrum and hence the intensity at some specific probe wavelength while the latter modulates the intensity of the relevant homogeneous spectral range.

In this study, the electronic phase relaxation time and the frequencies of the vibrational modes due to the wave-packet motion in the electronic excited state have been obtained. From our analysis of the dynamics of the mean distribution energy of the vibration energy, the rate of descent down the vibrational ladder is calculated for several modes. The vibrational phase relaxation rate is also calculated from the FWHM of the Fourier spectrum, and the electronic phase relaxation is obtained from the data in the negative time range when the probe pulse precedes the pump pulse.

The sample polymer studied is poly[o-TFMPA([o-(trifluoromethyl) phenyl]acetylene)] (hereafter abbreviated as PTTPA), and its molecular structure is shown in the inset of Figure 5b. From the real-time traces obtained in the experiment, as shown in Figure 5a, the two time constants were

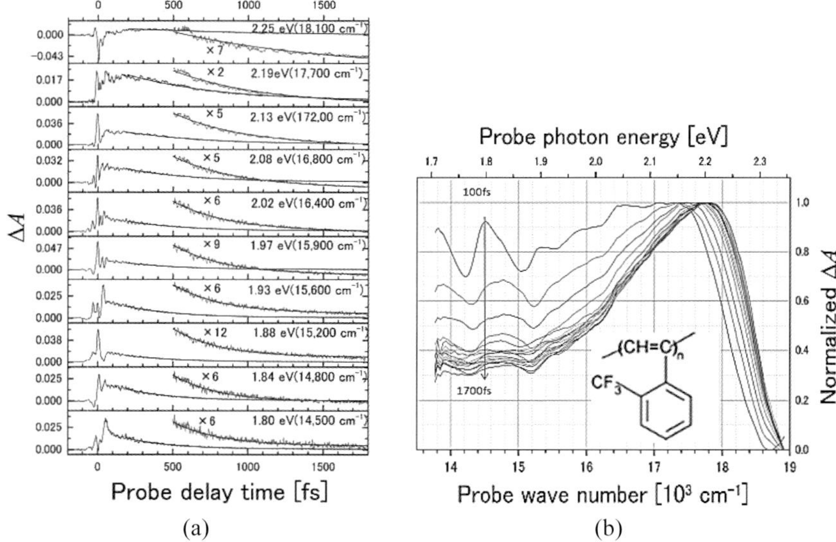

Figure 5. (a) Pump-probe delay time dependence of absorbance changes at ten typical probe photon energies from -200 to 1800 fs. Magnified curves from 500 fs with appropriate magnification factors are also shown to clarify the molecular vibrations. (b) Normalized time-resolved spectrum integrated for 100 fs delay time duration with the center delay times between 100 and 1700 fs with a 100-fs step.

determined to be $\tau_1 = 20\pm2$ fs and $\tau_2 = 320\pm50$ fs. The former corresponds to the ultrafast geometrical relaxation from free excitons to self-trapped excitons in PTTPA in the absence of a barrier between them. The latter, which is too short to be assigned to the population decay of STEs, namely the electronic population decay [32,33], is ascribed to the vibrational relaxation. Therefore, this blue-shift observed in the time-resolved spectra shown in Figure 5b can be explained in terms of the intra-chain thermalization process discussed below.

First, the assignment of the modulation must be mentioned. In the spectral range of positive ΔA, the initial state of the transition is the excited state, either free-exciton or STE. Then the modulation of ΔA, which is described as $\delta\Delta A$, is considered to be due to the wave packet in the excited state.

In the thermalization process, the vibrational quanta of modes with high vibrational frequencies are scattered, converted to low-frequency modes via dynamic vibrational mode coupling. This results in the reduction of the mean vibrational energy of the population distributed over vibrational levels with different quantum numbers and with many modes in the electronic excited state for the signal observed in the positive ΔA.

The blue shift is also found in Figure 5b. It is more clearly detected by the moment calculation discussed below and can be ascribed to the change in the induced absorption caused by that of the population distribution in the lowest excited state (i.e., the initial state of the observed transition in the induced absorption) without changing the energy position of the vacant higher excited state (the final state of the transition).

For a detailed study of the vibrational-energy relaxation process associated with the thermalization, the first moment of the induced absorption was calculated in the integration range of 529-726 nm (2.34-1.71 eV). As shown in Figure 6, the decay of the moment corresponding to the average transition energy is gradually reduced with the delay time t_D. This trend can be reproduced by the following fit to the calculated first moment,

$$\Delta E(t_D) = (\Delta E - \Delta E_0)\exp(-t_D/\tau_e) + \Delta E_0, \tag{11}$$

where the relaxation time, τ_e, the initial extra energy, ΔE, and the final energy after thermalization, ΔE_0, were taken as variable parameters. The parameters were found to be $\tau_e = 217\pm2$ fs, $\Delta E = 1.91\pm0.01$ eV, and $\Delta E_0 = 2.08\pm0.01$ eV.

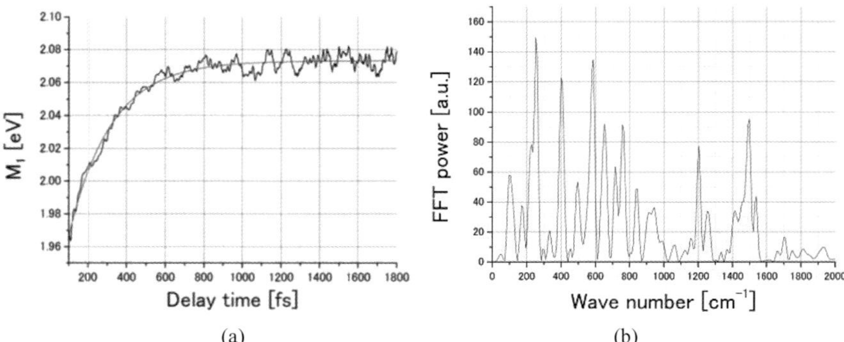

Figure 6. (a) Delay time dependence of the first moment (M1) of the induced absorption spectra calculated by integrating in the range of 529-726 nm (2.34-1.71 eV). Fitting to an exponential curve is also displayed. (b) FFT of the (a) curve after the fitting curve is subtracted from the experimental curve.

The FFT power spectrum of the first moment curve is shown in Figure 6b. In the spectrum there are six most intense vibrational modes with wavenumbers of 103, 171, 258, 383, 1201, and 1499 cm^{-1}. The Fourier power associated with the mean energy (first moment) reducing with time provides the strength of the vibronic coupling, which induces (mean) energy reduction. The modulations shown in the (mean) energy relaxation obtained by the first moment is the total effect of vibrational relaxation contributed from each vibrational mode. Therefore, the first moment can be expressed by the sum of main contributions from the abovementioned six modes as

$$M_1(t_\mathrm{D}) = \Delta E(t_\mathrm{D}) + \sum_i \Delta E_{vi} \exp(-t_\mathrm{D}/T_{2i}^{\mathrm{vib}})\cos(\omega_{vi}t_\mathrm{D}+\varphi_i), \quad (i=1,2,...,6) \quad (12)$$

Here, ω_{vi} is the vibrational frequency of each mode, T_{2i}^{vib} is the vibrational dephasing time, and φ_i is the initial phase. The individual contribution to the energy relaxation can be obtained from each of the six modes, out of all the modes, including those not well observed below the noise level using their relative peak values in the FT power spectrum in Figure 6b, in combination with the total amount of the energy relaxation in the excited state, obtained from the first moment of induced absorption. Using the individual contribution divided by the vibrational frequency of itself, the Huang-Rhys factors corresponding to the transition from the lowest to the higher excited states for these six modes can be determined as listed in Table 1.

Table 1 Lifetimes and Huang-Rhys factors of the vibrational modes determined from the first moment.

M1			
Peak Wavenumber [cm^{-1}]	Normalized FFT power	Lifetime τ_{M1vib} [ps]	Huang-Rhys factor
49	0.026	1.67	
103	0.492	1.69	0.25
171	1	1.96	0.15
219	0.17	1.42	
258	0.415	1.78	0.10
313	0.273	2.21	
383	0.765	1.15	0.11
1201	0.071	1.80	0.01
1252	0.028	1.63	
1355	0.016	1.75	
1424	0.013	1.21	
1499	0.089	1.26	0.01
1550	0.011	2.01	

Furthermore, we calculated the electronic dephasing time according to the data in the negative delay time, which are the perturbed free polarization decay terms representing the case when the probe pulse arrives earlier than the pump pulse without temporal overlap between them. The probe pulse generates electronic coherence in the sample with the duration of the electronic dephasing time. Then the intense pump field forms a grating, which interacts with another pump field to be diffracted into the probe direction, satisfying the causality. In the present case, the vibronic coupling expected to be strong in the conjugated electron system is the origin of the electronic spectrum of the ground state. Therefore, the polarization generated by the probe pulse that precedes the pump pulse is a vibronic transition, instead of pure electronic transition, which is associated with the transition between the ground vibrational level in the ground electronic state and the vibronically excited state, which is both excited vibrationally and electronically. The wave-packet formation in the ground state requires two fields of the pump pulse. Therefore, this signal increases with the delay time and the time constant T_2 and disappears quickly at $t=0$ [34].

The decay times of the signal in the negative time range are functions of the probe photon energy, as shown in Figure 7. The apparent lifetimes depend on the contribution of the coherent spike, which reduces the real degasing time. The longest among the observed values is estimated to be the closest to the true value 47±5 fs, around 1.91 eV in the figure. This is a reasonable duration of dephasing in condensed phase materials [35].

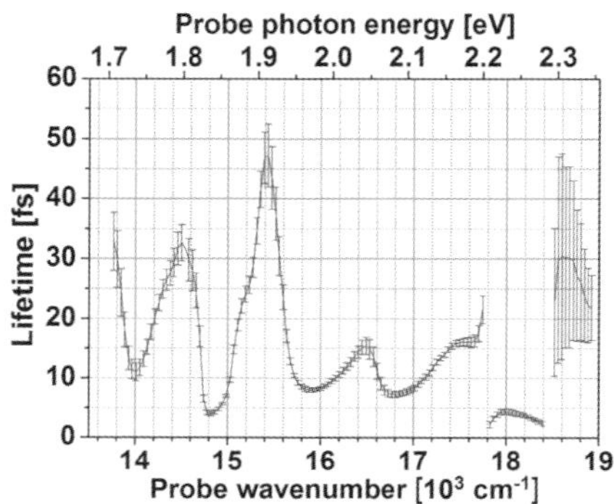

Figure 7. Decay times of the signal in the negative time range as a function of the probe photon energy.

By using the pump-probe data in the negative time range with 7 fs pulses, we have obtained the electronic phase relaxation time and the frequencies of the vibrational modes due to the wave-packet motion in the electronic excited sate. The absorbance change observed in the 'negative' time range has been used to estimate the electronic dephasing time at 47±5 fs. The Huang-Rhys factors corresponding to the transition from the lowest to the higher excited states for these six modes have been determined for the first time.

4.2. *Observation of Breather and Soliton in a Substituted Polythiophene*

Ever since the unusual wave propagation, soliton, was first discovered in 1844 [36], it has been identified in many fields of nonlinear physics [37-42] including water waves, sound waves, matter waves, and electromagnetic waves [43]. According to simulations performed using the Su-Schrieffer-Heeger (SSH) Hamiltonian [44], a photogenerated electron-hole $(e-h)$ pair evolves into a soliton-antisoliton pair $(s\bar{s})$ within 100 fs after photoexcitation due to barrier-free relaxation in a one-dimensional system. Matter-wave solitons have given rise to many interesting phenomena in the simplest conducting polymer, *trans-*

polyacetylene [45], including anomalous conductivity and huge optical nonlinearity [46]. The formation times of solitons in polyacetylene have been determined to be <150 fs [47]. Although the existence of a soliton in *trans*-polyacetylene is well known and has been extensively studied, there has been no systematic study of conjugated polymer systems besides *trans*-polyacetylene. This is probably due to the scarcity of polymers with a degenerate ground state.

It has also been theoretically predicted [48-51] that the soliton pair is spatially localized to form a dynamic bound state called a breather. The excess energy of the photogenerated $(e-h)$ pair over that of the soliton pair induces collective carbon-carbon (C-C) oscillations, namely the breather mode, due to electron-phonon coupling. The breathers predicted in Reference [52] have been observed in *trans*-polyacetylene [53], which was found to have a period of 44 fs and an extremely short lifetime of ~50 fs. However, there is not currently a consensus among researchers as to whether breathers are the primary photogenerated excitations, or how they affect the ultrafast vibronic dynamics [53-57].

We report the dynamics of solitons and breathers in a quinoid–benzenoid polythiophene (PHTDMABQ) with a degenerate ground state, the chemical structure of which is shown in Figure 8a. The degeneracy is due to the tautomerization of the inner structure of the polymer, and its repeat unit (two thiophene rings with both *cis* and *trans* configurations) is much larger than that of *trans*-polyacetylene (one single bond and one double bond). Given that there has been no other spectroscopic study of solitons except for *trans*-polyacetylene, the study of its dynamics and a comparison of any differences between it and *trans*-polyacetylene in terms of the repeat unit structure are of great interest. To the best of our knowledge, this is the first observation of the existence of a breather and solitons and their dynamics in a system other than trans-polyacetylene.

Three dominant vibrational modes, which peaked at 1111 ± 7, 1343 ± 7, and 1465 ± 7 cm^{-1} have been observed (P1, P2, and P3, respectively). Figure 8b shows the calculated dimensionless displacements Δ along the vibrational coordinates of the optimal geometries between the ground and excited states. This immediately enables us to identify them as intra- and inter-thiophene ring C-C and C=C stretching motions (see Figure 8b, top structures). The calculated vibrational frequencies (1345, 1533, and 1675 cm^{-1}) are overestimated by about 200 cm^{-1} compared to the experimental values, which is typical for semi-empirical calculations.

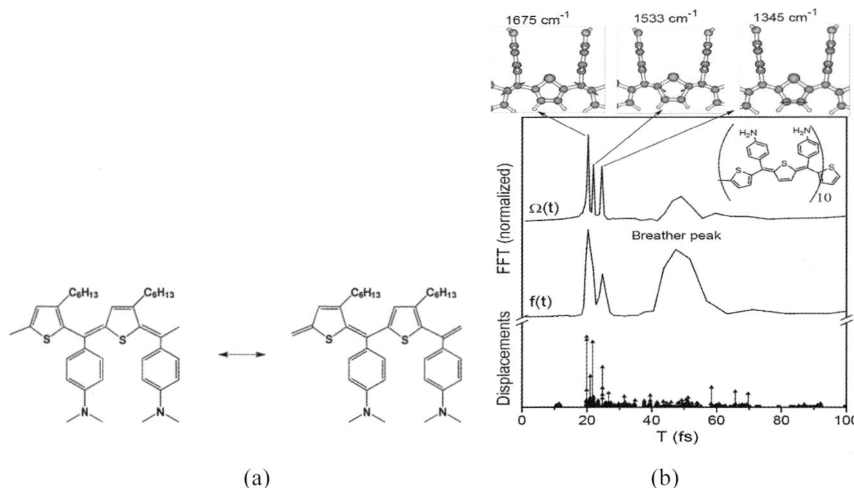

(a)   (b)

Figure 8. Two quantum mechanical resonant structures of (a) PHTDMABQ and (b) the excited state molecular dynamics simulations results. (b) shows the normalized Fourier spectra of the lowest dipolar allowed excited state transition energy $\Omega(t)$ and its respective oscillator strength $f(t)$ trajectories (top two plots), and the amplitudes of dimensionless displacements Δ (stick spectrum, bottom panel) along normal modes calculated in the oligomer with 10 repeat units, as shown in the inset. The three molecular structures at the top schematically show vibrational normal modes with frequencies strongly coupled to the electronic system, which lead to the formation of the breather excitation. These correspond to the C=C vibration and C-C stretches as schematically shown in the middle panel.

The requirement for the existence of solitons is a degenerate ground state structure, and as shown in Figure 8a, our quinoid-benzenoid polythiophene sample has a degenerate ground state because the two quantum-mechanical resonance structures have equivalent energies. The transient absorption spectra exhibit positive absorbance changes with photon energies smaller than 1.91 eV, which is attributed to the tail of the solitons not fully relaxing to the band-gap center, as is the case in trans-polyacetylene [53]. Therefore, the positive value indicates the increased contribution of induced absorption due to a soliton with a peak near the mid-gap, which is estimated to be around 1.4 eV.

Figure 9 shows the time-dependent spectra of the molecular vibrational modes. In addition to the three main modes, P1, P2, and P3, five more peaks appear as sidebands of the main peaks at 270, 500, 640, 1960, and 2200 cm⁻¹. These five modes are breather modes that are not visible in the full-time Fourier power spectrum or in the stationary resonance Raman spectrum because they have extremely broad widths due to their short phase-relaxation times including homogeneous and inhomogeneous broadening mechnisms. We also note that there are no detectable normal modes in this spectral region coupled to the electronic excitation, as evidenced by the lack of significant displacements calculated in Figure 8b. The average frequency difference between the three main bands (1111, 1343, and 1465 cm^{-1}) and their corresponding sidebands (270 and 1960cm^{-1} for 1111cm^{-1}, 500 and 2200cm^{-1} for 1343cm^{-1}, and 640cm^{-1} for 1465 cm^{-1}) is calculated to be 843$\pm$12 cm^{-1}, which is close to the experimentally observed value of about 760 cm^{-1} reported for polyacetylene [53], and theoretically expected values of 660-1000 cm^{-1} [49-52], indicating that the breather modulates the C-C stretching modes with a period of about 40 fs generating the sidebands.

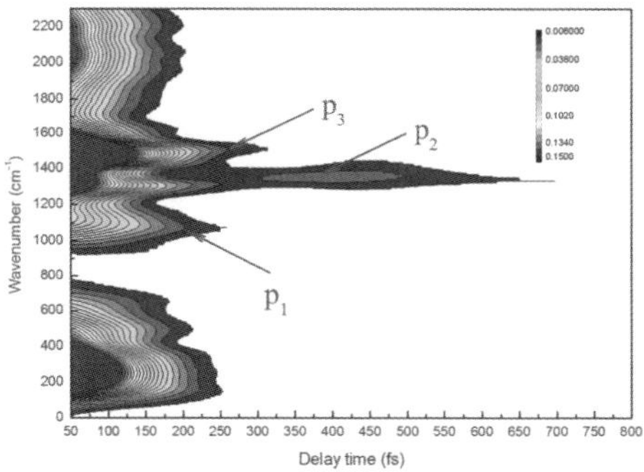

Figure 9. Contour maps of the two-dimensional Fourier power of the vibrational components obtained by spectrogram calculation for real-time data covering 680-690 nm.

The lifetimes of the sidebands were determined to be about 50, 32, and 40 fs for the sidebands of P1, P2, and P3, respectively. These short lifetimes also correspond to the shortest lifetime component of about 62 fs in the ΔA trace [59]. To analyze these sideband peaks from theoretical calculations, we compute the power spectra of 750 fs photoexcited trajectories of the excitation energy $\Omega(q,t)$ and oscillator strength $f(q,t)$ shown in Figure 8b [58-60]. Both plots show an additional broad peak centered around 50 fs, corresponding to a vibration in the range of 550 to 800 cm^{-1} that does not correspond to any of the vibrational normal modes. This peak is assigned to a nonlinear breather excitation that occurs due to the coupling of C-C vibrational motions. This vibrational excitation decays gradually over long time scales due to the dissipation of vibrational energy to internal vibrational degrees of freedom that are weakly coupled to the electronic system [58,60]. The power spectra of longer excited-state trajectories (not shown) demonstrate a diminishing breather peak. Our calculations estimate the breather lifetime to be about 0.2 ps without accounting for intermolecular dissipation channels (baths), which is considered to be in agreement with our experiment data of ~50 fs, in case we consider both intermolecular interactions and phonon energy leakage through the boundaries of conjugated segments (defects). These processes may reduce breather lifetime from 0.2 ps to ~50 fs [60]. As expected, the breather peak is more pronounced in the power spectrum of the oscillator strength (see Figure 9), because it is directly related to the modulation of the respective transition dipole moment [60].

We now discuss the effect of the differences in the sizes and structures of the repeat units. The repeat unit in the *trans*-polyacetylene is composed of one single bond and one double bond, while in PHTDMABQ it is composed of two thiophene rings. The lifetime of the breather in PHTDMABQ is even shorter than that in *trans*-polyacetylene. Since quinoid–benzenoid polythiophene has many more internal vibrational modes than *trans*-polyacetylene, we can conclude that the lifetime is not determined by the separation of soliton pairs from the originally generated site in the polymer chain, but by the energy dissipation to internal vibrational freedom.

In summary, using a polythiophene derivative with a degenerate ground state, we investigated the ultrafast dynamics that take place immediately after excitation. The breather lifetime was experimentally determined from the electronic spectral dynamics to be ~62 fs, and the lifetime of the breather was determined by the energy dissipation to internal vibrational freedom. Although extensive theoretical studies have been conducted, to the best of our knowledge, this is the first time that the modulation of the C-C single and double stretching modes by the breather have been experimentally observed other than in *trans*-polyacetylene.

5. Summary and Future Prospects

The benefits of developing ultrashort laser pulses in various scientific fields is demonstrated by their wide variety of applications, including defining frequency standards, taking distance measurements, and studying the ultrafast dynamics of chemicals, biological materials, and solid-state/condensed matter physics. Using sub-8-fs pulses, we found various new phenomena, such as dynamic mode coupling between molecular vibrational modes [61], a dynamic observation of the Duschinsky rotation [62], and a dynamic intensity borrowing effect [23-25]. Here, we have introduced several typical achievements obtained by ultrafast time-resolved spectroscopy.

First, we discussed the dynamic vibronic couplings in the vibrational real-time spectra of a polyacetylene derivative. Pump-probe spectroscopy was performed for the sample with a few cycle pulses of sub-5 fs duration. We obtained the electronic phase relaxation time and the frequencies of the vibrational modes due to the wave-packet motion in the electronic excited state, and estimated the electronic dephasing time to be 47 ± 5 fs. The Huang-Rhys factors corresponding to the transition from the lowest to the higher excited states for these six modes were also determined for the first time.

Second, we presented the ultrafast dynamics just after excitation in a substituted polythiophene with degenerated ground state. We succeeded in observing how the breather modulates the C-C stretching modes with a period of ~40 fs generating the sidebands for a period of ~40 fs. The breather lifetime was experimentally determined using the electronic spectral dynamics at ~62 fs, and the lifetime of the breather was determined by the energy dissipation to internal vibrational freedom. To the best of our knowledge, this is the first observation of the existence of a breather and solitons and their behavior in systems other than trans-polyacetylene.

Stable ultrashort visible pulses and simultaneous detection in the broadband spectral region enabled us to perform highly reliable time-resolved vibration spectroscopy. This advantage allows us to apply the methodology of ultrafast dynamics study even for materials with measurement constraints, such as biomaterials. As a future prospect of this work, we can extend the application of this ultrafast broadband spectroscopy to the real-time observation of vibration modes in the electronic ground and excited states of other unstudied condensed matter such as molecules, polymers, and biopolymers. In this field, it is thought that there are still many more interesting phenomena that have not yet been observed. For example, observation of the real-time vibration spectrum confirms the transition state during which the chemical reaction proceeds from its initial state to products proceeding via intermediates. Thus, the real-time observation of molecular vibration in the transition state will elucidate the whole pictures of chemical reactions.

Acknowledgments

The authors would like to express thanks to Drs. A. Shirakawa, A. Baltuska, A. Yabushita, Z. Wang, Y. Wang, and Mr. T. Iiyama for their collaboration. This work was partly supported by the National Science Council of the Republic of China, Taiwan (NSC 98-2112-M-009-001-MY3), and a grant from the Ministry of Education, Aiming for Top University (MOE ATU) Program at National Chiao-Tung University (NCTU). A part of this work was performed under the joint research project of the Institute of Laser Engineering, Osaka University under Contract No. A3-01.

References

1. F. Schotte, J. Soman, J. S. Olson, M. Wulff, and P. A. Anfinrud, *J. Struct. Biol.* **147**, 235 (2004).
2. B. J. Siwick, J. R. Dwyer, R. E. Jordan, and R. J. D. Miller, *Science* **302**, 1382 (2003).
3. R. Srinivasan, J. S. Feenstra, S. T. Park, S. Xu, and A. H. Zewail, *Science* **307**, 558 (2005).
4. T. Kobayashi, T. Saito, and H. Ohtani, *Nature* **414**, 531 (2001).
5. A. Shirakawa, I. Sakane, M. Takasaka, and T. Kobayashi, *Appl. Phys. Lett.* **19**, 2268 (1999).
6. A. Baltuska, and T. Kobayashi, *Appl. Phys. B* **75**, 427 (2002).
7. A. Baltuska, T. Fuji, and T. Kobayashi, *Opt. Lett.* **27**, 306 (2002).
8. A. Shirakawa, and T. Kobayashi, *Appl. Phys. Lett.* **72**,147 (1998).
9. A. Shirakawa, I. Sakane, and T. Kobayashi, *Opt. Lett.* **23**,1292 (1998).
10. R. Danielius, A. Piskarskas, P. Di Trapani, A. Andreoni, C. Solcia, and P. Foggi, *Opt. Lett.* **21**, 973 (1996).
11. P. Kukura, D. W. McCamant, R. A. Mathies Ann, *Rev. Phys. Chem.* **58**, 461-488 (2007)
12. D. W. McCamant, P. Kukura, and R. A. Mathies, *Appl. Spectrosc.* **57**, 1317 (2003).
13. Time-Resolved Vibrational Spectroscopy, Springer Proceedings in Physics Vol. 4, edited by A. Laubereau and M. Stockburger, Springer-Verlag, Berlin, 1985.
14. Time-Resolved Vibrational Spectroscopy V, Springer Proceedings in Physics Vol. 68, edited by H. Takahashi, Springer-Verlag, Berlin, 1991.
15. Time-Resolved Vibrational Spectroscopy VI, Springer Proceedings in Physics Vol. 74, edited by A. Lau, F. Siebert, and W.Werncke, Springer-Verlag, Berlin, 1993.
16. J. Du, T. Kobayashi, *Chem. Phys. Lett.* **481**, 204 (2009).
17. I. Iwakura, A. Yabushita, T. Kobayashi, *J. Am. Chem. Soc.* **131**, 688 (2009).
18. A. Yabushita, T. Kobayashi, *Biophys. J.* **96**, 1447 (2009).
19. T. Kobayashi, Z. Wang, and T. Otsubo, *J. Phys. Chem.* **A 111**, 12985 (2007).
20. J. B. Renucci, W. Richter, M. Cardona, and E. Schöstherr, *Phys. Status Solidi* **b 60**, 299 (1973).
21. A. T. N. Kumar, F. Rosca, A. Widom, and P. M. Champion, *J. Chem. Phys.* **114**, 701 (2001).
22. N. Ishii, E. Tokunaga, S. Adachi, T. Kimura, H. Matsuda, and T. Kobayashi, *Phys. Rev.* **A 70**, 023811 (2004).
23. H. Kano, T. Saito, and T. Kobayashi, *J. Phys. Chem.* **A 106**, 3445 (2002).
24. H. Kano, T. Saito, and T. Kobayashi, *J. Phys. Chem.* **B 105**, 413 (2001).
25. H. Kano, and T. Kobayashi, *J. Chem. Phys.* **116**, 184 (2002).

26. T. Kobayashi (Ed.), Relaxation in Polymers, World Scientific, Singapore, 1993.

27. A.J. Heeger, S. Kivelson, J.R. Schrieffer, and W.P. Su, *Rev. Mod. Phys.* **60**, 781 (1988).

28. T. Kobayashi,M. Yoshizawa, U. Stamm, M. Taiji, and M. Hasegawa, *J. Opt. Soc. Am. B* **7**, 1558 (1990).

29. T. Kobayashi, M. Yoshizawa, T. Masuda, and T. Higashimura, *IEEE J. Quantum Electron.* **28**, 2508 (1992).

30. T. Kobayashi, T. Teramoto, V.M. Kobryanskii, and T. Taneichi, *Synth. Met.* **159**, 1751 (2009).

31. J. Du, and T. Kobayashi, *Chem. Phys. Lett.* **481**, 204 (2009).

32. T. Kobayashi, A. Shirakawa, H. Matsuzawa, and H. Nakanishi, *Chem. Phys. Lett.* **321**, 385 (2000).

33. T. Kobayashi, M. Hirasawa, Y. Sakazaki, H. Hane, *Chem. Phys. Lett.* **400**, 301 (2004).

34. C.H. Brito Cruz, J.P. Gordon, P.C. Becker, R.L. Fork, and C.V. Shank, *IEEE J. Quantum Electron.* **24**, 261 (1988).

35. T. Kobayashi, J. Du, W. Feng, K. Yoshino, *Phys. Rev. Lett.* **101**, 037402 (2008).

36. J. S. Russell, in Report of the Fourteenth Meeting of the British Association for the Advancement of Science, 311 (Murray, London, 1844).

37. N. Manton, and P. Sutcliffe, Topological Solitons (Cambridge Univ. Press, Cambridge, 2004).

38. L. Khaykovich, F. Schreck, G. Ferrari, T. Bourdel, J. Cubizolles, L. D. Carr, Y. Castin, C. Salomon, *Science* **296**, 1290 (2002).

39. K. E. Strecker, G. B. Partridge, A. G. Truscott, and R. G. Hulet, *Nature* **417**, 150 (2002).

40. M. Nakazawa, E. Yamada, and H. Kubota, *Phys. Rev. Lett.* **66**, 2625 (1991).

41. W. P. Su, J. R. Schrieffer, and A. J. Heeger, *Phys. Rev. Lett.* **42**, 1698 (1979).

42. A. M. Kosevich, B. A. Ivanov, and A. S. Kovalev, *Phys. Rep.* **194**, 117 (1990).

43. G. I. Stegeman, and M. Segev, *Science* **286**, 1518 (1999).

44. W. P. Su, and J. R. Schrieffer, *Proc. Natl. Acad. Sci.* **USA 77**, 5626 (1980).

45. H. Shirakawa, *Rev. Mod. Phys.* **73**, 713 (2001).

46. H. Naarmann, and N. Theophilou, *Synth. Met.* **22**, 1 (1987).

47. C. V. Shank, R. Yen, R. L. Fork, J. Orenstein, and G. L. Baker, *Phys. Rev. Lett.* **49**, 1660 (1982).

48. D. K. Campbell, *Nature* **432**, 455 (2004).

49. M. Sasai, and H. Fukutome, *Prog. Theor. Phys.* **79**, 61 (1988).

50. S. R. Phillpot, A. R. Bishop, and B. Horovitz, *Phys. Rev.* **B 40**, 1839 (1989).

51. S. Block, and H.W. Streitwolf, *J. Phys. Condens. Matter* **8**, 889 (1996).

52. A. R. Bishop, D. K. Campbell, P. S. Lomdahl, B. Horovitz, and S. R. Phillpot, *Phys. Rev. Lett.* **52**, 671 (1984).

53. S. Adachi, V. M. Kobryanskii, and T. Kobayashi, *Phys. Rev. Lett.* **89**, 027401 (2002).

54. G. S. Kanner, Z.V. Vardeny, G. Lanzani, and L. X. Zheng, *Synth. Met.* **116**, 71 (2001).

55. K. M. Gaab, and C. J. Bardeen, *J. Phys. Chem.* **B 108**, 4619 (2004).

56. G. Lanzani, G. Cerullo, C. Brabec, and N.S. Sariciftci, *Phys. Rev. Lett.* **90**, 047402 (2003).

57. R. Lécuiller, J. Berréhar, C. Lapersonne-Meyer, and M. Schott, *Phys. Rev. Lett.* **80**, 4068 (1998).

58. S. Tretiak, A. Saxena, R. L. Martin, and A. R. Bishop, *Proc. Natl. Acad. Sci. USA* **100**, 2185 (2003).

59. J. Du, Z. Wang, W. Feng, K. Yoshino, and T. Kobayashi, *Phys. Rev.* **B 77**, 195205 (2008).

60. S. Tretiak, A. Piryatinski, A. Saxena, R. L. Martin, and A. R. Bishop, *Phys. Rev.* **B 70**, 233203 (2004).

61. T. Saito, and T. Kobayashi, *J. Phys. Chem.* **A**, **106**, 9436 (2002).

62. T. Fuji, T. Saito, and T. Kobayashi, *Chem. Phys. Lett.*, **332**, 324 (2000).

MULTIDIMENSIONAL OPTICAL SPECTROSCOPY USING A PUMP-PROBE CONFIGURATION: SOME IMPLEMENTATION DETAILS

ZHENGYANG ZHANG and HOWE-SIANG TAN
Division of Chemistry and Biological Chemistry,
School of Physical and Mathematical Sciences,
Nanyang Technological University, 21 Nanyang Link,
Singapore 637371

We describe the experimental implementation of 2D and 3D optical spectroscopy using a pulse shaper assisted pump-probe configuration. We use a conventional pump-probe transient absorption spectroscopy experiment as a basis for our description. Details pertaining to hardware and computer control procedures are provided.

1. Introduction

There are several ways of experimentally implementing two-dimensional and three-dimensional optical spectroscopy experiments (2DOS and 3DOS, respectively) [1]. They can be classified as non-collinear [2], collinear [3] and partially collinear [4] beam geometry configurations.

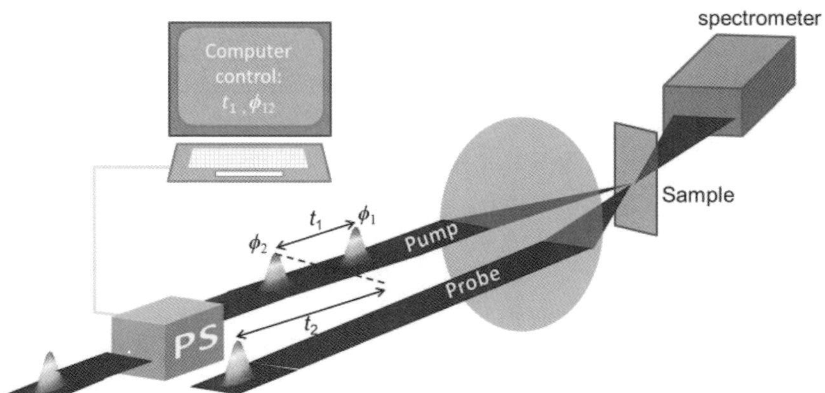

Figure 1. Schematic of our pump-probe configuration 2D optical spectrometer. PS, Pulse Shaper

In this article we focus on describing the pump-probe configuration [1] method, an example of a partially collinear configuration. The schematic of a third-order 2DOS setup is depicted in Fig. 1. Briefly, the first two excitation pulses are collinear, while the third excitation pulse interacts non-collinearly with the sample. As the emitted signal is in the same direction as the third excitation pulse, the third pulse acts also as a local oscillator that heterodyne-detects the signal. The first two collinear pulses are termed "pump" pulse group while the third pulse is termed the 'probe' pulse. The pump pulse group can typically be created using a pulse shaper or other related methods [5], with controllable interpulse phase $\phi_{12} = \phi_1 - \phi_2$ and interpulse delay t_1. Using a mechanical translational stage, the probe pulse arrives at a delay t_2 after the pump pulse group. This t_2 delay can be viewed as analogous to the pump-probe delay in a typical transient absorption experiment. Terms commonly used for t_2 are waiting time T_w and population time T. The data collection is akin to collecting a conventional pump-probe transient absorption signal, the transient absorption signal, $S^{(3)}$ however is now dependent on the parameters of the pump pulse group,

$$S^{(3)}(\phi_{12}, t_1, t_2, \omega_3) \tag{1}$$

To obtain a 2D spectrum at waiting time t_2, $S^{(3)}$ is collected, scanning through t_1. Afterwhich, a process termed phase cycling is then applied before a Fourier transform is applied about t_1. Phase cycling is required in order to obtain the required signal that is free from spurious signals. In phase cycling, the desired 2D signals can be retrieved by the weighted summation of data collected from using different interpulse phases ϕ_{12}. Phase cycling can be easily implemented using a pulse shaper to select different pathways by systematically changing the relative interpulse phase, which is cycled over 2π radians in a number of equally spaced steps. We have shown that to get the rephasing, non-rephasing and purely absorptive 2D signals, a 3×1 phase cycling scheme is needed [6]. For example, using the phase cycling schemes

$$S_R^{1 \times 3}(t_1, t_2, \omega_3) \propto S^{(3)}\left(0, t_1, t_2, \omega_3\right) + S^{(3)}\left(\tfrac{-2\pi}{3}, t_1, t_2, \omega_3\right)e^{i2\pi/3}$$
$$+ S^{(3)}\left(\tfrac{-4\pi}{3}, t_1, t_2, \omega_3\right)e^{i4\pi/3} \tag{2}$$

or

$$S_{NR}^{1 \times 3}(t_1, t_2, \omega_3) \propto S^{(3)}\left(0, t_1, t_2, \omega_3\right) + S^{(3)}\left(\tfrac{-2\pi}{3}, t_1, t_2, \omega_3\right)e^{-i2\pi/3}$$
$$+ S^{(3)}\left(\tfrac{-4\pi}{3}, t_1, t_2, \omega_3\right)e^{-i4\pi/3} \tag{3}$$

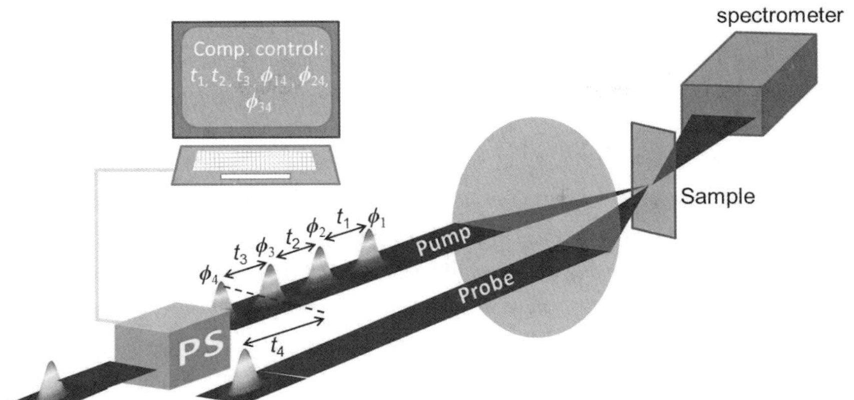

Figure 2. Schematic of our pump-probe configuration 3D optical spectrometer. PS, Pulse Shaper

purely absorptive 2D spectra at t_2, $S(\omega_1, t_2, \omega_3)$ can be obtained after Fourier transform about t_1 of either $S_R^{1\times3}$ or $S_{NR}^{1\times3}$. If we further ensure causality (Refer to [6] for full explanation), a rephasing and non-rephasing 2D spectrum can be obtained from Eq. (1) and Eq. (2), respectively.

This pump-probe configuration can be used without any change in the physical setup to perform fifth-order 3DOS. The schematic is shown in Fig. 2. The first four collinear pulses forms the 'pump' group and are produced by a pulse shaper with controllable interpulse phases ϕ_{14}, ϕ_{24}, and ϕ_{34} and interpulse delays t_1, t_2 and t_3. Using a mechanical translational stage, the probe pulse comes at a delay t_4 after the pump pulse group. Analogous to the 2DOS scheme presented above, this t_4 delay can be view as analogous to the pump-probe delay in a typical transient absorption experiment. In a general fifth-order 3DOS, there is two waiting/population times t_2 and t_4. The transient absorption signal $S^{(5)}$ obtained as a function of the parameters of the pump pulse group

$$S^{(5)}(\omega_5; t_1, t_3, \phi_{14}, \phi_{24}, \phi_{34}, t_2, t_4) \tag{4}$$

As in 2DOS, phase cycling is necessary to obtain meaningful 3D spectra. For 3DOS, it was shown that a $3\times3\times3\times1$ phase cycling scheme where the first three pulses are each cycled over phases ϕ_{14}, ϕ_{24}, and $\phi_{34} = 0$, $2\pi/3$ and $4\pi/3$ is most useful. The full detail of these phase cycling schemes are discussed in ref [7].

In the next section, we present in detail the experimental setup and computer programming procedure to implement the phase cycling schemes and to acquire 2D and 3D spectra.

2. Some Experimental Details

From the experimental schematic of Fig. 1 and Fig. 2, one can see that the difference between our 2D/3DOS setup with a conventional pump-probe transient absorption is a pulse shaper inserted in the path of the pump pulse. We therefore only concentrate our description of the implementation of the control and data acquisition procedure of the pulse shaper, and spectrometer. The experimental setup for the multidimensional optical spectrometer is as follows.

The pulse shaper used in this experiment is an acousto-optic programmable dispersive filter (AOPDF) [8] (Dazzler, Fastlite) . Briefly, the pulse shaper sends shaped acoustic waves into a TeO_2 crystal where the incoming light will be shaped at different parts of the crystal. The phase-matched shaped pulses will exit with an orthogonal polarization, and at a small diffracted angle, to the incoming light field. By customizing the acoustic waveforms, we can obtain the desired shaped pulses that are required for performing multidimensional optical spectroscopy. The waveform generator can store 128 shaped waveforms on its random-access memory (RAM), and 1024 shaped waveforms in its flash memory. We shall first describe the procedures for loading the waveforms for 2D phase cycling follow by 3D phase cycling.

The relative phases required for a 3×1 three-step phase cycling scheme in 2DOS, as shown in Fig. 1, are 0, $2\pi/3$ and $4\pi/3$. For every increment of the interpulse delay step size Δt_1, the waveforms to generate a two pulse train with the respective three relative phases are stored in the AOPDF. The storage sequence will be $\left[(0, 2\pi/3, 4\pi/3; n\Delta t_1), n = \{0, 1, 2, ..., m-1\} \right]$, where m is the total number of steps taken.

To decide Δt_1 for the first coherence time t_1, we have to consider the Nyquist sampling limit of the experiment. In principle, the Nyquist sampling limit is half the period of an oscillation. The Nyquist limit can be varied by changing the frame of reference of the oscillation. In a collinear setup using an interferometer, the reference frequency between two pulses is at 0 Hz. This means the oscillations of the coherence, or free induction decay (FID), decays with a high frequency and many data points are needed to sample the full decay. To reduce the number of data points needed, a partial rotating frame is chosen by shifting the reference frequency toward the absorbance frequency. For instance, to study the Q_y transition of chlorophyll a (Chl a) in methanol at 451

THz, the reference frequency is set at 420 THz, and the Nyquist limit is thus 16 fs. As the FID of the coherence signal at 451 THz completely decays at approximately 150 fs, 23 data points are needed to sample the delay t_1 from 0 to 154 fs in step sizes of 7 fs.

With a 3×1 phase cycling scheme, in total 69 waveforms are stored in the waveform generator of the AOPDF. Each waveform is loaded twice consecutively. With the same pump waveform, by collecting data consecutively with and without the probe light, scattered light can easily be subtracted from the collected signals. The optical arrangement will be described later. The AOPDF is triggered by the laser, and it changes every waveform at a repetition rate of 1 kHz. The waveforms are repeated in a loop until the end of the experiment.

The AOPDF has two electronic outputs that generate TTL signals, labeled as S1 and S2. The output S1 generates a TTL signal when the first stored waveform is loaded, and it triggers a data acquisition device (NI USB-6212 BNC, National Instruments) to initiate data collection from the first waveform. For the Chl a experiment described earlier, the TTL signal repetition rate generated from S1 is at 7.2 Hz. The second output S2 generates TTL signals at the rate where different waveforms are changed. Due to consecutive loading of identical waveforms as mentioned earlier, signals from S2 has a rate of 500 Hz. An optical chopper is connected to S2, and chops the probe light at 500 Hz.

The close-circuit detector (CCD) camera (PIXIS-100B, Princeton Instruments), with an active pixel dimension of 500×100, is connected to the spectrometer (Princeton Instruments, Acton-SP2500), and triggered by the laser at 1 kHz. The CCD camera is split into two sections of 500×50, where one section detects the reference beam and the other detects the probe beam. The reference beam is split from a small portion of the probe beam, and does not interact with the pump light. The probe light fluctuation can be reduced by normalizing the probe light to the reference light, hence reducing noise in the signal.

To obtain the signal for a certain population time t_2, the motorized translational stage is delayed accordingly. During data collection, the camera collects and stores 20 sets of signals in a single run on the RAM of the computer, before transferring the data to the hard-disk memory. For better signal-to-noise ratio, typically 1000 or more data sets are collected and averaged.

All data processing is done with a home-written MATLAB code. Briefly, scattered light is subtracted from the signal by taking the difference between the un-chopped and chopped probe light. The signal is averaged, and normalized with the reference light to remove probe light fluctuation. By treating the data with the mathematical operations required for a 3×1 phase cycling scheme [6, 9, 10], spurious signals are removed from the data. The data set is padded up to twice the number of delay step size m, and Fourier transformed along delay t_1, to retrieve well-resolved features in a 2D spectrum [1].

For 3DOS, the data collection and processing is similar to 2DOS. A four pulse train is used for 3DOS as shown in Fig. 2, and a $3 \times 3 \times 3 \times 1$ 27-step phase cycling scheme is required to remove spurious signals. The pulse shaper scans the delays t_1, and t_3, and also the relative phases ϕ_{14}, ϕ_{24}, and ϕ_{34}. For every increase in delay step sizes Δt_1 and Δt_3, the waveforms to create a four pulse train with relative phases from the 27-step phase cycling scheme are saved. The waveform sequences are stored where the full delay t_1 is scanned before an increment of Δt_3. The data is collected in a partially rotating frame, where the reference frequencies for both the coherence decay in t_1 and t_3, set by the pulse shaper, are the same. Both Δt_1 and Δt_3 have the same step size, determined by the Nyquist sampling limit. The population time t_2 is controlled by the pulse shaper, while population time t_5 is set by a mechanical translational stage.

Apart from the computer programming, the physical experimental setup and the electronics triggering are the same for both 2D/3DOS. Due to specific operating procedures of the pulse shaper software, the shaped waveforms for phase cycling in 3DOS can only be saved in the RAM of the AOPDF. As the total number of waveforms required for phase cycling in 3DOS is much greater than the RAM of the AOPDF, the sequence of waveform is broken down into block sizes equal or smaller than 128 waveforms. The collected data parts are combined and processed with a home-written MATLAB code. The MATLAB data processing code for 3DOS is similar to 2DOS, and the only difference to obtain a 3D spectrum is the required $3 \times 3 \times 3 \times 1$ phase cycling mathematical data treatment and a Fourier transform along an additional delay t_3.

3. Conclusions

We have provided a description of our 2D/3DOS spectrometer based on a pulse shaping assisted pump-probe configuration. We explain it in a way that will help a researcher who wishes to upgrade an existing pump-probe spectroscopic setup to a 2D/3D optical spectrometer.

Acknowledgments

This work is supported by the Singapore National Research Foundation (NRF-CRP5-2009-04) and the Singapore Agency for Science, Technology and Research Science and Engineering, (A*STAR SERC grant no. 102-149-0153). Z.Z. thanks the Nanyang President's Graduate Scholarship for support.

References

1. P. Hamm, and M. T. Zanni, *Concepts and Methods of 2D Infrared Spectroscopy* (Cambridge University Press, 2011).
2. M. Khalil, N. Demirdöven, and A. Tokmakoff, J. Phys. Chem. A **107**, 5258-5279 (2003).
3. P. Tian, D. Keusters, Y. Suzaki, and W. S. Warren, Science **300**, 1553-1555 (2003).
4. S. H. Shim, and M. T. Zanni, Phys. Chem. Chem. Phys. **11**, 748-761 (2009).
5. D. Brida, C. Manzoni, and G. Cerullo, Opt. Lett. **37**, 3027-3029 (2012).
6. Z. Zhang, K. L. Wells, E. W. J. Hyland, and H.-S. Tan, Chem. Phys. Lett. **550**, 156-161 (2012).
7. Z. Zhang, K. L. Wells, M. T. Seidel, and H. S. Tan, J. Phys. Chem. B (2013).
8. P. Tournois, Opt. Commun. **140**, 245-249 (1997).
9. H.-S. Tan, J. Chem. Phys. **129**, 124501 (2008).
10. S. Yan, and H. S. Tan, Chem. Phys. **360**, 110-115 (2009).

HIGH-SENSITIVITY ULTRAFAST TRANSIENT ABSORPTION SPECTROSCOPY OF ORGANIC PHOTOVOLTAIC DEVICES

ALEX J. BARKER, KAI CHEN, SHYAMAL K.K. PRASAD and JUSTIN M. HODGKISS

School of Chemical and Physical Sciences, Victoria University of Wellington

and

MacDiarmid Institute for Advanced Materials and Nanotechnology
Wellington, New Zealand

The design of effective materials for efficient organic photovoltaic cells requires developing a detailed photophysical model of the processes that link photon absorption to photocurrent collection. Transient absorption spectroscopy offers the potential to do so, but its value depends on the ability to carry out measurements with sensitivity of better than 10^{-5} and from femtosecond to microsecond timescales. In this article, we describe the transient absorption spectroscopy tools that we have developed specifically for probing charge photogeneration and recombination in organic photovoltaic cells, and we illustrate their implementation with several examples.

1. Introduction

The past decade has seen significant growth in our understanding of the rich photophysics of organic semiconductor films and a corresponding improvement in organic photovoltaic (OPV) cell power conversion efficiencies.[1,2] The frontiers of our understanding have been continuously expanded by improvements in spectroscopic methods; time-resolved spectroscopy can now be carried out with sufficient sensitivity and time resolution to comprehensively map out kinetic models connecting species on vastly disparate time scales,[3] and new types of multiple pulse measurements have successfully identified extremely short-lived states that play a key role in photocurrent generation.[4]

Transient absorption (TA) spectroscopy is ideally suited to probing the physics of photocurrent generation in OPVs. The variety of different optical excitations involved (e.g., singlet and triplet excitons, charge-transfer states, and polarons) all present spectroscopic signatures throughout the visible-infrared regions as well as sensitivity to their environment (e.g., polymer morphology, temperature). As depicted in Figure 1, transitions between these states can occur on time-scales ranging from tens of femtoseconds for charge photogeneration in polymer-fullerene OPVs[5] to microseconds and beyond for charge recombination.[3] Moreover, since free charges can recombine bimolecularly, the recombination dynamics are dependent on excitation density. Likewise, excitations existing earlier in the photophysical cascade are also subject to bimolecular reactions at sufficiently high excitation densities, including exciton-exciton annihilation or exciton-charge annihilation.[6] The importance of bimolecular reactions motivates the need to employ excitation intensity dependence as a standard tool in OPV spectroscopy and also constrains the excitation intensity that should be used to resolve inherent monomolecular processes. For example, an excitation fluence on the order of only 1 μJ cm^{-2} produces an initial excitation density of only 10^{17} cm^{-3} for a typical OPV film on the order of 100 nm thick and with an absorption coefficient of 2×10^4 cm^{-1}. This excitation density is on the upper end of the steady-state excitation density expected for an OPV device operating under solar illumination,[2,3] yet the TA signal is expected to be only 2×10^{-4}, assuming TA cross-sections on the order of 2×10^{-16} cm^2. This calculation, and similar consideration of faster bimolecular exciton annihilation reactions, immediately dictates the need for TA sensitivity better than 10^{-5} when studying photocurrent generation in OPV films.

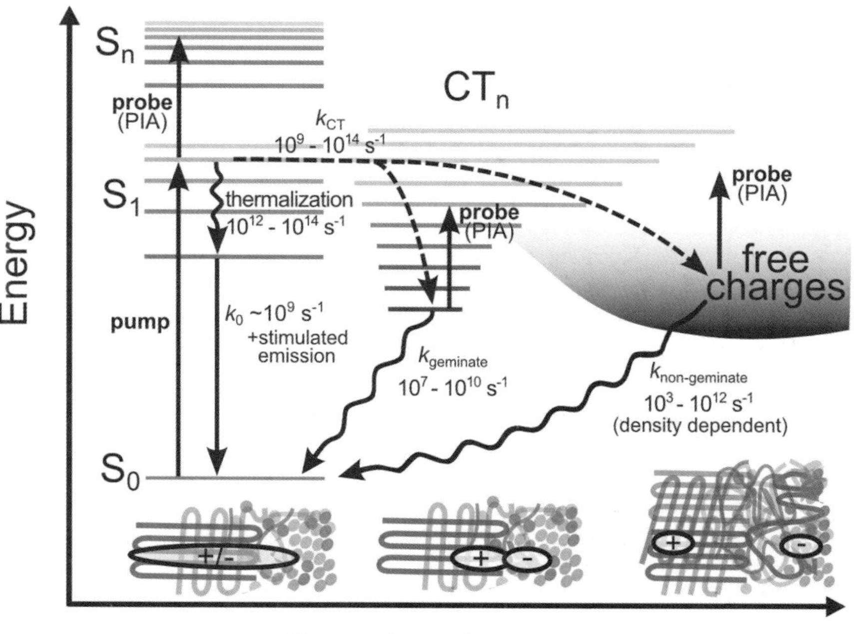

Figure 1. Photophysical scheme showing the relevant processes and characteristic timescales for photocurrent generation in OPV films.

In principal, sensitivity can easily be improved simply by extending data acquisition times. However, more elegant and efficient means of achieving high sensitivity become extremely valuable for high throughput measurements that resolve the dependence of dynamics on a wide range of initial conditions like temperature and excitation intensity. As we will demonstrate, a further benefit of high sensitivity is the ability to resolve the TA spectra of secondary excitations generated by re-exciting an already small population of transient states. An important application of these experiments is probing the short-lived non-thermalized charge-transfer states that are found to precede the formation of free charge carriers.

With those challenges in mind, the purpose of this article is to give an overview of the apparatus and methods that we employ to acquire extremely high sensitivity TA data capable of resolving processes over femtosecond to microsecond timescales, and to present several case studies from our investigations of photocurrent generation in OPVs in which the benefits of high sensitivity and broad time resolution are highlighted.

2. Experimental Details

TA spectroscopy employs pairs of pump (excitation) and probe pulses that overlap in the sample. The time- and wavelength-dependent changes in the transmission of the probe pulse induced by the excitation pulse are extracted from the ground state transmission spectrum by modulating the excitation (usually at $\omega/2$, where ω is the probe pulse repetition rate) and calculating the differential transmission;

$$\frac{\Delta T}{T}(\lambda, t) = \frac{T_{pump\,on}(\lambda,t) - T_{pump\,off}(\lambda)}{T_{pump\,off}(\lambda)} \tag{1}$$

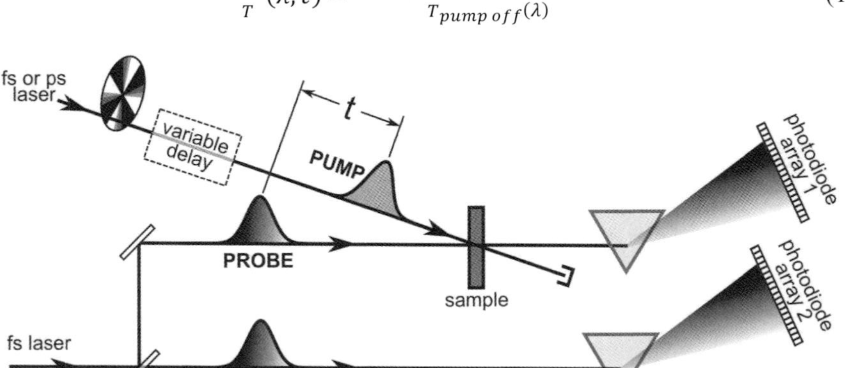

Figure 2. The main components of a typical pump-probe transient absorption spectrometer.

where T is time- and wavelength-dependent probe pulse transmission. In the small signal limit ($\Delta T/T \ll 1$), $\Delta T/T$ is proportional to both excitation density and sample thickness.[7,8] Given that the TA signal is derived from the intensity differences between pairs sequential probe pulses, it is clear that the sensitivity towards real pump induced signals depends on the ability to minimize the effect of other fluctuations between sequential probe pulses.[9]

2.1. *Excitation light sources and pulse timing*

Our TA spectrometer is based on a commercial Ti:Sapphire amplifier[a] that generates 100 fs, 1.3 mJ pulses at a repetition rate of 3 kHz. This relatively high peak pulse energy allows us to drive several non-linear processes simultaneously, specifically two optical parametric amplifiers which function as tunable light sources for excitation and probing of the sample. The choice of a 3 kHz repetition rate over the more common 1 kHz systems allows us to collect a larger number of datapoints in a given time, increasing sample throughput and/or effective sensitivity *via* averaging.

Our primary source of excitation light is a commercial travelling wave optical parametric amplifier[b] (TOPAS) capable of producing narrowband (<1.3 times transform-limit) 100 fs pulses over a tunable range from 1150 – 2600 nm. Frequency doubling and mixing stages at the TOPAS output extend this capability, allowing us to access a total range of excitation wavelengths spanning 240 – 2600 nm.

The delay of the pump pulse relative to the probe pulse is varied using a broadband retroreflector mounted on an automated linear translation stage. As the positioning repeatability of the stage equates to an uncertainty in delay of less than 10 fs, the time resolution of our system is limited only by the pulse width of our pump and probe pulses.

As delays longer than ~3.5 ns exceed the range of our mechanical delay stage, we excite using the second harmonic of an electronically triggered Q-switched laser (0.7 ns pulse width[c]). This laser is synchronized with the femtosecond laser via an electronic delay generator[d] triggered from the previous femtosecond pulse to collect data from 1 ns to arbitrarily long pump-probe delays (although for delays longer than 300 μs this requires operating the amplifier at a reduced repetition rate). The combination of these two pump-

---

[a] Spectra-Physics Spitfire Pro, with Mai Tai SP oscillator and Empower pump
[b] TOPAS-C, Light Conversion Ltd
[c] Advanced Optical Technology Ltd. AOT-YVO-25QSP
[d] Stanford Research Systems DG535

probe delay configurations spans the femtosecond to millisecond time ranges of interest to OPVs with the same detection system and without any time gaps.[3,10,11]

The choice of pump-probe delays to be used for a given experiment is based on the need to capture all photophysical processes under investigation with sufficient time resolution, while minimizing the time required to run the experiment. Photophysical processes in OPVs are found to span a wide range of timescales, therefore we typically employ a logarithmic distribution of time points to generate data that can be fitted with an equal weighting on each time scale.[9] The main exception to this is for regions where significant probe chirp exists, where linearly spaced time points aid in the implementation of a chirp-correction algorithm (see Section 3.1).

2.2. *Probe light sources*

To generate broadband visible probe pulses, we employ a homebuilt noncollinear optical parametric amplifier (NOPA).[12] The visible NOPA provides intense and stable light from 520–760 nm. If the experiment requires broader and higher energy probe spectra, we generate unamplified white light supercontinuum as far as 300 nm in a thin CaF_2 window.[9] The ability to probe at near-infrared (NIR) energies is especially important as low-bandgap polymers that absorb into the NIR emerge as promising materials for widespread OPV implementation. We currently generate NIR supercontinuum in a 3 mm sapphire window using an intense 1200 nm pulse from the TOPAS. However these unamplified light sources typically result in greater noise levels due to inherently lower shot-to-shot stability than the NOPA, and to an inability to fill the photodiode bins in our detector to an optimal level. We note that variations on our visible NOPA are able to generate more stable and intense light in the NIR region,[12] at the expense of bandwidth and tunability.

2.3. *Light detection*

The effect of probe pulse fluctuations can be minimized by employing a reference channel to monitor probe intensities on a shot-to-shot basis. Prior to interaction with the sample, the probe beam is split using a broadband beamsplitter. The reflected portion is sent through the sample, overlapping with the pump beam, while the transmitted portion is used as a reference beam.[8] After passing through the sample, the vertically displaced probe and reference beams are dispersed by a short focal length spectrometer that is selected to achieve high throughput while maintaining sufficient resolution for the broad electronic transitions observed in OPV materials. We use a fibre-coupled grating based spectrometer[e] when probing with the NOPA. A free-space coupled prism spectrometer is used when probing with unamplified light, as the higher throughput (especially at shorter wavelengths) and ability to adjust the dispersion to optimize light intensity per pixel minimizes digital readout noise as a result of low light intensity.

The probe and reference spectra are recorded using a pair of 256-pixel linear silicon photodiode arrays[f], which are read out at the laser repetition rate of 3 kHz. We can optionally switch to a linear InGaAs array for probing NIR energies. Our analogue-to-digital converter has four additional analog inputs, which are used to monitor pulse intensities via additional single channel photodetectors. The speed of our read out system enables shot-to-shot detection of complete probe and reference spectra – a capability that exceeds most CCD cameras and opens the door to several noise reduction methods described below.

2.4. *System automation*

The brain of our transient absorption spectrometer is a modular system of LabVIEW virtual instruments that communicate with each other *via* National Instruments' proprietary Datasocket Server protocol. The tasks performed by each module are depicted in Figure 3.

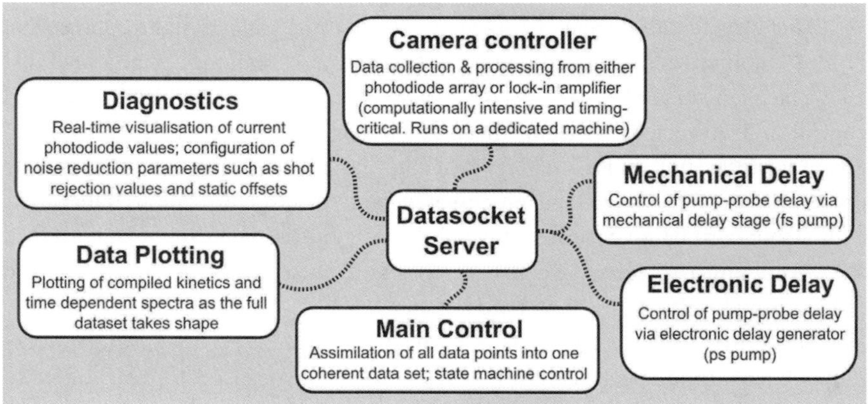

Figure 3. Schematic showing automation software connections.

The goal of our data acquisition system is to provide shot-to-shot detection at 3 kHz with as high a duty cycle as possible, i.e., minimum processing time

---

[e] Princeton Instruments Acton SP150
[f] Entwicklungsbuero Stresing

between data collection cycles to capture the maximum amount of data in a given time. It is crucial that the software is able to keep up with this rapid rate of data collection, as we have found that digital noise can be introduced to the data as a result of timing and memory problems. To ensure optimum performance, the LabVIEW code uses standard "best practice" measures such as memory preallocation, an event-driven architecture, and appropriate prioritization of resources within the operating system. As the separate LabVIEW modules can communicate *via* Ethernet using the Datasocket Server, we dedicate one computer to the main data collection and processing role, and a second computer is used for all other roles such as user interface and data presentation.

2.4.1. *Noise reduction during data collection*

The most significant source of noise in our system is due to wavelength-dependent fluctuations in the intensity of our pump and probe light. Non-linear processes such as those taking place in our NOPA and TOPAS are extremely sensitive to beam alignment and mode quality, as well as laboratory temperature and humidity. Therefore, in addition to implementing tight environmental stability controls, we enclose beam paths prior to non-linear processes to minimize unnecessary fluctuations due to air currents and particles passing through the beam.

The wavelength-dependent intensity fluctuations in our light sources are typically comprised of several components. Slower, 'breathing' oscillations that take place on the order of seconds can be minimized through environmental control and corrected for using the methods described below. The remaining noise is fast shot noise, inherent to the laser itself and an unavoidable consequence of the quantized processes central to laser operation. This noise exhibits no correlation between sequential shots, and as such cannot be corrected for. Averaging is our best weapon against shot noise, with the signal-to-noise ratio scaling by $\sqrt{N}$ with our sample number N.

Several thousand probe shots are typically collected per pump-probe delay point. Each probe spectrum is divided by the reference spectrum in order to correct for shot-to-shot variations in probe shape.[8] An external photodiode collects scattered pump light, allowing us to correlate fluctuations in pump intensity with each collected spectrum. Additionally, the pump photodiode acts as a trigger for the $\Delta T/T$ calculation, labeling spectra "pump on" or "pump off". Each batch of "pump on" and "pump off" spectra are separately averaged and then used to calculate $\Delta T/T$ as in Equation 1. This approach is less susceptible to noise from occasional large probe fluctuations than if $\Delta T/T$ values are calculated individually prior to averaging.

As the signal from the pump photodiode provides us with a measure of relative pump intensity for each collected spectrum, we are able to define a window of acceptable intensities for the experiment. Spectra taken when the pump intensity is outside the window are rejected and do not contribute to the averaged spectrum for each time point. This is particularly valuable when pumping with the TOPAS, which is generally stable (within specification) but exhibits occasional slow fluctuations that persist for longer than the several thousand shots comprising each time point and are therefore not effectively averaged out.

For a typical high-sensitivity measurement, we conduct several runs through the entire time series before implementing the post-processing measures outlined in the next section. Finally, we have developed a new method that exploits chirped pulses in order to acquire very high sensitivity kinetics, which is described in the first case study below (Section 3.1).

2.4.2. *Noise reduction after measurement: Data processing*

Once the measurement is complete, the collected data is processed in MATLAB. The results of each pump-probe delay scan are compiled, and compared to check for effects from photodegradation and experimental drift, before being averaged into a single wavelength- and pump-probe delay-dependent array of $\Delta T/T$ values.

The next step is to apply a transformation to correct for the "chirp" of our system – the wavelength-dependent value of pump-probe delay corresponding to temporal overlap of our pump and probe pulses in the sample, $t_0(\lambda)$, where the wavelength dependence is primarily due to the net group-velocity dispersion of the broadband probe pulse in our experiment. For each experimental configuration, a single run is taken at high pump intensity, such that a strong and fast (within instrument response) TA signal or alternatively, coherent artifact[13] is observed at t_0. Points are manually entered along the artifact using a graphical interface, allowing us to fit a 3rd degree polynomial corresponding to the net chirp of our system. The dataset is temporarily upsampled (linear 1-D interpolation of the data for each wavelength value), the appropriate temporal offset is applied for each wavelength, and the chirp-corrected dataset is downsampled back to a common set of pump-probe delays.

Any static features (that may exist, for example, from scattered pump light incident on the photodiode array) are removed by subtracting from the entire dataset the average of the spectra collected at $t < t_0$. A smoothing filter is then typically applied to the individual spectra, as the effective wavelength spacing of our photodiode pixels is less than the wavelength resolution of our spectrometer, which is optimized for light throughput.

Finally, other signal recovery or data fitting techniques may be applied as appropriate for the experiment in order to determine the ensemble of lifetimes and spectral components contributing to the observed signal, and to deconvolute these from the instrument-response function. Techniques often used include singular value decomposition, global fitting, genetic fitting algorithms, and maximum-entropy methods.

2.5. *Measuring beam profiles*

In order to determine the spatial distribution of pump and probe beam energy density, we place the exposed CMOS detector from a common webcam[g] in the sample plane of our experiment. Using the published pixel size and pitch of the CMOS elements, gaussian fits of the data give us the FWHM of each beam for the purposes of estimating excitation density within the sample, or for any data fitting that requires knowledge of incident excitation density.

3. Case Studies

3.1. *Ultra-sensitive studies of charge photo-generation in P3HT:PCBM thin film blends*

As one of the most commonly studied OPV blends, P3HT:PCBM thin films provide an excellent benchmark for us to demonstrate the high sensitivity of our transient absorption spectrometer. The non-linearities exhibited by P3HT:PCBM with respect to excitation density provide a clear motivation for spectroscopists to develop techniques for resolving the dynamics of charge photogeneration at continually lower excitation intensities. Exciton-exciton annihilation,[14] exciton-charge annihilation,[6] and bimolecular charge recombination[2,3] all result in excitation density dependent TA data.[3,15]

The main sources of noise in a typical TA experiment are the wavelength-dependent fluctuations in our probe pulses, fluctuations in pump intensity, and read-out noise from the photodiode array and read-out circuitry. After taking appropriate measures to minimize noise from these sources (respectively: the

---

[g] Microsoft LifeCam Cinema 720p HD Webcam, which uses an OmniVision OV9712 image sensor with 3.04 μm pixel pitch

use of a reference probe, pump intensity-based shot-rejection, and a sufficiently bright probe source to take advantage of the full dynamic range of our detection system), the best approach to improving sensitivity is to maximize the number of samples N for the purposes of averaging. In an ideal situation, our signal-to-noise ratio will be proportional to $\sqrt{N}$.

For a given system, the maximum N that can be collected within a measurement period t is given by $N = \omega x t / 2$, where ω is laser repetition rate, x is the number of detection elements (in this case, the 256 pixels on our probe photodiode array) and the factor of $1/2$ due to the fact that two laser shots are required to generate one $\Delta T/T$ value. Many reasons for not attaining this ideal N are obvious, for example pauses in sampling while the delay stage is moving or the acquisition software is busy. However, other reasons are less obvious.

Many TA features in organic semiconductors are spectrally broad, in some cases over 100 nm. A kinetic trace for such a feature is typically generated by integrating the signal over the x detection pixels associated with the relevant wavelength range. While this should multiply our signal-to-noise ratio by $\sqrt{x}$ compared to the single pixel kinetic trace, we do not usually see such an improvement as wavelength-correlated probe noise results in constructive additions from probe fluctuations. One way to remove these correlations is to introduce significant chirp to our probe pulse, for example by the insertion of a strongly dispersive element such as a glass block into the beam path. If sufficient chirp is present, then after the chirp-correction algorithm the values of $\Delta T/T$ for each pixel at a given pump-probe delay will have been collected from separate laser shots. This method effectively removes the effect of correlated noise due to probe pulse fluctuations, and improves the signal-to-noise ratio for our kinetic trace by approximately $\sqrt{x}$ when integrating over x pixels.

Figures 4c) and d) show the growth of the photoinduced absorption feature in P3HT:PCBM (probed from 650–760 nm, dispersed across 55 pixels on the photodiode array) after excitation at 480 nm. For the measurement shown in Fig. 4d) a 10 cm thick, high refractive-index glass block was inserted in the probe beam path prior to interaction with the sample. Both measurements took approximately 30 minutes to collect, and (other than the glass block) were collected and processed identically.

We assess the noise contribution to our signal by fitting a monoexponential growth curve to the data and calculating the standard deviation of the distribution of residuals. We find that noise from the chirped method has a standard deviation of 3.2×10^{-6}, compared with 1.6×10^{-5} for the traditional measurement – an improvement in the signal-to-noise ratio of a factor of 5. This is approaching the improvement of 5.6 that we would expect as a result of the $\sqrt{x}$ relation for x uncorrelated measurements (in this case, over 55 pixels) after correcting for the number of shots averaged in each case.

While chirped probe pulses enable us to obtain extremely sensitive kinetics, there is an associated increase of noise in the spectral domain. This is because different wavelengths of each spectrum now originate from different laser shots, which are subject to the same fluctuations that we removed from the kinetic data. We find a practical solution to this is to collect data for spectra separately from kinetics, at fewer delay positions (enabling high sample number per spectrum) and with a minimum of dispersive optical elements in the beam path. The result is shown in Fig. 4b), which required 14 minutes to collect and nicely compliments the kinetic traces described earlier.

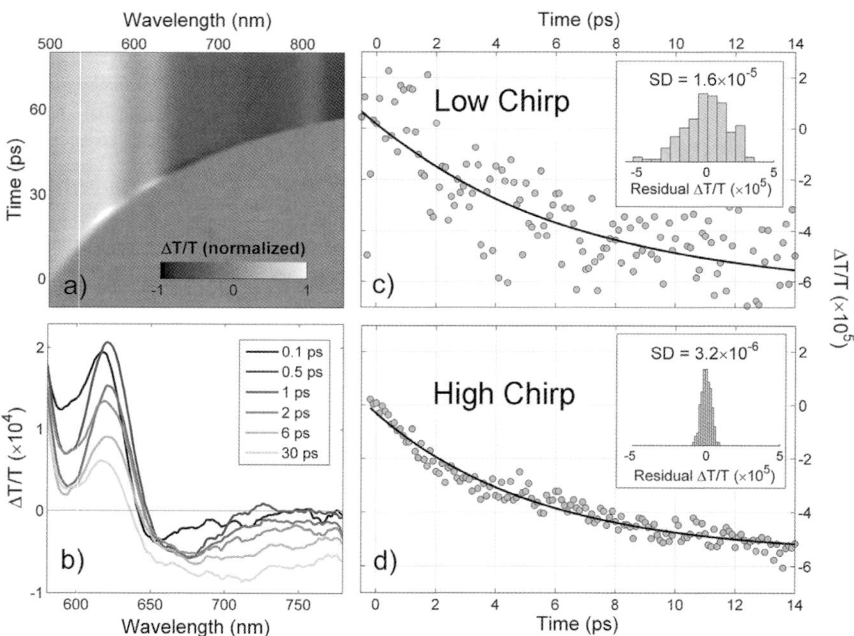

Figure 4. TA of charge photogeneration dynamics in a P3HT:PCBM thin film. **a)** TA surface collected with an extremely chirped probe pulse: NOPA light is stretched over ~ 60 ps after a single pass through a 10 cm high refractive index glass block. **b)** Unchirped TA spectra at 6 time points only (64,000 shot pairs per spectrum, with a 10 nm wide smoothing filter applied after measurement. Total measurement took 14 minutes to collect). **c)** Charge growth (650-760 nm) under low pump fluence measured using all-reflective (unchirped) probe pulse (7,000 shot pairs per time point) and **d)** chirped probe pulse (4000 shot pairs per time point). Measurements **c)** & **d)** each took approximately 30 minutes, as the second measurement requires acquisition of additional time points that are moved out of the observation window by the chirp correction algorithm.

There is some degradation of our temporal instrument response function as a result of the increased chirp per detection element within the spectrometer. However this can be managed through selection of an appropriate amount of chirp (glass thickness) for the dynamics to be resolved, and/or the use of a prism-based spectrometer where spectral resolution can be adjusted.

The net result of our noise reduction efforts are shown in Fig. 4, where TA data taken under extremely low pump fluence reveals the broadband dynamics of diffusion-limited charge generation over several picoseconds in the absence of exciton annihilation and bimolecular recombination reactions. Visible in Fig. 4c) are spectra showing bleaching of the ground state below 650 nm, and photogenerated polaron absorption, overlapping with short-lived exciton-based stimulated emission at wavelengths greater than 650 nm (with associated kinetics shown in Figs. 4c) and d)).

3.2. *Temperature and excitation-density dependent charge recombination in P3HT-PCBM blends*

The charge photogeneration studies presented in Fig. 4b) show that after ~ 10 ps, the spectrum is dominated by charges that exhibit a photoinduced absorption signature beyond 660 nm. We can cleanly resolve charge recombination dynamics by monitoring the decay of this signal in different blends and under varying excitation conditions. Charge recombination can occur from sub-nanosecond to beyond microsecond timescales,[3] hence motivating the need to exploit the electronically delayed pump-probe configuration described in Section 2.1. By resolving timescales relevant to macroscopic charge dynamics in devices, this technique bridges the gap between ultrafast optical spectroscopy and slower electronic measurements such as open-circuit voltage decay.

We chose to probe the polaron absorption using the fundamental harmonic from our amplifier at 800 nm, which has the benefit of being more stable than either of our broadband light sources. Light at 800 nm is equally likely to be absorbed by bound and free charges,[3] allowing us to track the decay of the total charge population. To fully explore the range of intensity dependence, we need to be able to resolve signals at extremely low pump intensities. In the most efficient blend at 290 K, the minimum pump power density was 2 μ J/cm^2, inducing a maximum $\Delta T/T$ of only 1.6×10^{-4} for the relevant trace in Fig. 5.

The dependence of charge recombination rates on excitation density in P3HT:PCBM blends reveals two distinct decay paths that contribute to the observed decay rate: a monomolecular component, which scales linearly with excitation density and is attributed to bound charge pairs recombining

48

geminately, and a second component which scales non-linearly with excitation density and is attributed to free charge pairs recombining bimolecularly.[3]

Transport of free charges in organic semiconductors typically takes place by way of a thermally activated hopping mechanism.[16] It follows that as we remove heat from the system, charge mobility – and by extension, the rate of bimolecular recombination – should decrease. In order to investigate this, we mounted spin-coated thin film samples in a closed-cycle helium cryostat, allowing us to undertake transmission-mode TA at temperatures as low as 10 K.

In agreement with charge recombination measurements in neat P3HT films conducted by Paquin et al,[17] we find that at sufficiently low temperatures ($\leq$ 25 K, see Fig. 5) there is no dependence of recombination rate on charge density, indicating that all charge transport – and therefore bimolecular recombination – has been frozen out of the system. This measurement allows us to isolate the distribution of monomolecular recombination tunneling rate constants and the thermal activation energies.

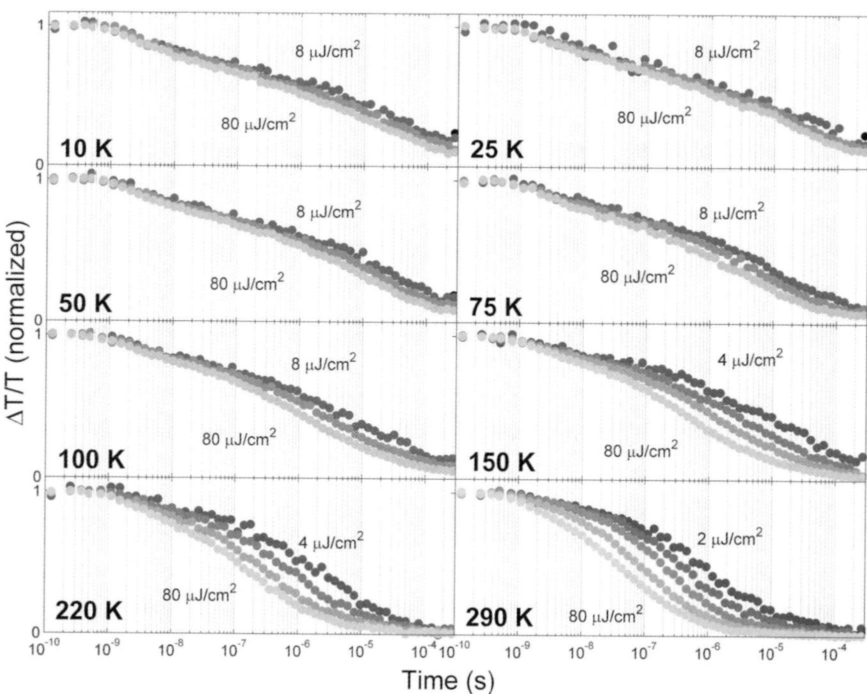

Figure 5. Temperature and excitation density dependent charge recombination measurements for a thin film blend of P3HT:PCBM. Low noise TA techniques enable the collection of a large number of small signal datasets in a reasonable amount of time.

The low-noise capabilities of our TA spectrometer were crucial in collecting this dataset, not only for the ability to capture signals at extremely low pump fluence, but also for the ability to conduct all measurements in a very short period of time. This improved efficiency enabled us to collect the large intensity- and temperature-dependent dataset shown in Figure 5 in a single 8 hour lab session. This improved efficiency of data collection and ability to avoid the compromises of splitting measurements over several sessions (e.g., sample and experimental stability) is an extremely welcome feature in our experimental setup.

3.3. *Probe saturation and observation of higher order perturbations*

While the insight gained from measuring the dependence of TA dynamics on pump intensity is now well established, the dependence of signal amplitude on probe intensity is rarely discussed. The signal should be independent of probe intensity in the weak-probe limit; however, ultrafast lasers present the potential to go beyond the Beer-Lambert regime with a strong perturbative probe pulse.

In order to investigate the dependence of TA signal amplitude on probe intensity, we investigated the singlet exciton absorption in films of the conjugated polymer F8BT blended in an inert polystyrene matrix to minimize interchain effects. We directly probed the photoinduced absorption of the singlet exciton at 800 nm[18] using the fundamental harmonic of the amplifier, which combined with the measures described in Section 2 allowed us to obtain very stable and repeatable signal amplitudes. Unlike most TA studies, which focus on dynamics (i.e., *relative* signal intensity over time) and qualitative spectral shapes, this investigation challenges our ability to make quantitative and repeatable measurements of absolute signal intensities that can be compared with separate measurements.

Figure 6a) shows a reduction in signal amplitude as the energy density of the probing light is increased. We attribute this to a saturation effect, where light within the leading part of the probe pulse excites a significant fraction of excitations from the first singlet excited state (S_1, see Fig. 1) into the higher manifold of excited states (S_n), effectively reducing the population of excitations available in S_1 to absorb the remainder of the probe pulse. The dependence of the absolute signal magnitude on probe intensity can provide us with useful information about the absorption cross-section of S_1. However, we must place limits on the amount of sequential absorption[19] taking place due to the probe pulse acting directly on the manifold of excited state S_n.

We investigate such contributions using a 3-pulse experiment where initial excitation resonant with the ground state generates a population of S_1 at a chopping frequency of $\omega/2$, as earlier. Second, an intense 800 nm push pulse, chopped at $\omega/4$ replicates the excitation from S_1 to S_n due to the strong probe beam in the earlier experiment. Finally, a broadband probe pulse (in this case from the NOPA) at low intensity overlaps with the first two beams at the sample and is dispersed onto the photodiode array to measure the secondary differential transmission spectrum induced by the push pulse.

$$\Delta\left(\frac{\Delta T}{T}\right) = \left[\left(\frac{\Delta T}{T}\right)_{pushed} - \left(\frac{\Delta T}{T}\right)\right] \bigg/ \left(\frac{\Delta T}{T}\right) - \left(\frac{\Delta T}{T}\right)_{pump\ off,pushed} \tag{2}$$

If $\Delta T/T$ is the signal due to first pump alone, then by taking the normalized difference between the TA signal with and without push pulse we obtain $\Delta(\Delta T/T)$, the differential transmission spectrum of S_n as a result of the push pulse acting on S_1. The final term is to remove any effect of direct excitation of the ground state sample multi-photon absorption of the push pulse. It is essential to achieve excellent signal-to-noise ratios in this measurement, as we are looking at a small change between two spectra that are themselves already differential measurements, thus the effect of noise is amplified. This measurement also highlights the benefit of collecting all combinations of pulses under multiple chopping frequencies with shot-to-shot detection.

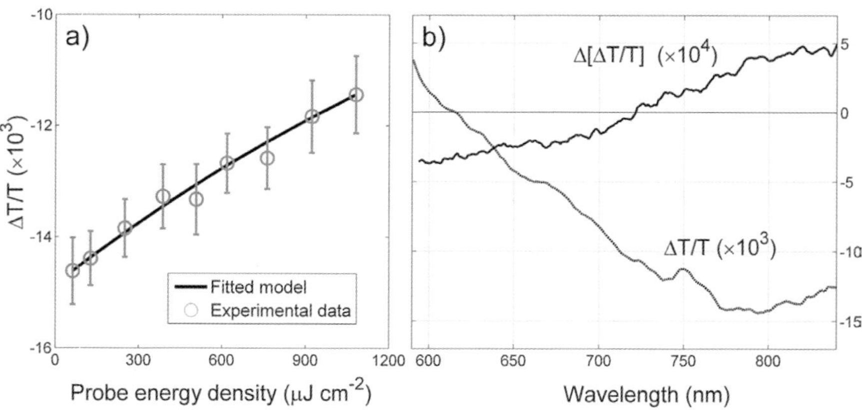

Figure 6. TA saturation effects for an F8BT:polystyrene thin film. **a)** Observed $\Delta T/T$ signal can be saturated under an intense 800 nm probe pulse. **b)** Standard TA spectra and 3-pulse TA spectra at a pump-probe delay of 6 ps and an 800 nm push pulse arriving 1 ps before the probe. The $\Delta T/T$ spectrum is dominated by the photoinduced absorption of the S_1 state, while the positive $\Delta(\Delta T/T)$ signal induced by the 800 nm push pulse shows that it depletes the S_1 population.

The spectra in Fig. 6b) show $\Delta(\Delta T/T)$ for F8BT, with a pump-probe delay of 6 ps and the push pulse arriving 1 ps before the probe pulse. The positive $\Delta(\Delta T/T)$ signal indicates that at wavelengths greater than 720 nm, the higher manifold of states S_n are less absorptive than the S_1 state. Thus, the apparent bleaching of the S_1 absorption around 800 nm confirms that the observed reduction in signal intensity with increased probe intensity in Fig. 6a) can be accounted for by saturating the strong absorption of the S_1 state.

The ability to measure secondary TA spectra of re-excited species offers exciting possibilities in the spectroscopy of photocurrent generation in OPVs. Recent investigations have shown that free charge carrier generation is preceded by delocalized charge-transfer states that can be re-accessed with an infrared push pulse.[4] These states are not isolated via conventional TA spectroscopy because they are extremely short-lived, yet their formation is convoluted with preceding excitonic processes following excitation. With sufficient sensitivity to quantitatively measure secondary TA spectra of re-excited species, we are now positioned to examine the properties of non-thermalized excited states and other transient states existing further along the photophysical cascade.

4. Conclusions

TA spectroscopy is arguably the most incisive experimental tool available for establishing photophysical models. In spite of a clear demand for TA spectroscopy to understand the mechanisms of photocurrent generation in OPV cells, its application is compromised by severe constraints on signal sensitivity and time resolution. As described in this article, we have developed the capability to measure charge photogeneration and recombination dynamics in OPV thin films with the requisite sensitivity of better than 10^{-5} and from femtosecond to microsecond timescales. The examples selected illustrate the benefits of designing data acquisition and processing methods around the best available laser sources and detectors that can be read out on a shot-to-shot basis. We demonstrate that chirped probe pulses can be exploited to significantly suppress the effects of laser pulse fluctuations, and that our exceptional sensitivity can be used to measure higher order perturbations when transient states are re-excited. Having proven these techniques for OPV materials, we expect that many of these techniques will prove useful in other demanding applications of TA spectroscopy.

52

References

1. J.-L. Brédas, J. E. Norton, J. Cornil, and V. Coropceanu, *Accounts of Chemical Research* **42**, 1691 (2009).
2. T. M. Clarke and J. R. Durrant, *Chemical Reviews* **110**, 6736 (2010).
3. I. A. Howard, R. Mauer, M. Meister, and F. Laquai, *Journal of the American Chemical Society* **132**, 14866 (2010).
4. A. A. Bakulin, A. Rao, V. G. Pavelyev, P. H. M. van Loosdrecht, M. S. Pshenichnikov, D. Niedzialek, J. Cornil, D. Beljonne, and R. H. Friend, *Science (New York, N.Y.)* **335**, 1340 (2012).
5. G. Grancini, M. Maiuri, D. Fazzi, a. Petrozza, H.-J. Egelhaaf, D. Brida, G. Cerullo, and G. Lanzani, *Nature Materials* **11**, 1 (2012).
6. J. M. Hodgkiss, S. Albert-Seifried, A. Rao, A. J. Barker, A. R. Campbell, R. A. Marsh, and R. H. Friend, *Advanced Functional Materials* **22**, 1567 (2012).
7. J. Cabanillas-Gonzalez, G. Grancini, and G. Lanzani, *Advanced Materials* **23**, 5468 (2011).
8. V. I. Klimov and D. W. McBranch, *Optics Letters* **23**, 277 (1998).
9. U. Megerle, I. Pugliesi, C. Schriever, C. F. Sailer, and E. Riedle, *Applied Physics B* **96**, 215 (2009).
10. S. Westenhoff, I. a Howard, J. M. Hodgkiss, K. R. Kirov, H. a Bronstein, C. K. Williams, N. C. Greenham, and R. H. Friend, *Journal of the American Chemical Society* **130**, 13653 (2008).
11. J. M. Hodgkiss, A. R. Campbell, R. A. Marsh, A. Rao, S. Albert-Seifried, and R. H. Friend, *Physical Review Letters* **104**, 177701 (2010).
12. C. Manzoni, D. Polli, and G. Cerullo, *Review of Scientific Instruments* **77**, 023103 (2006).
13. M. Lorenc, M. Ziolek, R. Naskrecki, J. Karolczak, J. Kubicki, and A. Maciejewski, *Applied Physics B: Lasers and Optics* **74**, 19 (2002).
14. M. Stevens, C. Silva, D. Russell, and R. H. Friend, *Physical Review B* **63**, 1 (2001).
15. H. Ohkita, S. Cook, Y. Astuti, W. Duffy, S. Tierney, W. Zhang, M. Heeney, I. McCulloch, J. Nelson, D. D. C. Bradley, and J. R. Durrant, *Journal of the American Chemical Society* **130**, 3030 (2008).
16. H. Bässler and A. Köhler, *Topics in Current Chemistry* **312**, 1 (2012).
17. F. Paquin, G. Latini, M. Sakowicz, P.-L. Karsenti, L. Wang, D. Beljonne, N. Stingelin, and C. Silva, *Physical Review Letters* **106**, 197401 (2011).
18. S. Gélinas, J. Kirkpatrick, I. a Howard, K. Johnson, M. W. B. Wilson, G. Pace, R. H. Friend, and C. Silva, *The Journal of Physical Chemistry. B* (2012).
19. X. Zhang, Y. Xia, R. H. Friend, and C. Silva, *Physical Review B* **73**, 245201 (2006).

TRANSIENT ABSORPTION DATA ANALYSIS BY SOFT-MODELLING

I.A. HOWARD,* H. MANGOLD, F. ETZOLD, D. GEHRIG and F. LAQUAI

Max-Planck-Institute for Polymer Research
55128 Mainz, Germany
** E-mail: ian.howard@mpip-mainz.mpg.de*
http://www.mpip-mainz.mpg.de/11529/Organische_Optoelektronik

This chapter describes techniques for transient absorption data analysis, that is the analysis of the observed data matrix to determine the number of excited-state species, the evolution of the concentration of each excited-state species' population with time, and the $\Delta T/T$ spectrum of each excited-state species over the wavelength region measured. It is shown that the critical step in data analysis is not finding an accurate mathematical decomposition of the data matrix, but rather finding the actual (physical) decomposition of the data matrix $\mathbf{C_a}$ and $\mathbf{S_a}$ from within the possibly infinite set of equivalently valid mathematical decompositions. When detailed knowledge of the system of kinetic equations along with their boundary conditions is available, global analysis can be used to determine the physical decomposition. However, normally such specific information is not available. In this case, we demonstrate how a soft-modelling technique can be used to determine the bands of decompositions in which the physical decomposition must lie. Depending on the observations and the known physical constraints, this may lead to a single very constrained band, in which case the physical decomposition can be precisely determined. In other cases, information can be obtained as to what additional experimental information would be most effective to reduce the size of the solution bands, and most effectively hone in on the physical decomposition. The clarity with which the range of possible solutions appear in this soft-modelling technique avoids the potential pitfalls for misassignment of the physical decomposition present when global analysis is attempted with insufficient (or insufficiently robust) *a priori* knowledge.

Keywords: Transient Absorption, Data Modelling, Global Analysis, Soft-Modelling, Organic Solar Cells

1. Introduction

The result of modern transient absorption experiments is a data matrix consisting of m rows, each of which is the n long transient absorption spectrum (where n corresponds to the number of illuminated pixels on the linear array detector behind the spectrograph) measured at the mth time delay. The data matrix expresses $\Delta T/T$ as a function of wavelength and time, $\mathbf{D}(m,n) = \Delta T/T(t, \lambda)$, and is often plotted as a surface against these two variables. This is shown in the first panel of Fig. 1.

The goal of transient absorption data analysis is to determine the number of excited-state species, the evolution of the concentration of each excited-state species' population with time, and the $\Delta T/T$ spectrum of each excited-state species over the wavelength region measured. As long as the spectrum of each excited-state species does not change with time, analysis of the transient absorption data corresponds to finding a bilinear decomposition of the measured data matrix with a given number, p, of excited-state species: $\mathbf{D} = \mathbf{CS} + \mathbf{E}$, where $\mathbf{C}$ is an $(m \times p)$ matrix in which the ith column corresponds to the population of the ith species at each of the m measured timepoints, $\mathbf{S}$ is a $(p \times n)$ matrix in which the ith row represents the transient absorption spectrum of the ith species (at the n wavelengths of the measured data), and $\mathbf{E}$ is an $(m \times n)$ matrix that represents the difference between the measured data and its idealized description (given by $\mathbf{CS}$) due to experimental noise. This is graphically illustrated in the first panel of Fig. 1. It is important to note that the number of excited-state species, p, that are required to describe the experimental data corresponds to the rank of the data matrix. The matrix multiplication $\mathbf{CS}$ can be written as the series from $i = 1 \ldots p$ of the outerproducts between the ith column of $\mathbf{C}$ and the ith row of $\mathbf{S}$, $\mathbf{CS} = \sum_1^p \mathrm{c}(i) \otimes \mathrm{s}(i)$, where $\mathrm{c}(i)$ and $\mathrm{s}(i)$ are the ith column of the concentration matrix and the ith row of the spectral matrix respectively. Physically speaking, this means that the total transient absorption surface can be thought of as a sum of rank 1 surfaces, each given by the outer product of the concentration profile of an excited-state species with that species' transient absorption spectrum. This is again shown graphically in the last two panels of Fig. 1.

Following from the results of the last paragraph, the goal of transient absorption data analysis is to factor the observed data matrix into the product of two matrices, which to reiterate are $\mathbf{C}$, an $(m \times p)$ matrix in which the ith column gives the population of the ith excited-state species as a function of time, and $\mathbf{S}$, a $(p \times n)$ matrix in which the ith row gives the spectrum of the ith excited-state species as a function of wavelength. Finally, p is the number of excited-state species that contribute to the measured transient absorption surface. We will now for the moment assume that we can find (based on mathematical techniques) a decomposition $\mathbf{CS}$

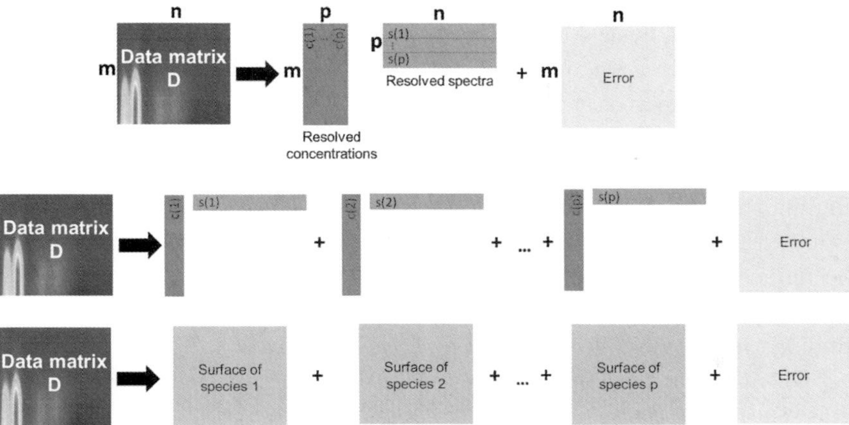

Fig. 1. The upper panel shows a bilinear decomposition of a transient absorption data surface $(m \times n)$ into resolved concentration $(m \times p)$ and spectra $(p \times n)$ profiles plus an error matrix $(m \times n)$. This means that the data matrix can be represented by the summation of the outer products of the individual concentration and spectra profiles of the p components, shown schematically in the lower panels.

that accurately describes the measured data, and consider whether or not this decomposition is unique. As it turns out, it is easy to show that the decomposition is not mathematically unique, which will mean that mathematical techniques alone are insufficient to determine the physically true $\mathbf{C}$ and $\mathbf{S}$ matrices (that we will label with the subscript a, for actual, as $\mathbf{C_a}$ and $\mathbf{S_a}$). Equation 1 shows that any invertible matrix $\mathbf{T}$ and its inverse can be introduced between $\mathbf{C}$ and $\mathbf{S}$.

$$\mathbf{D} = \mathbf{CS} = (\mathbf{CT})\left(\mathbf{T}^{-1}\mathbf{S}\right) = \mathbf{C}_{new}\mathbf{S}_{new} \tag{1}$$

As the product of $\mathbf{T}$ and its inverse is the identity matrix, their introduction does not change the result of the matrix multiplication at all. As matrix multiplication is associative we can calculate the matrix product as $(\mathbf{CT})(\mathbf{T}^{-1}\mathbf{S})$, as also shown in Eq. 1. The term in the first brackets $(\mathbf{CT})$ can be considered a new concentration matrix, $\mathbf{C_{new}}$, wherein each of the new columns (concentration profiles) is a mixture of the old columns (concentration profiles) determined by the elements of $\mathbf{T}$. Similarly, the second bracketed term, $(\mathbf{T}^{-1}\mathbf{S})$, can be considered a new spectra matrix, $\mathbf{S_{new}}$, in which each of the rows (spectra) is a combination of the old rows (spectra) based on the elements in the inverse of $\mathbf{T}$. As we know that the product of $\mathbf{C_{new}}\mathbf{S_{new}}$ describes the data identically well as the product $\mathbf{CS}$

we immediately see that the decomposition is not mathematically unique. Neglecting any physical constraints, we see that the real physical solutions $\mathbf{C_a}$ and $\mathbf{S_a}$ lie in a set of solutions that is infinitely large (given there are an infinite number of choices for $\mathbf{T}$). So based on this knowledge, we can now understand that the critical step in transient absorption data analysis boils down not to finding a decomposition of the data matrix, but rather finding the actual (physical) decomposition of the data matrix $\mathbf{C_a}$ and $\mathbf{S_a}$ from within the theoretically infinite set of equivalently valid mathematical decompositions.

In order to achieve the goal introduced in the preceding paragraph of finding the physical decomposition of the measured data matrix $\mathbf{C_a}$ and $\mathbf{S_a}$ which reveal the concentration evolutions and spectra of the excited-state species present, constraints based on physical insights into the system measured must be introduced. In the next two sections we will consider two techniques for introducing such physical constraints, and using them to search out the physically meaningful decomposition. The first technique requires very specific knowledge of the system under study, namely the functional form of the rate equations for all excited-state species. Due to the necessity for such specific knowledge, we call this approach hard-modelling. In the second section we introduce a technique that uses more general knowledge, such as the requirement that concentrations be positive, along with insight gained from the mathematical analysis of the measured data surface (such as local-rank information) to search for the physical decomposition. We call this second technique soft-modelling, as it can be applied in general to transient absorption data, with little or no prior knowledge of the reacting species or their kinetics. For both cases we consider the question of estimating how certain one is that the true physical $\mathbf{C_a}$ and $\mathbf{S_a}$ were found. For the second case, we introduce the technique we have developed in order to quickly and powerfully examine the uniqueness and physicality of the found solution in the case of a two-excited-state decomposition wherein one spectrum can be constrained from the data (an example of such a case is shown in the demonstrations section, and actually often occurs in energy conversion reactions in which excited-state species of very disparate lifetimes are involved).

2. Global analysis

The first technique we consider to find the physically meaningful matrix decomposition $\mathbf{C_a}$ and $\mathbf{S_a}$ is global analysis. This technique has a long history in the analysis of transient absorption data, and a good review of its current status was recently presented by van Thor and coworkers.[1]

In broad strokes, the most common implementation of this technique can be described in the following fashion. Firstly, the functional form of the population flow between the (known number of) excited-state species along with the boundary conditions of the initial population of each species must be known *a priori*. A general form of these equations is given in Eq. 2, where C_i is the concentration of the ith excited-state species, and F_i is a function that governs the dynamics of the ith population, which takes in a vector of parameters $(\vec{p}_i)$ and the populations. An explicit example, from a common application of TA spectroscopy, that is, the analysis of photophysical processes in organic solar cells, is a scheme of kinetic equations that describe an initial population of excitons that either decay to the ground-state or form a pair of free charges which can then bimolecularly recombine. This is shown in Eq. 3. In this example F_1 is the decay to the ground state and the transfer to charge pairs given by $-k_1 C_1 - k_2 C_1$, $p_1 = (k_1, k_2)$, F_2 is the growth of charge pairs from the exciton population and their decay due to bimolecular recombination $k_2 C_1 - k_3 C_2{}^2$, and $p_2 = (k_2, k_3)$. The global fit is performed by using a non-linear least squares minimization routine (like *lsqnonlin* in Matlab) to vary the parameter vectors $\vec{p}_1 \ldots \vec{p}_p$ in order to find a minimum of the least squares residual between the model and the data. The model is calculated for each set of input parameters in a two step process. First the concentration matrix is found by finding the numerical solutions to the population equations (using a solver such as ode15s in Matlab) with the current parameter set. Once the concentration matrix (for the given step) is known, the corresponding best possible spectra matrix is quickly determined using a linear least squares approach (like a matrix division in Matlab). The model is then computed by the multiplication of the concentration and spectra matrices and the residual at this step calculated. The minimization routine continues to adjust the input parameters until a minimum difference exists between the model and data, at which point $\mathbf{C_a}$ and thereby also $\mathbf{S_a}$ are found (given the population equations and initial conditions are indeed correct).

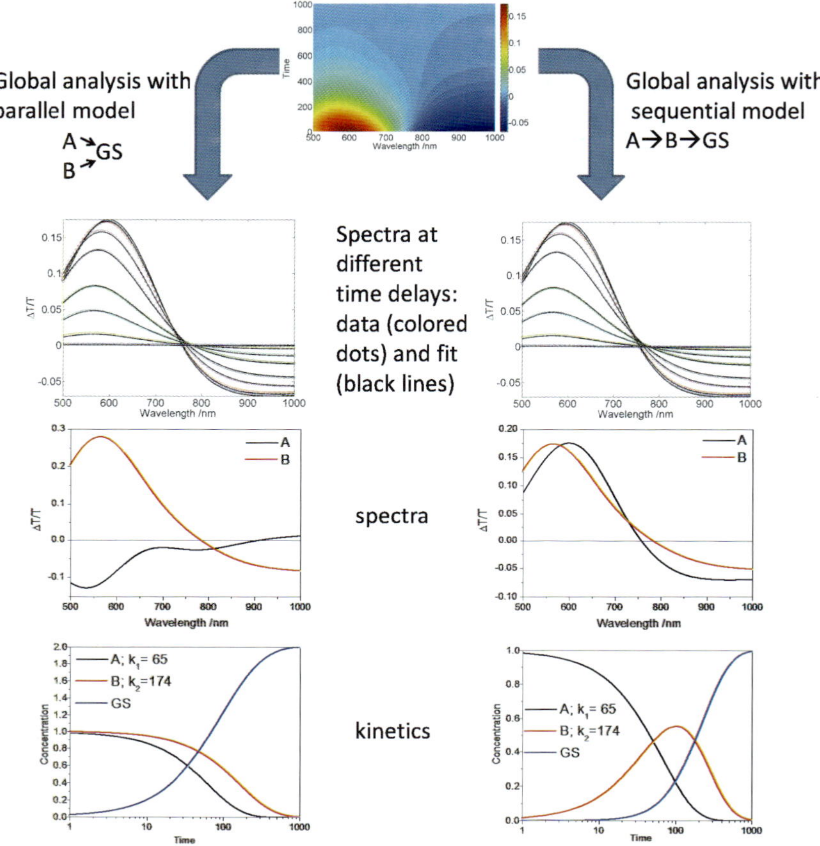

Fig. 2. (Color online) Result of global analysis on a simulated dataset. On the left-hand side the data matrix is fitted with a parallel model with two components A and B which both decay monoexponentially with the respective decay constants k_1 and k_2. The right-hand side presents the fit with a sequential model with A and B also exhibiting monoexponential decay with k_1 and k_2. The first panel on each side shows a comparison between the data (colored dots) and the fit (black lines) at time delays of 1, 10, 50, 100, 200, 300, 500 and 1000. Both models fit the data equally well. The $\Delta T/T$ spectra of the two components are shown in the second panel, the corresponding kinetics in the lower panel. GS denotes the ground state where $\Delta T/T=0$ for all wavelengths.

$$\frac{d}{dt}C_1 = F_1(\vec{p}_1, C_1, C_2, \ldots, C_p)$$

$$\frac{d}{dt}C_2 = F_2(\vec{p}_2, C_1, C_2, \ldots, C_p)$$

$$\vdots$$

$$\frac{d}{dt}C_p = F_p(\vec{p}_p, C_1, C_2, \ldots, C_p) \tag{2}$$

$$\frac{d}{dt}C_1 = -k_1 C_1 - k_2 C_1$$

$$\frac{d}{dt}C_2 = k_2 C_1 - k_3 C_2{}^2 \tag{3}$$

2.1. *Global analysis applied to a simple example*

In order to illustrate the global analysis technique, a simple example of a simulated dataset is considered here. It is assumed that two species are present in the dataset and that the decay of a species is proportional to its concentration. Two different models are applied to analyze the data: In the parallel model both species A and B are present from the start with the initial condition $C_A(t = 0) = C_B(t = 0) = 1$ while in the sequential model only A is present in the beginning and B is formed from A (initial condition: $C_A(t = 0) = 1$; $C_B(t = 0) = 0$). The following schemes of kinetic equations describe these simple models: Parallel model:

$$A \xrightarrow{k_1} GS$$

$$B \xrightarrow{k_2} GS$$

$$\frac{d}{dt}C_A = -k_1 C_A$$

$$\frac{d}{dt}C_B = -k_2 C_B \tag{4}$$

Sequential model:

$$A \overset{k_1}{\to} B \overset{k_2}{\to} GS$$

$$\frac{d}{dt}C_A = -k_1 C_A$$

$$\frac{d}{dt}C_B = k_1 C_A - k_2 C_B \tag{5}$$

The results of the global analysis with the rate equations from Eq. 4 and 5 are presented in Fig. 2. Both parallel and sequential model fit the data equally well as can be seen from the solid lines overlaying the data points in Fig. 2. This means that both decompositions $\mathbf{CS}$ describe the data matrix $\mathbf{D}$ with the same accuracy. So in order to know that the decomposition obtained by global analysis is the physically correct decomposition into $\mathbf{C_a}$ and $\mathbf{S_a}$ one must be completely certain that the rigorously correct system of equations was chosen, and that the boundary conditions applied were correct!

In conclusion, global analysis is a useful tool to decompose an experimental data matrix. However, to obtain the physically meaningful decomposition into $\mathbf{C_a}$ and $\mathbf{S_a}$ it is crucial that the correct set of kinetic equations and boundary conditions are used. As seen in this simple example, it is often the case that more than one set of rate equations (and also more than one set of initial conditions) can lead to decompositions that give a high quality fit of the data. Therefore, the fact that a given kinetic model leads to a good fit of the data after global analysis is, in and of itself, not adequate evidence that the kinetic model is correct, or that the decomposition is indeed $\mathbf{C_a}$ and $\mathbf{S_a}$.

In the next section we will investigate a data analysis approach that does not require such specific *a priori* knowledge of the population kinetics. It also provides an approach to examining the range of decompositions that accurately reproduce the observed data and satisfy the known physical constraints. This allows the experimenter to determine the bands of reasonable decompositions in which the physically true $\mathbf{C_a}$ and $\mathbf{S_a}$ must lie. In certain cases these bands can be quite narrow, and $\mathbf{C_a}$ and $\mathbf{S_a}$ can be precisely determined.

3. Multivariate curve resolution

Here we investigate how a physically meaningful decomposition can be approached by soft modelling using constraints that can be obtained from basic physical knowledge and mathematical investigation of the observed

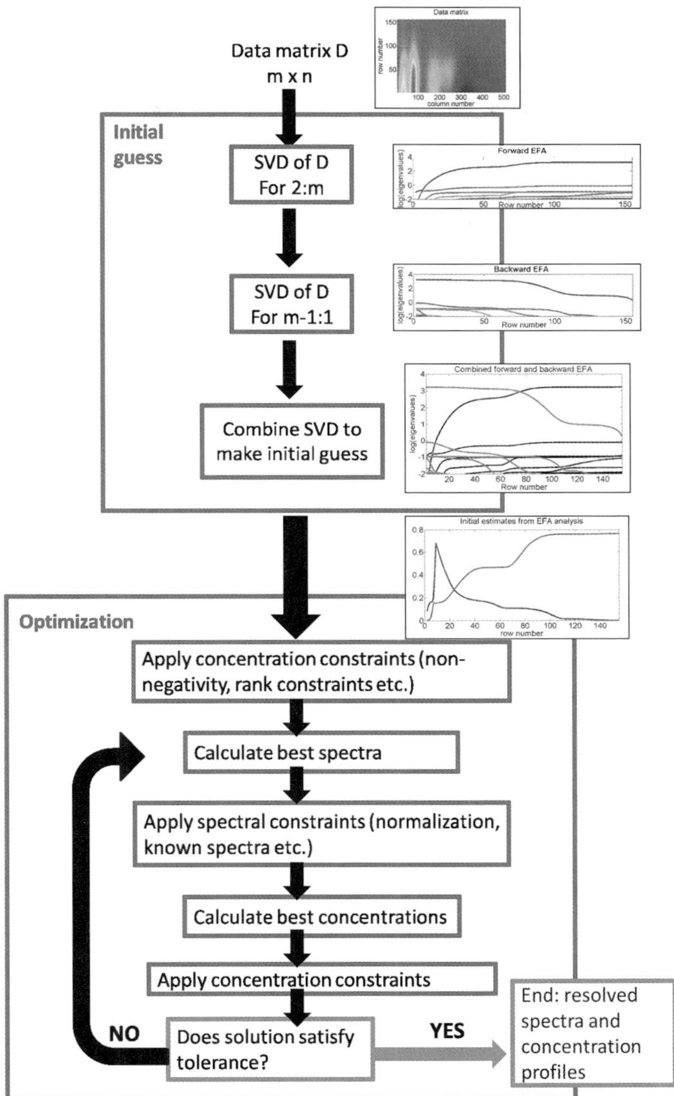

Fig. 3. Schematic flowchart of the analysis done with MCR. The initial guess is obtained with evolving factor analysis, in doing a singular value decomposition on gradually enlarged datasets in the forward and backward direction. The optimization procedure involves the application of constraints and an iterative linear least square algorithm.

data structure. This section presents the application of a multivariate curve resolution (MCR) method developed by Tauler and coworkers[2–5] to transient absorption data.

3.1. *Initial guess*

A schematic flowchart of the steps of the MCR process is shown in Fig. 3, in the following these steps will be explained in detail. First, information about the number of components and an initial guess of the concentration profiles is needed. This is done with evolving factor analysis (EFA), which is based on singular value decomposition (SVD). The singular value decomposition of a matrix $\mathbf{D}$ is a mathematical technique that factors any matrix into the product of three matrices:

$$\mathbf{D} = \mathbf{USV}^* \tag{6}$$

where $\mathbf{U}$ is a $(m \times m)$ matrix, $\mathbf{S}$ is a diagonal $(m \times n)$ matrix, and $\mathbf{V}^*$ (conjugate transpose of $\mathbf{V}$) is a $(n \times n)$ matrix. The diagonal elements of $\mathbf{S}$ are the singular values which are the square roots of the eigenvalues of $\mathbf{DD}^*$ arranged in decreasing order. The number of significant values in S gives the rank of $\mathbf{D}$. The columns of $\mathbf{U}$ and $\mathbf{V}$ are the left and right singular vectors of $\mathbf{D}$ which are the eigenvectors of $\mathbf{DD}^*$ and $\mathbf{D}^*\mathbf{D}$ respectively. In evolving factor analysis we will use SVD to do something similar to estimating matrix rank, so we will be concerned only with the values along the diagonal of $\mathbf{S}$.

The basis of evolving factor analysis is to do SVD analysis on a sequence of submatrices and use the singular values to estimate how populations of excited-states evolve with time. To start with, an SVD of a submatrix containing the first two columns of the data matrix is carried out and the singular values plotted as points with their value as the y-value and an x-value of 1. Then SVD is done on the first 3 columns, and the singular values again plotted but this time with an x-value of 2. This procedure of adding one more column, taking the SVD, then plotting the singular values at the next integer on the x-axis is repeated until there are no more columns to add. Lines are then drawn connecting the biggest singular value at every x point, the second-largest, third-largest, etc. This is called the forward-going EFA, and the increase of the second, third, etc., singular value with x (or timepoint) is indicative of the grow in of additional excited-state species at the times when the respective singular value line increases. In order to complete the EFA, a second graph is made in the backward-going direction. This is made by starting with only the last column, and adding one column to the left after each iteration. Looking at when the second, third, etc. component increase in this plot shows when excited-state species disappear from the data matrix. In order to make an approximate guess of the concentration profiles for a starting position of the MCR-ALS algorithm, it is assumed that the species first present in the data are the first to disappear. Looking at the EFA and the motion of singular value lines relative to the noise level an estimate of the number of excited-state species, p, is made. The initial concentrations are then estimated by iterating i from 1 to p and taking the ith component of the forward-going EFA until it intersects with the $(p\text{-}i\text{+}1)$th component of the backward-going EFA from which time the $(p\text{-}i)$th component of the backward-going EFA is taken to give the excited state concentration. From EFA we can also identify if a region at late time exists where one component is sufficient to describe the data matrix. If only one singular value is necessary to describe the data matrix for a long time in the backward-going EFA, then it can be concluded that only a single species accounts for the absorption left at long-time. Such information is very useful, as will be shown later.

3.2. *Optimization process*

This initial guess for the concentration profiles is then used in the optimization process: First, the constraints are applied to the concentration profiles. Possible constraints are that the concentrations should always be positive (non-negativity constraint), have only one peak (unimodality constraint) or rank constraints which can be obtained from EFA. With these concentration profiles, the best spectra are calculated in Matlab with a right matrix division $\mathbf{S}^T = \mathbf{C}/\mathbf{D}$. This is done with a least squares calculation which is equivalent to finding the pseudoinverse of the concentration $(\mathbf{C}^+)$: $\mathbf{S}^T = \mathbf{C}^+ \mathbf{D}^*$. The calculated spectra can be adjusted by constraints such as normalization, or known spectra of one or a number of excited-states. From there the concentration profiles are calculated again using the least squares method. Subsequently, the constraints are applied to the concentration profiles and it is assessed whether the solution satisfies the tolerance. If this is not the case, the next fitting round starts by calculating the spectra. If the tolerance is satisfied, then the current spectra and concentration profiles are shown as they adequately describe the data and meet all of the physical constraints. These are then a decomposition that is physically reasonable, however we do not yet know whether or not this decomposition has given us $\mathbf{C_a}$ and $\mathbf{S_a}$. Depending on the soft constraints and the data these concentration and spectra may be practically unique (ignoring scale uncertainty) or it may still be possible to find some invertible matrices, see Sec. 1, which mix the components of the solution found by the MCR-ALS procedure and result in new concentration and spectra profiles that still satisfy the physical constraints. We will examine some

64

methods for considering the uniqueness of the solution found by MCR-ALS in the coming sections. This will develop the understanding of whether the solution gives reliable insight into the true physical processes or whether a band of possible processes could all describe the experimental observations equally well.

3.3. *Data augmentation*

Before we turn to determining the uniqueness of the MCR-ALS solution, we will present a method that is broadly useful for making the solution as unique as possible. This technique relies on the fact that often some reaction rates that affect the transient absorption data surfaces are non-linear. This means that they occur at different rates when different initial population densities are created after excitation with different pulse fluences. Thus, by taking transient absorption surfaces for the same sample at a series of different excitation fluences, we can create an 'augmented' transient absorption surface by placing the individual surfaces on a common wavelength axis and an extended time axis as shown in Fig 4. Performing MCR-ALS on this augmented data set is possible, the spectra of the species are the same for all of the individual experiments, and the concentration profiles although different are found as augmented vectors wherein the unique evolutions of the species for the different fluences are appended to each other (just as the surfaces were). Usually, the data is normalized before augmentation to weight variances during optimization more equally. The fact that the various concentration profiles for each experiment must all simultaneously satisfy the soft constraints imposes a greater constraint on this overall augmented solution. This constraint means that the range of possible solutions is decreased, and that more precise insight into the physical decomposition can be gained.

3.4. *Exploring the linear transformation ambiguity*

As introduced in Sec. 1 we can create new spectra and concentration profiles that are identically mathematically valid by taking linear superpositions of the found concentration and spectra vectors using an invertible matrix $\mathbf{T}$. This is repeated in Eq. 7. In this section we present some methods for exploring how many $\mathbf{T}$ matrices exist that mix the concentration and spectra vectors found by MCR-ALS and result in new vectors which still satisfy the known soft constraints. We do not, as of yet, know of a general solution to this problem for an arbitrary number of components. However, we

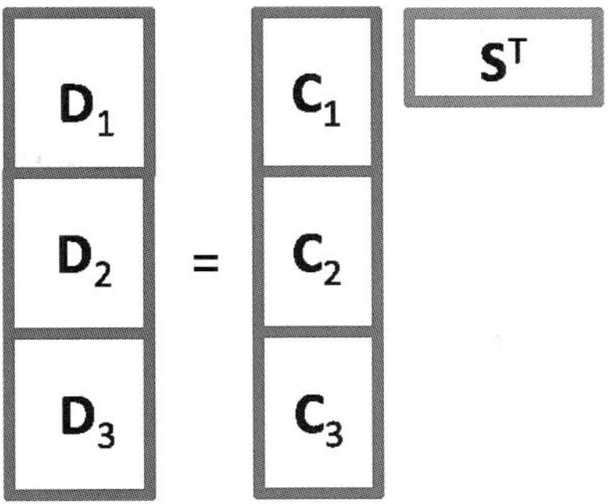

Fig. 4.   Column-wise augmentation of datasets merging into one data matrix which is represented by common spectra and individual concentration profiles.

present results that are useful for commonly occurring cases where only two excited-state species are present. These special cases cover a large number of energy conversion reactions.

$$D = CS = (CT)\left(T^{-1}S\right) = C_{new}S_{new} \qquad (7)$$

Let us now consider the form of the linear transformation matrix T. Specifically, we will see that it is useful to split T into the product of a mixing matrix M and a scaling matrix F, both of which must be invertible. The scaling matrix is a diagonal matrix with constants along its trace, $F_{ii} = a_i$, and zero off-diagonal elements. The mixing matrix then has a trace of ones, and off-diagonal elements are each a constant. The scaling matrix represents a scaling of a concentration profile by a given constant, and the associated spectra by the inverse of the same constant. This clearly leaves the surface created from the outerproduct of that spectra and concentration profile unchanged. In order to settle the scale uncertainty, some absolute information regarding the species cross section or population boundary conditions are necessary. However, we are often interested in establishing the normalized concentration evolution and spectral shape (if just as a first step). Therefore, in the first step we are not interested in the scaling matrix, but in the mixing matrix (that actually changes the normalized concentration and spectral profiles). We can learn about the uniqueness of the (normalized) solution with respect to the mixing matrix, without considering the exact scaling. In the following subsections, the practical utility of this separation will be demonstrated.

3.4.1. *Linear transformation with two populations*

If the transient absorption surface can be described by two excited-state species then $\mathbf{T}$ must be a 2×2 matrix with a non-zero determinant. It can be expressed in terms of $\mathbf{M}$ and $\mathbf{F}$ as shown in Eq. 8. The inverse of $\mathbf{T}$ can also be expressed in terms of $\mathbf{M}$ and $\mathbf{F}$ as shown in Eq. 9. Eq. 10 illustrates how the data surface is generated from the obtained concentration and spectral profiles, and how an equivalent surface can be generated by new matrices given by the the product of the original vectors and the mixing matrix (and its inverse). In this case, the scaling vectors are multiplied together and give the identity matrix. In order to compare solutions, we generally renormalize the spectra after rotation to have unit length, and also make the appropriate rescaling of the concentration profile. This allows convenient visual comparison of the mixed vectors, whose forms are shown in Eq. 11.

Examining the form of the mixed vectors shown in Eq. 11, we can make some general observations. First, in order that the determinant is non-zero $bc \neq 1$. If both b and c are large, then $\vec{C_1}$ is switched with $\vec{C_2}$ and $\vec{S_1}$ with $\vec{S_2}$. This is a trivial case. As b and c both approach 1 (or -1), then the concentration profiles become very similar, and the spectra approach a mirroring of each other through the x-axis. With the similarity of the concentrations, this parameter choice leads to the most parallel-like solutions. In order to judge the physicality of these solution, constraints regarding the shape of the spectra are useful. For example if the states have a common absorption bleach, observing when this feature stops looking similar in the two spectra is useful. For the case of approaching -1, then positivity of the concentration profiles is also useful. To make solutions more sequential-like, rotations can be found that push the initial population of one concentration profile towards zero, and the other towards a higher level at early time. Depending on which population profile one wishes to push toward zero, b and c can be selected by looking at the initial conditions of the resolved $\vec{C_1}$ and $\vec{C_2}$. A standard analysis procedure is to calculate the mixed vectors over a mesh of b and c values and highlight the regions where the non-negativity constraints are satisfied. This reveals all the possible concentration and spectra profiles which can describe the observations. From this, one gains a detailed insight into how certain one can be regarding the concentration evolution and spectra of given components.

$$\mathbf{T} = \mathbf{MF} = \begin{pmatrix} 1 & b \\ c & 1 \end{pmatrix} \begin{pmatrix} a & 0 \\ 0 & d \end{pmatrix} \tag{8}$$

$$\mathbf{T}^{-1} = \mathbf{F}^{-1}\mathbf{M}^{-1} = \begin{pmatrix} \frac{1}{a} & 0 \\ 0 & \frac{1}{d} \end{pmatrix} \frac{1}{1-bc} \begin{pmatrix} 1 & -b \\ -c & 1 \end{pmatrix} \tag{9}$$

$$\mathbf{CS} = \left(\vec{C}_1, \vec{C}_2 \right) \begin{pmatrix} \vec{S}_1 \\ \vec{S}_2 \end{pmatrix} = \vec{C}_1 \otimes \vec{S}_1 + \vec{C}_2 \otimes \vec{S}_2 = \mathbf{Surf}_1 + \mathbf{Surf}_2 = \ldots$$

$$\ldots = \mathbf{CTT}^{-1}\mathbf{S} = \mathbf{CMFF}^{-1}\mathbf{M}^{-1}\mathbf{S} = \ldots$$

$$\ldots = \left(\vec{C}_1 + c\vec{C}_2, \, b\vec{C}_1 + \vec{C}_2 \right) \begin{pmatrix} \vec{S}_1 - b\vec{S}_2 \\ -c\vec{S}_1 + \vec{S}_2 \end{pmatrix} \frac{1}{1-bc} \tag{10}$$

$$\vec{C}_{1new} = \vec{C}_1 + c\vec{C}_2$$
$$\vec{C}_{2new} = b\vec{C}_1 + \vec{C}_2$$
$$\vec{S}_{1new} = (\vec{S}_1 - b\vec{S}_2)\frac{1}{1-bc}$$
$$\vec{S}_{2new} = (-c\vec{S}_1 + \vec{S}_2)\frac{1}{1-bc} \tag{11}$$

In energy conversion reactions, it is not uncommon that one species is much longer-lived than the others. For example, when an exciton population (with an un-quenched lifetime on the sub-nanosecond scale) creates long-lived charges, we know that the signal left after a nanosecond comes exclusively from the long-lived charges. That means we know exactly what the (normalized) spectrum of the charges is. This knowledge of the spectral shape of a single component significantly simplifies the examination of the linear mixing ambiguity. First of all, when we do the MCR-ALS analysis, we can constrain the spectral shape of one of the species (let us assume that we constrained the first species to represent the charges). Then, we know that $\vec{S}_1$ is the correct spectrum, so that b must equal zero. This has the immediate consequence that the population evolution of the other species, that is $\vec{C}_2$, is precisely known! This rather surprising result can be very useful, as we will see in the coming sections. On the other hand, the spectra of the other species $\vec{S}_2$, and the population evolution of the species with the known spectra, $\vec{C}_1$, can still be changed by mixing. The validity of the mixed solutions can be in this case checked by running a linear sweep of c from positive to negative values. This will be demonstrated in the next section.

68

Such an analysis can be extended for greater number of components, although it is clear that as the number of components grows the parameter space for mixed solutions quickly expands. For simplicity, we will introduce this technique with some examples of two-state systems.

3.5. *MCR applied to a simple example*

In order to illustrate the process of analyzing a data matrix with MCR-ALS the simulated dataset which was analyzed in Sec. 2.1 with global analysis is decomposed and investigated with MCR-ALS here. Figure 5 shows a flowchart of the analysis process: First, forward and backward evolving factor analysis are carried out, from which two components are identified to be sufficient to explain the dataset as their eigenvalue is distinctly larger than the eigenvalue of the other components. The initial guess is taken from EFA as described in Sec. 3.1 and the ALS optimization is started, with the constraint that the concentrations should be non-negative. From EFA we know that the matrix is rank 1 at longer times, therefore only one species (A, shown in green) is known to be left in the end. Therefore we can introduce an additional constraint, namely that the spectrum of species A is known, as it is the spectrum which is found in the data matrix at long time delays. From the optimization process as depicted in the lower panel of Fig. 3 we obtain the concentration profiles of the species A and B and the before unknown spectrum of species B.

However, the analysis is not complete at this point even though the quality of the fit in terms of residuals is very good. As discussed in the previous section, we have to investigate the ambiguity due to possible linear transformations. In the current example we found that the matrix becomes essentially rank 1 after 200 ps. This means that we know the spectrum of the long-lived component with certainty (it is the spectra measured at long times) and only have a single parameter c that influences the rotation. Fig. 6 shows how the variation of this parameter affects the concentration and spectral profiles. We see that by choosing positive values of c the solution is rotated to a more parallel-like one, where the populations tend to monotonically decay. Rotation too far in this direction leads to spectral shapes that are mirror-images of each other, and are likely to be physically unreasonable as either induced absorption or absorption bleach is often expected for both species in given wavelength ranges. Rotation with negative c values leads to more sequential-like models. The limit of negative rotation is easy to find as a concentration profile starts to go negative at this threshold value. In this way, a clear band of possible factorizations (in which the

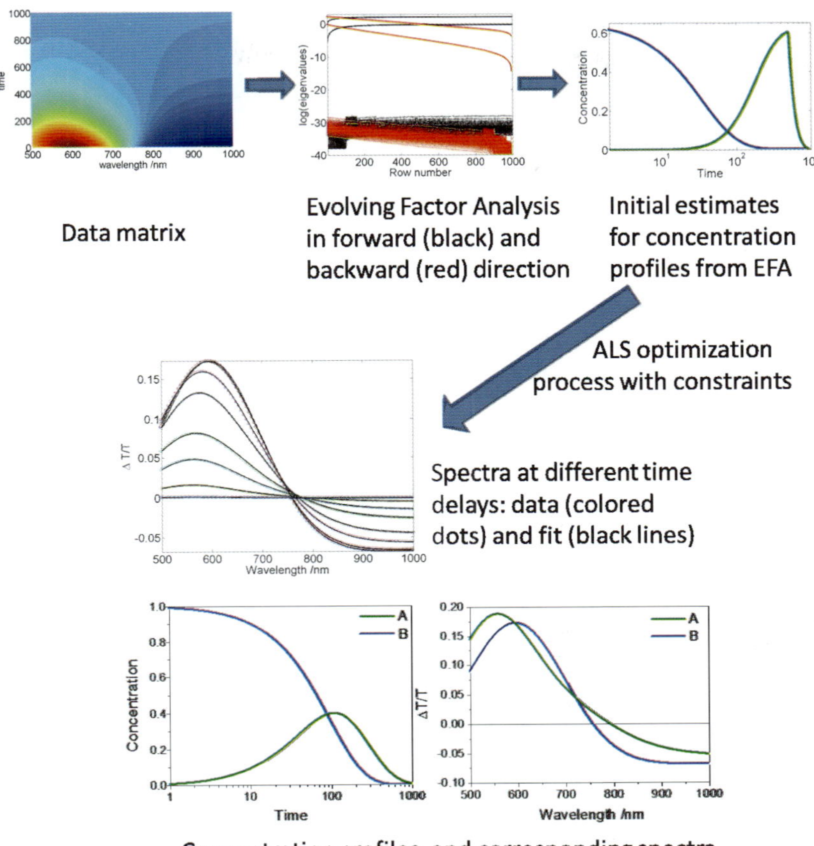

Fig. 5. (Color online) MCR-ALS analysis applied to a simulated dataset. EFA gives the initial estimates for the concentration profiles. The ALS algorithm incorporating constraints yields the spectra and concentration profiles shown in the bottom of the panel. The graph in the middle presents a comparison between the data (colored dots) and the fit (black lines) at time delays of 1, 10, 50, 100, 200, 300, 500 and 1000.

true physical factorization will be contained) can be simply generated by this soft modelling approach.

3.6. *General conclusions regarding hard and soft modelling*

In this section we have examined two methods for extracting concentration and spectral profiles from transient absorption data. In the section on global analysis by hard modelling we saw that good fits to the data surface could be generated using a system of parametrized rate equations. It is vitally

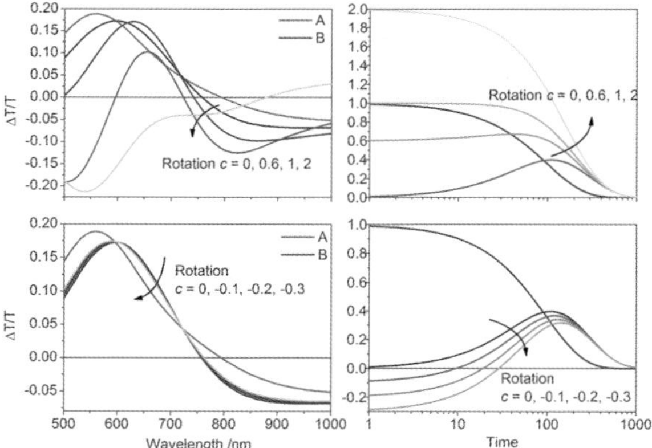

Fig. 6. Determining the rotational ambiguity of the MCR-ALS analysis applied to a simulated dataset. The upper panel shows how positive c values rotate the solution to more parallel-like models. Rotation to the positive eventually leads to spectra that become almost mirror images of each other. These are rarely likely to be physically-reasonable solutions. Rotations with negative values of c are shown in the bottom panel. These move the decomposition to a more sequential-like solution. In the current case, the resolved solution was the most sequential possible as we see that further rotation leads to negative populations.

important to note though, that many different systems of equations could fit a given dataset very well. So the simple observation that a given rate equation system yields a good fit in a global fitting approach is not sufficient evidence that the concentration profiles so-obtained are the physically correct ones! Global fitting only extracts the physically correct factorization if the system of rate equations are correct, this must be carefully considered and physically justified. Basing the argumentation of physical correctness on the goodness of the fit must be approached with extreme caution in the case of global analysis. Soft modelling, on the other hand, finds a factorization which obeys some known soft constraints, such as non-negativity in the concentration profiles, and known spectral information if available. In the same way that many different systems of rate equations can lead to good global fits, the concentration and spectra profiles found by MCR-ALS are usually not unique. However, the uniqueness of the MCR-ALS solution can be explored by mixing the found solution with linear transformation matrices as described in Sec. 3.4. This allows the researcher to see the full range of possible solutions that satisfy the most rigorously known constraints. From the solutions it is often possible to make further assumptions, for example relating to spectral form, which can lead to a better constrained estimate of the factorization. In any case, we feel that a general advantage of the soft-modelling technique is that the linear transformation uncertainty is clearly apparent throughout the analysis, and false 'certainty' in the factorization (often an unwanted by-product of global analysis) can thus be avoided. If the data surface can only be explained by a unique factorization, it can be found by soft-modelling. If additional insight is needed to find the physical decomposition, soft-modelling can provide clear insight into exactly how additional information can further constrain the solution bands. It provides a useful analytic tool to guide experimental effort and rigorously determine the valid physical decomposition of transient absorption surfaces.

4. Demonstrations of soft-modelling applied to the analysis of transient absorption surfaces of organic solar cells

4.1. *Separating exciton and charge dynamics during charge generation*

4.1.1. *Absorption and devices*

A rational design strategy to increase the efficiency of organic solar cells is to absorb a bigger part of the visible spectrum. As C_{60}-fullerene derivatives don't contribute strongly to the absorption in the sun's emission spectrum and C_{70}-derivatives are very expensive there is a need for alternative acceptors that fulfill requirements like high absorption coefficients, good photo stability, easy production and/or decent charge transport properties. One suitable class are perylene diimides (PDI).[6]

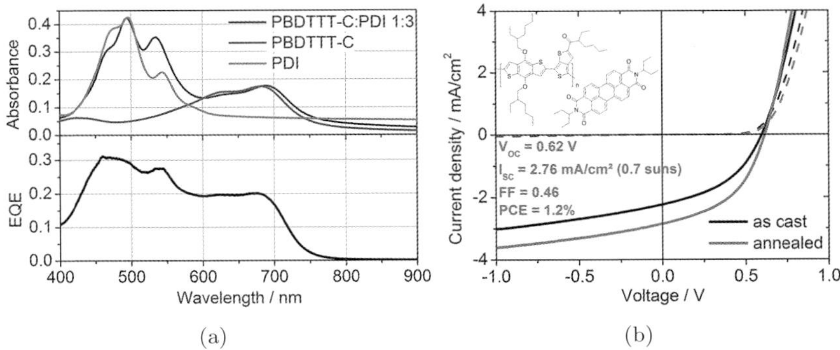

Fig. 7. (a) Upper panel: absorption data of PBDTTT-C, PDI and a blend of a 1:3 ratio of PBDTTT-C:PDI. Lower panel: EQE spectrum of a 1:3 blend. (b) IV-characteristics of a 1:3 blend using an PEDOT:PSS/ITO/Blend/Ca/Al structure.

For this case study a combination of a low-bandgap polymer, namely PBDTTT-C,[7] as donor and N,N'-bis-(1-ethylpropyl)-perylene-3,4,9,10-tetracarboxidiimide (PDI) as acceptor material were used (see Fig. 7(b)). This material combination exhibits complementary absorption spectra which is also reflected in the EQE spectrum that shows contributions of the donor as well as the acceptor (see Fig. 7(a)). In a solar cell this blend gives a power conversion efficiency of about 1.2 %.

4.1.2. *Transient absorption data and MCR analysis*

The TA data in the upper panel of Fig. 8(b) show the evolution with delay time of the transient absorption (TA) in the near infra-red of a PBDTTT-C:PDI 1:3 blend. In this near infra-red region only induced absorptions of excited-states are seen. At early times the spectral shape is strongly dominated by the photo-induced absorption of the exciton on the polymer (this can be established by comparison with a pure polymer film). At later times the spectra evolves until at long time, (~ 3 ns), the spectral shape is dominated by charge-induced absorption. The charge spectrum matches that observed in quasi-steady state photo-induced absorption measurements. The top panel of Fig. 8(b) shows the evolution of the induced absorption signal at two specific wavelength regions. The peak at 1425 nm is dominated by the exciton population, and the peak at 1100 nm is at later times dominated by charge-induced absorption. However, at early times the shoulder of the exciton-induced absorption strongly overlaps this charge peak, so the raw kinetics of the transient absorption signal do not reveal the population evolution of the charge population. As the growth of the charge signal is related to a partially diffusion-limited process, writing a physically meaningful system of rate equations is non-trivial. However, we will see that using the 'soft' insights into the solution that we already have, we can use MCR analysis to simply and precisely determine the physical factorization of the data surface in this case. Using non-negativity constraints, and the known spectra of the charges only, we obtain the concentration and spectra profiles from MCR-ALS shown in the bottom panels of Fig. 8(a). Now we will investigate whether this factorization is unique.

4.1.3. *Linear transformation ambiguity and summary*

The obtained solution needs to be investigated further with respect to rotational ambiguity. The known charge spectrum has been applied to population 1, so b is known to be zero. Rotation to negative values of c is not

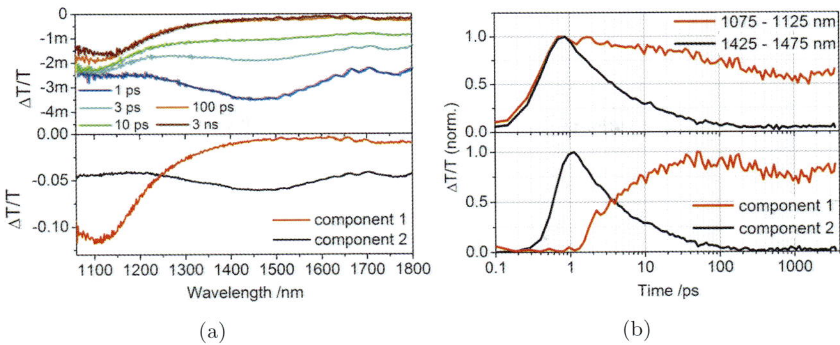

Fig. 8. (Color online)(a) Upper panel: TA spectra of a PBDTTT-C:PDI (1:3) blend excited at 650 nm. Lower panel: components obtained by MCR analysis of the shown TA data. (b) Upper panel: kinetics tracked at 1075–1125 nm (black, charge absorption) and 1425–1475 nm (red, exciton absorption). Lower panel: concentration profiles of the components shown in (a), showing that charges are not present from the beginning but evolve from excitons.

possible as this immediately violates the constraint of non-negative concentrations. An example of this is shown in the upper panels of Fig. 9 for a rotation with $c = -0.2$. The rotation to positive values is also very restricted, as the exciton spectrum starts becoming positive after small rotations. As there can be no absorption bleach or stimulated emission from the exciton in this wavelength region, a positive value in the spectrum is not physically reasonable. Several values for positive rotations are shown in the bottom panel of Fig. 9. Starting with the unrotated spectra and kinetics ($c = 0$) we can see how the exciton spectra and charge kinetics can be changed over a limited range of rotations. Therefore, we can find with soft-modelling the kinetics and spectra of the exciton and charge populations with a high degree of certainty based on simple and robust physical insights. The uncertainty in the evolution of the charge population on the short-timescale can be clearly quantified and related to uncertainty in the exciton induced-absorption spectrum. Naturally if we were to incorporate the measured exciton-induced absorption this ambiguity would be fully removed, but in this example we wished to present how unambiguous a factorization can be even when some physical insight into the system is absent.

4.2. *Analysing triplet formation mechanisms, specifically the role of nongeminate recombination*

Over the past years organic solar cells experienced a continuous improvement in performance. One of the best studied systems so far

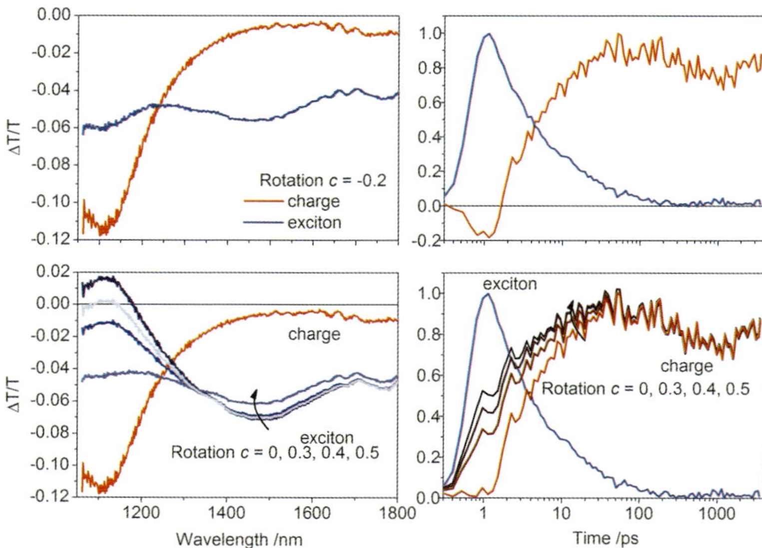

Fig. 9. Rotation of the results obtained from MCR analysis and presented in the lower panels of Fig. 8. Spectra of component 1 (charges) and component 2 (excitons) are shown on the left-hand side, the corresponding concentration profiles on the right-hand side. Rotation with negative c leads to negative concentrations, and therefore non-physical solutions. Rotation with positive c works for a short while until unphysical positive components in the exciton spectra emerge. From this analysis we can see the limited range of kinetics and spectra in which the physically true value must lie.

is the polythiophene/methanofullerene blend P3HT/PCBM for which power conversion efficiencies around 5 % were reported.[8] The limiting factor in P3HT, however, is the rather poor overlap with the solar spectrum leaving solar photons above the optical bandgap, which contain approximately 50 % of the total optical power, unused. Instead of choosing a stronger red-absorbing acceptor, several approaches were undertaken to push the donor polymer absorption further to the low energy spectral region to guarantee more efficient photon harvesting. The bridged bithiophene PCPDTBT,(poly[2,6-(4,4-bis-(2-ethylhexyl)-4H-cyclopenta[2,1-b;3,4-b']-dithiophene)-alt-4,7-(2,1,3-benzothiadiazole)), was one of the first low bandgap polymers used in OPV with an onset of photocurrent generation starting beyond 900 nm.[9] Although only moderate efficiencies around 3 % were reported in the beginning, improvement of the blend morphology by adding a high boiling point co-solvent during processing let to further improvements yielding power conversion efficiencies exceeding 5 %.[10] Hou et al. then exchanged the carbon bridgehead atom with a silicon atom which produced a more favorable morphology and increased mobility even without co-solvents, partially caused by an enhanced inter-chain packing.[11–13] Despite the broadened absorption in these low bandgap polymers compared to P3HT, the increase in power conversion efficiency is still only moderate, and therefore PSBTBT and PCPDTBT are interesting candidates for time-resolved optical spectroscopy to gain insight into the loss mechanisms limiting device performance.

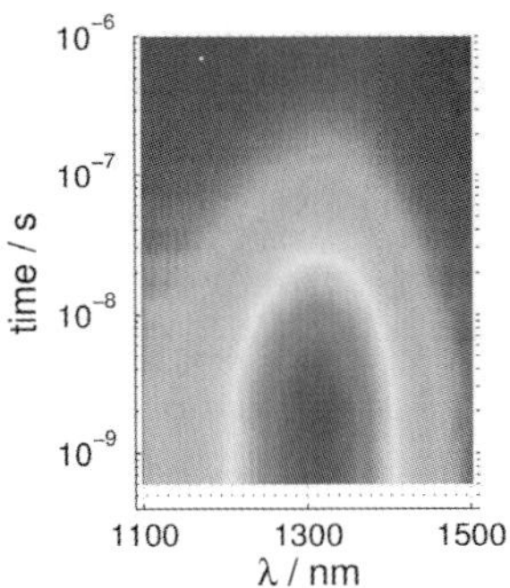

Fig. 10.   Normalized transient absorption of PSBTBT/PC$_{70}$BM.

In this case study ns -μs transient absorption data of PSBTBT/PC$_{70}$BM in a 1:1.5 ratio are presented and analyzed using the MCR-ALS method. Figure 10 shows the typical transient surface with a long-lived feature around 1300 nm. Species with associated lifetimes extending to the μs-regime are usually photo-generated charge carriers, i.e. fully dissociated electron-hole pairs, long-lived interfacial charge-transfer states, or triplet states. Due to the good photovoltaic performance this long-lived peak is intuitively assigned to charge-carriers as has previously been shown in literature for the carbon bridged polymer.[14–17] However, MCR-ALS analysis shows that this assignment may be ambiguous. The evolving factor analysis performed on the PSBTBT/PC$_{70}$BM blend film, Fig. 11, indicates the contribution of two species to the NIR feature throughout the whole time range. This shift in population between excited-state species with rather similar spectra manifests itself as a subtle spectral shift in the measured surface shown in Fig. 10. Therefore, MCR-ALS was performed with two species for an augmented matrix consisting of six different excitation densities with

76

the only restriction of non-negativity. It returned two strongly overlapping
spectra with long-lived concentration profiles, as shown in Fig. 12 as the
thick lines. Due to the long-lived character of both species further analysis becomes difficult because local rank restrictions do not allow the direct
assignment of the spectra like in the preceding case. Hence, the solution
contains a high degree of rotational ambiguity. For demonstration purposes
the higher energy species was fixed and the other rotated, resulting in large
spectral changes and corresponding alterations in the concentration profiles
rendering any conclusion about the relative contributions of the individual
species useless (see Fig. 12). This analysis reveals a significant potential
pitfall of global fitting. Because so many different concentration profiles
can fit this surface, a great variety of rate equation systems can be written
which would accurately describe the data surface after global fitting has
found the related spectra. However, what may not be clear after this global
fit, is that another system of equations could equally well fit the data (with
new spectra found). So for a meaningful global analysis of the data, a set
of equations must be found that are rigorously known to be correct, other-
wise the presented analysis is essentially meaningless. In this soft modelling
approach the significant uncertainty becomes immediately apparent, which
helps erroneous assignment of the physical decomposition to be avoided. It
allows the experimenter to see more accurately how much information regarding population evolutions and spectra is contained within experimental
observations, and aids the consideration of which additional experimental
information would be useful to increase the certainty with which these parameters can be extracted.

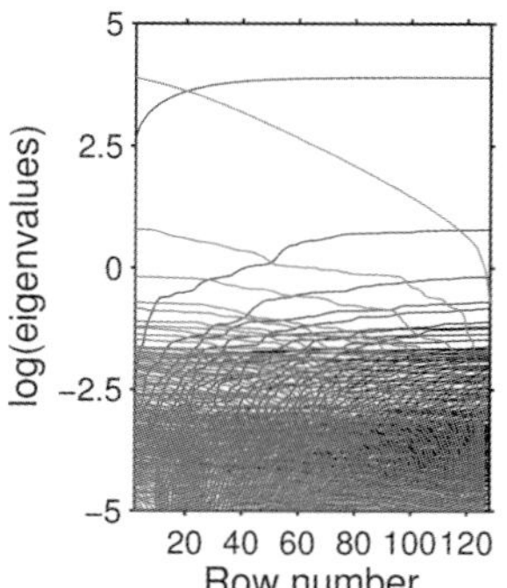

Fig. 11.   EFA of transient data. Black lines indicate the forward direction EFA, red lines
are backward direction.

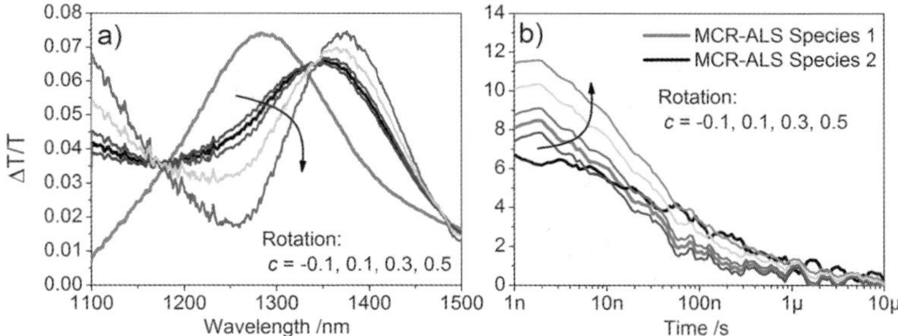

Fig. 12. MCR-ALS analysis of PSBTBT/PC$_{70}$BM together with rotational ambiguity. a) The initial spectral solution of the MCR analysis for species 1 (thick red line) and species 2 (thick black line). The rotated spectra of species 2 are plotted in blue colors. b) Concentration profiles corresponding to the spectra shown on the left. The rotated concentration profiles for species 1 are plotted in red colors.

5. Conclusion

In this short chapter we have discussed the general aim of transient absorption data analysis, namely the decomposition of an observed data matrix into the physically true concentration evolutions and spectra of a given number of excited states. In order to do this matrix factorization we have discussed global analysis, and a soft-modelling technique followed by an analysis of the linear transformations of the obtained factorization. When a system of equations for the population flows is not rigorously known, this second method has very significant value. It can be used to find the bands of factorizations that satisfy the known physical properties of the system. These bands accurately present how much information can be extracted from the measured data (using all the known physical properties of the system). Through the demonstrations presented, we hope to have illustrated how this analytical procedure can be of use in rigorously interpreting measurements and also guiding experimental design.

Acknowledgment

FL and IH acknowledge the Max Planck Society for funding of a Max Planck Research Group and Max Planck Scholarship, respectively. FE acknowledges the Max Planck Graduate Center with the University of Mainz (MPGC) and the Deutsche Forschungsgemeinschaft (DFG) in the framework of the priority program SPP1355 for financial support. DG thanks the Fonds der Chemischen Industrie for providing a Kekule scholarship.

References

1. L. J. G. W. van Wilderen, C. N. Lincoln and J. J. van Thor, *PLoS ONE* **6** (2011).
2. A. de Juan, M. Maeder, M. Manuel and R. Tauler, *Chemometrics and Intelligent Laboratory Systems* **54**, 123 (2000).
3. J. Jaumot, R. Gargallo, A. de Juan and R. Tauler, *Chemometrics and Intelligent Laboratory Systems* **76**, 101 (2005).
4. A. de Juan and R. Tauler, *Analytica Chimica Acta* **500**, 195 (2003).
5. A. de Juan and R. Tauler, *Critical Reviews in Analytical Chemistry* **36**, 163 (2006).
6. V. Kamm, G. Battagliarin, I. Howard, W. Pisula, A. Mavrinskiy, C. Li, K. Muellen and F. Laquai, *Adv. Energy Mater.* **1**, 297 (2011).
7. J. Hou, H. Chen, S. Zhang, R. Chen, Y. Yang, Y. Wu and G. Li, *Journal of the American Chemical Society* **131**, 15586 (2009).
8. M. D. Irwin, D. B. Buchholz, A. W. Hains, R. P. H. Chang and T. J. Marks, *Proceedings of the National Academy of Sciences* **105**, 2783 (2008).
9. D. Muehlbacher, M. Scharber, M. Morana, Z. Zhu, D. Waller, R. Gaudiana and C. Brabec, *Advanced Materials* **18**, 2884 (2006).
10. J. Peet, J. Y. Kim, N. E. Coates, W. L. Ma, D. Moses, A. J. Heeger and G. C. Bazan, *Nat Mater* **6**, 497 (2007).
11. J. Hou, H.-Y. Chen, S. Zhang, G. Li and Y. Yang, *Journal of the American Chemical Society* **130**, 16144 (2008).
12. M. Morana, H. Azimi, G. Dennler, H.-J. Egelhaaf, M. Scharber, K. Forberich, J. Hauch, R. Gaudiana, D. Waller, Z. Zhu, K. Hingerl, S. S. van Bavel, J. Loos and C. J. Brabec, *Advanced Functional Materials* **20**, 1180 (2010).
13. H.-Y. Chen, J. Hou, A. E. Hayden, H. Yang, K. N. Houk and Y. Yang, *Advanced Materials* **22**, 371 (2010).
14. I. W. Hwang, C. Soci, D. Moses, Z. Zhu, D. Waller, R. Gaudiana, C. J. Brabec and A. J. Heeger, *Advanced Materials* **19**, 2307 (2007).
15. F. C. Jamieson, T. Agostinelli, H. Azimi, J. Nelson and J. R. Durrant, *The Journal of Physical Chemistry Letters* **1**, 3306 (2010).
16. D. Di Nuzzo, A. Aguirre, M. Shahid, V. S. Gevaerts, S. C. J. Meskers and R. A. J. Janssen, *Advanced Materials* **22**, 4321 (2010).
17. F. Etzold, I. Howard, N. Forler, D. Cho, M. Meister, H. Mangold, J. Shu, M. Hansen, K. Mullen and F. Laquai, *Journal of the American Chemical Society* **134**, 10569 (2012).

INFRARED ULTRAFAST OPTICAL PROBES OF PHOTOEXCITATIONS IN Π-CONJUGATED POLYMERS/FULLERENE BLENDS FOR PHOTOVOLTAIC APPLICATIONS

C.-X. SHENG[1,2], U. HUYNH[1] and Z.V. VARDENY[1,†]

[1] Department of Physics & Astronomy, University of Utah,
Salt Lake City, Utah 84112, USA
[2] School of Electronic and Optical Engineering,
Nanjing University of Science & Technology, Nanjing 210094, China

We used the transient and cw photomodulation spectroscopies for studying the photoexcitations dynamics in blends of fullerene with two typical π-conjugated polymers, namely regio-regular poly (3-hexyl thiophene) (RR-P3HT) with self-organized π-stacked two-dimensional lamellae, and 2-methoxy-5-(2'-ethylhexyloxy) poly(p-para-phenylene-vinylene) (MEH-PPV) with amorphous nanomorphology, in a broad spectral range from 0.13 to 2.7 eV. In (1:1) weight ratio of C_{60}-doped MEH-PPV film, the intrachain excitons dissociate into polarons via an electron transfer reaction from the polymer chain (donor) onto the C_{60} molecule (acceptor) within 200 fs. In blends of RR-P3HT/fullerene [6,6]-phenyl-C61-butyric acid methyl ester (PCBM) with (1.2:1) weight ratio, we found that exciton dissociation proceeds in two steps: first, within a couple of ps the excitons generated in the polymer domains populate charge transfer complex states at the RR-P3HT/fullerene interfaces; this is followed by charge transfer ionization into delocalized charge polarons in the polymer and fullerene constituents within ~20 ps. We also report on the occurrence of ultrafast quantum interference anti-resonances between photoinduced infrared-active vibrations and the delocalized polaron band in RR-P3HT/PCBM blends, that lack in MEH-PPV/C_{60}; this shows the delocalization character of the photogenerated charges that contribute to the photocurrent in P3HT/PCBM blend. Our findings indicate that the film morphology plays a crucial role in carrier photogeneration in donor-acceptor blends.

1. Introduction

The charge photogeneration mechanism in π-conjugated polymer (PCP) based organic solar cells is one of the most important questions that is still debated. Studying this process is expected to be useful in the pursuit of more efficient organic solar cells. One of the more efficient experimental techniques for

---

[†] To whom correspondence should be addressed; e-mail: val@physics.utah.edu

investigating the charge photogeneration mechanism in D-A blends has been the transient and continuous wave (cw) photoinduced absorption (PIA) spectroscopies [1-20]. Among various organic photovoltaic materials based on PCPs, two of them have been extensively studied in the last decade. One is the blend of poly[2-methoxy,5-(2′-ethyl-hexyloxy)-p-phenylene-vinylene] (MEH-PPV) and fullerene, which forms amorphous film [6]. The other is the blend of regio-regular poly (3-hexylthiophene) (RR-P3HT) and [6,6]-phenyl-C61-butyric acid methyl ester (PCBM) [4] with self-organized π-stacked two-dimensional (2D) lamellae of the polymer domains in the films [5]. In disordered PCP such as MEH-PPV a single charge carrier added onto the polymer chain forms a spin-1/2 polaron, with two allowed optical transitions, P_1 and P_2 below the optical gap (Fig. 1). For highly ordered PCP such as RR-P3HT film, self-organization of the polymer chains results in the formation of lamellae structure perpendicular to the film substrate; such 2D lamellae sheets have strong interchain interaction due to the short interlayer distance of the order of 3.8 Å [5]. An interaction model introduced before [5, 7] shows that the interchain interaction in RR-P3HT lamellae splits the intrachain localized polaron levels so that the delocalized polaron (DP) levels are shallower; this results in energy shift of the allowed optically transitions, namely DP_1 and DP_2 depicted in Fig. 1 [5, 7].

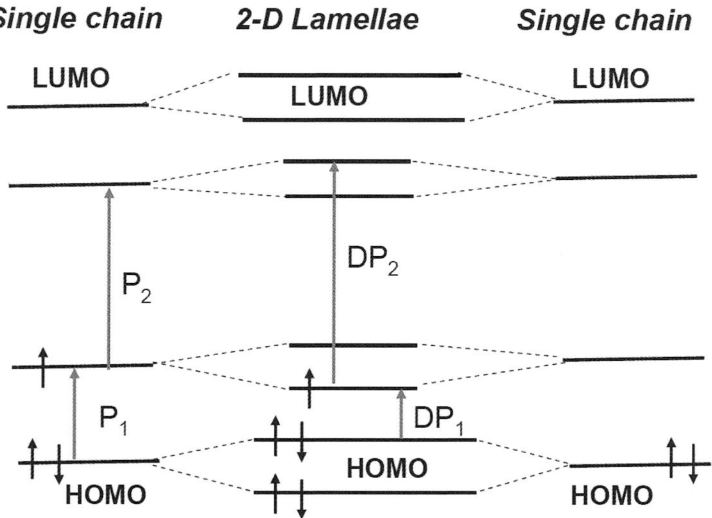

Figure 1. A model for the energy levels and related allowed optical transitions of intrachain localized polarons and interchain delocalized polarons.

Although there are still many unanswered questions [21-24], the polarons in polymer/fullerene blends are formed from the dissociation of photogenerated intrachain excitons in the polymer chains. However, the photoinduced absorption band of excitons in the visible spectral range partially overlaps with the P_2 and DP_2 bands[6, 7, 15]. Therefore, to clear out the role of excitons and polarons in the earliest time following photon absorption, ps transient spectroscopy measurement in the mid-IR spectral range are crucial. This is especially true for the new D-A blends that involve π-conjugated copolymers having low optical gap.

In this work we summarize our research of photoexcitations in blends of fullerene with these two typical PCPs, namely, 2-methoxy-5-(2'-ethylhexyloxy) poly(phenylene-vinylene) (MEH-PPV) and regio-regular poly (3-hexylthiophene) (RR-P3HT), using transient photomodulation (PM) spectroscopy in a broad spectral range from 0.13 to 2.7 eV with 150 femtosecond time resolution [6, 11, 15]. Our measurement technique is sensitive to non-equilibrium excitations in all states.

2. Experimental

For measuring the transient photoexcitation response in the fs to ns time domains, we used the fs two-color pump-probe correlation technique with linearly polarized light beams. The transient PM signal, $\Delta T/T(t)$ is the fractional change, ΔT in transmission, T, which is negative for PA, and positive for photobleaching (PB). Two fs Ti:sapphire laser systems were utilized to cover the broad probe photon energy that we have used; (i) a low power (energy/pulse ~0.1 nJ) high repetition rate (~80 MHz) [15] laser system for the mid-IR spectral range; and (ii) a high power (energy/pulse ~10 μJ) low repetition rate (~1 kHz) laser system for the near-IR/visible spectral range [6].

For the low intensity system in the mid-IR range both signal and idler output of an optical parametric oscillator (Tsunami, Opal, Spectra Physics) were used respectively as probe beams in the spectral interval $\hbar\omega$(probe) ranging from 0.55 to 1.05 eV. In addition a difference frequency set up based on a nonlinear optical crystal was used to extend the probe spectral range from ~0.13 to ~0.43 eV. For the high intensity system a white light super-continuum was generated having $\hbar\omega$(probe) in the range from 1.2 to 2.7 eV. The transient PM spectra from the two laser systems were normalized to each other in the near-IR/visible spectral range, where $\hbar\omega$(probe) from the low power laser system was doubled. The pump and probe beams were carefully adjusted to get complete spatial overlap on the film, which was kept under dynamic vacuum [6].

The steady state PM spectrum was obtained using a standard cw setup [7]. For excitation we used a cw Ar^+ laser pump beam at $\hbar\omega_L=2.5$ eV that was modulated at frequency f; and an incandescent tungsten/halogen lamp as the probe. The PM spectrum was measured using a lock-in amplifier referenced at f, a monochromator, and various combinations of gratings, filters, and solid-state photodetectors spanning the spectral range $0.3 <\hbar\omega(\text{probe})< 2.3$ eV. To cover the range $\hbar\omega(\text{probe})<0.3$ eV we used a FTIR spectrometer. The Ar^+ laser beam was modulated with a shutter, and the transmission spectrum of the film was measured with the excitation beam on and off. About 5000 scans for $T(\text{off})$ and $T(\text{on})$ (where $T(\text{on})$ and $T(\text{off})$ is the obtained transmission spectrum with the excitation laser beam on and off, respectively) were recorded, and subsequently $\Delta T/T$ spectrum was calculated using the relation $[T(\text{on})-T(\text{off})]/T(\text{off})$. The PM spectra obtained with both set ups were normalized to each other in the $\hbar\omega(\text{probe})$ range of 0.3-0.4 eV.

MEH-PPV, RR-P3HT, C_{60} and PCBM were purchased from ADS Inc.; they were used without further purification. The mixing ratio of the RR-P3HT/PCBM blend was 1.2:1 by weight; this ratio was chosen since it gives the highest power conversion efficiency in OPV cells application [25]. The mixing ratio of the MEH-PPV/C_{60} was 1:1 by weight. The films were drop cast onto CaF_2 substrates from dilute toluene solution. The films preparation was done in nitrogen atmosphere in a glovebox.

3. Results and discussion

Fig. 2(a) shows the cw PM spectrum of a MEH-PPV/PCBM film at 80 K; it contains two PA bands, namely P_1 and P_2 which are due to intrachain polarons. Fig. 2(b) shows the cw PM spectrum of a RR-P3HT/PCBM film at 80 K; it contains four PA bands. Two PA bands (DP_1 at 0.09 eV and DP_2 at 1.8 eV) are due to 2D delocalized polarons (DP) in the RR-P3HT ordered domains (lamellar structure with enhanced interchain coupling [5]); whereas the other two PA bands (P_1 at 0.35 eV and P_2 at 1.25 eV) are due to localized intrachain polarons (LP) in the disordered P3HT regions of the film [5, 7]. The interaction model depicted in Fig. 1 shows that the interchain interaction in the RR-P3HT lamellae splits the intrachain DP levels, resulting in energy shift of the DP allowed optically transitions [5] respect to those of LP. This may explain the reason that DP_1 is red shifted respect to P_1, whereas DP_2 is blue shifted respect to P_2 [5].

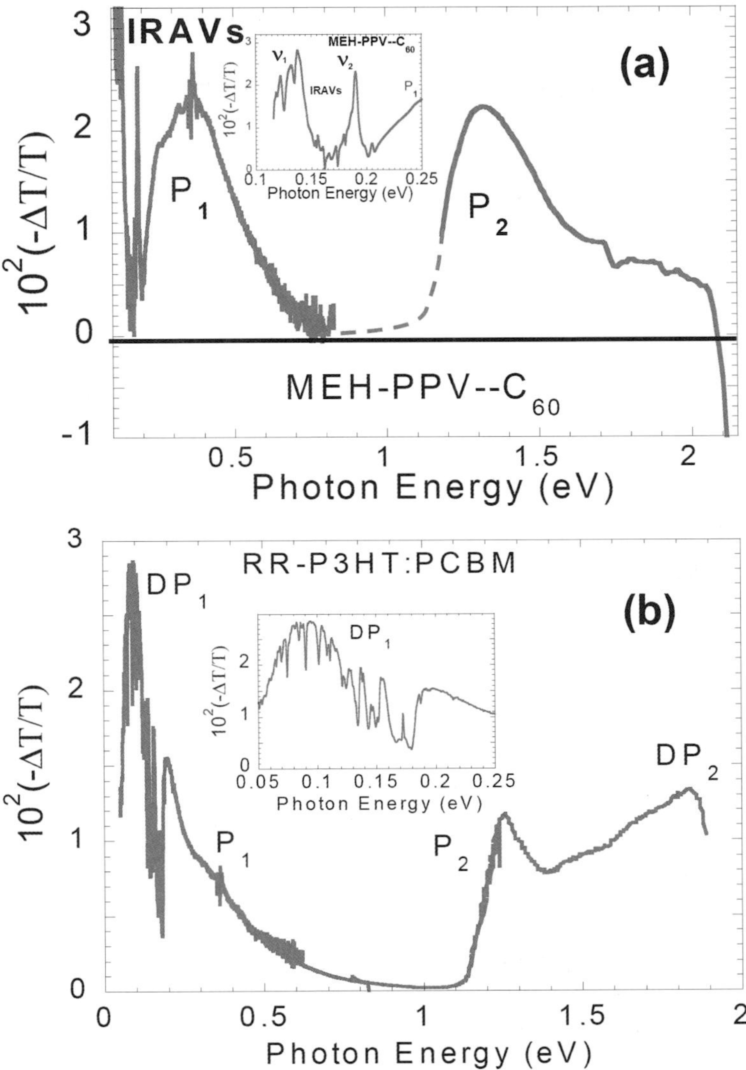

Figure 2. Cw PA spectrum of MEH-PPV/C_{60} (a) and RR-P3HT/PCBM (b) blend films at 80 K. Various PA bands are assigned where DP_1 and DP_2 denote delocalized polarons, and P_1 and P_2 denote localized intrachain polarons. Photoinduced IRAVs ν_1 and ν_2 are assigned in (a) inset. The inset in (b) shows the DP_1 band in more detail.

The sharp dips (or anti-resonances) superimposed on the DP_1 band (shown more clearly in Fig. 2(b) inset) are due to photoinduced infrared-active vibrations (IRAVs), which usually appear as PA lines in less ordered films such as MEH-PPV)/C_{60} blend (Fig. 2(a)) [6]. The neutral polymer chain has a set of Raman active A_g-type vibrations that are strongly coupled to the electronic bands via the e-p coupling [26]. These vibrations have been dubbed '*amplitude modes*' (AM) [27]. The AM model has had a spectacular success in explaining the resonant Raman scattering (RRS) dispersion as well as photoinduced and doping induced IRAVs in PCPs [26, 28]. Because the Raman frequencies are much smaller than the optical gap, and the corresponding IRAV frequencies are much smaller than the energies of photoinduced and doping induced electronic bands, previous applications of the AM model were based on the adiabatic approximation [26-28]. However this approximation fails in the case of ordered PCPs such as RR-P3HT/PCBM blends, since the IR-active lines overlap with the electronic transition in this case [5, 7, 15]. Therefore the cw photoinduced Fano-type anti-resonances (AR) have been explained using the non adiabatic amplitude mode model and charge density wave conductivity band, which indicates the existence of a continuum band in RR-P3HT/PCBM blend films [5].

Fig. 3(a) shows the transient PM spectra of a 1:1 MEH-PPV/C_{60} blend with 400 nm excitation. The $t = 0$ PM spectrum contains three PA bands peaked at 0.4 eV (P_1), 1.5 eV (P_2), and 1.0 eV (PA_1), which are respectively due to polarons and excitons on the polymer chains; and an accompanying PB band above 2.2 eV. The dissociation process of exciton in this film occurs within 200 fs (Fig. 3(a) inset), in agreement with transient measurements done in other laboratories [29]. The inset to Fig. 3(a) shows that P_1 polaron and PB (at 2.5 eV) increase with life time about ~9.5 ps (black line) *on the expense* of PA_1 exciton decay; so that at $t \sim 10$ ps only the polaron PA bands P_1 and P_2 remain in the PM spectrum, in contrast to pristine MEH-PPV [11]. This shows that with above gap excitation the intrachain excitons *separate into polarons* via an electron transfer reaction from the polymer chain (donor) onto the C_{60} molecule (acceptor). This is the traditional photoinduced charge transfer mechanism postulated to occur in the blend [30]. In Fig. 3(b) we show that P_1 band has a much longer lifetime compared with PA_1 of excitons, as well as with P_1 lifetime in pristine MEH-PPV [11].

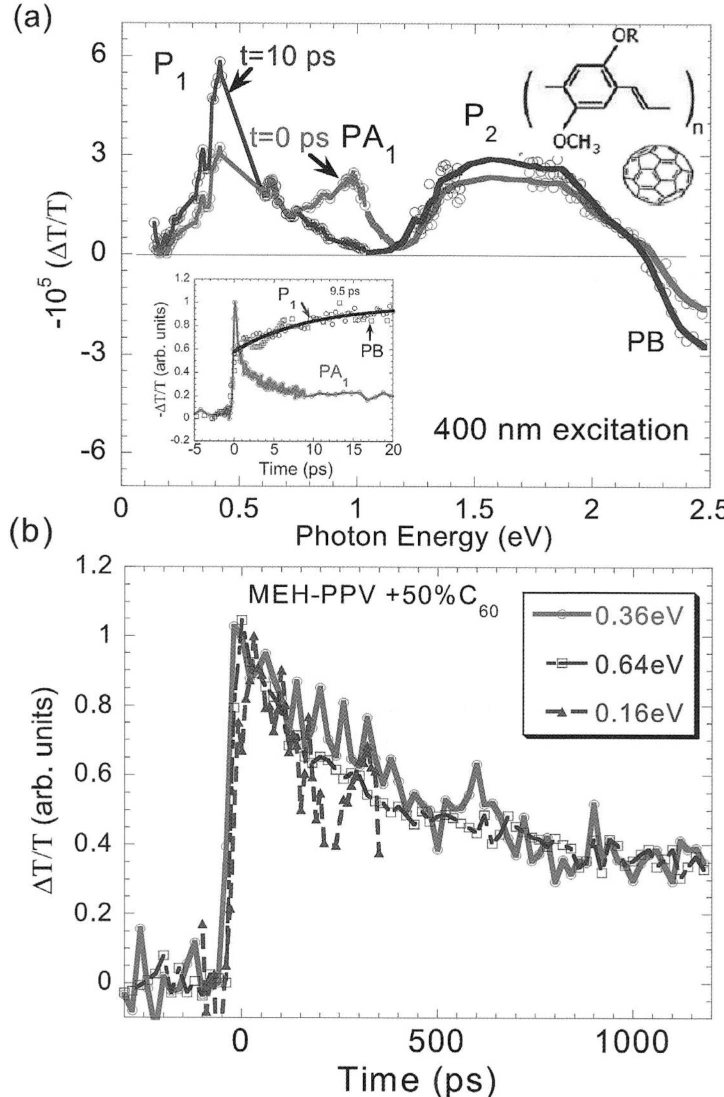

Figure 3. (a) The transient PM spectrum of MEH-PPV/C$_{60}$ (1:1) blend film generated using above gap excitation at t=0 (red) and 10 ps (blue). The polaron PA bands P$_1$ and P$_2$, exciton PA$_1$, and PB band are assigned. The insets in (a) show the ultrafast dynamics of the various bands up to 15 ps, and the polymer and fullerene backbone structures, respectively. (b) PA decay dynamics at various probe energies.

Although the DP excitation in the lamellae of RR-P3HT/PCBM blend has been recognized for playing an important role in the optoelectronic response of organic devices [31], the mechanism for the DP photogeneration in RR-P3HT/PCBM is still debated. One model proposes that the exciton photoexcitation in the polymer constituent dissociates within hundreds of femtoseconds [31], while the electron transfers to the PCBM molecule, and the remaining hole resides in the polymer chain. However whether the hole excitation in the polymer domain is LP or DP type is still unclear. It has been also suggested [21] that the ultrafast charge separation in polymer/fullerene blends occurs before localization of the primary excitation to form a bound exciton. In contrast, evidence for the existence of a charge transfer complex (CTC) state in the interfaces between the polymer and PCBM domains, and its role as an intermediate state in the DP photogeneration have been also reported [1, 6, 9, 32]. Several techniques such as electro-absorption, below-gap pump excitation PM, as well as photoluminescence [6, 9, 32] have been applied to study the CTC in polymer/fullerene blends. For further studying the CTC role in the charge photogeneration in the blend we need to investigate the photoexcitation ultrafast dynamics in the mid-IR spectral range.

Fig. 4(a) and (b) present the transient PM spectra of RR-P3HT/PCBM blend film measured at t = 0 ps and t = 50 ps, from 0.13 eV to 1 eV and from 1.25 eV to 2.25 eV, respectively. The t=0 PM spectrum contains five PA bands: PA_1 at ~0.95 eV, P_1 at ~0.35 eV, and DP_1 band below 0.3 eV in the mid-IR range; P_2 at ~1.4 eV and DP_2 at 1.9 eV (seen in Fig. 4(b)) in the near-IR range. PA_1 band was previously assigned to optical transitions related to the photogenerated singlet intrachain excitons [8, 10]; it is seen that it decays within about 20 ps (Fig. 5(a)). Consequently the transient PM spectrum at t = 50 ps contains only four bands, namely DP_1, P_1, P_2 and DP_2 (in increasing photon energy order); consistent with the cw PM spectrum (Fig. 2(b)). However, the relative intensities of the PA bands are very different in the cw and transient spectra. This indicates that the various PA bands have different dynamics.

PA_1 dynamics shown in Fig. 5(a) cannot be fit using a single exponential decay; a better fit is achieved using two exponentials, namely $A_1\exp(-t/\tau_1)+A_2\exp(-t/\tau_2)$ with $\tau_1 = 2.6$ ps, $\tau_2 = 22$ ps, respectively. Compared to the fast decay of excitons in MEH-PPV/C_{60} blends, the relatively large τ_1 value is surprising because the exciton dissociation was expected to be on a time scale less than 1 ps [9, 22]. We thus conclude that the delayed dynamics involves

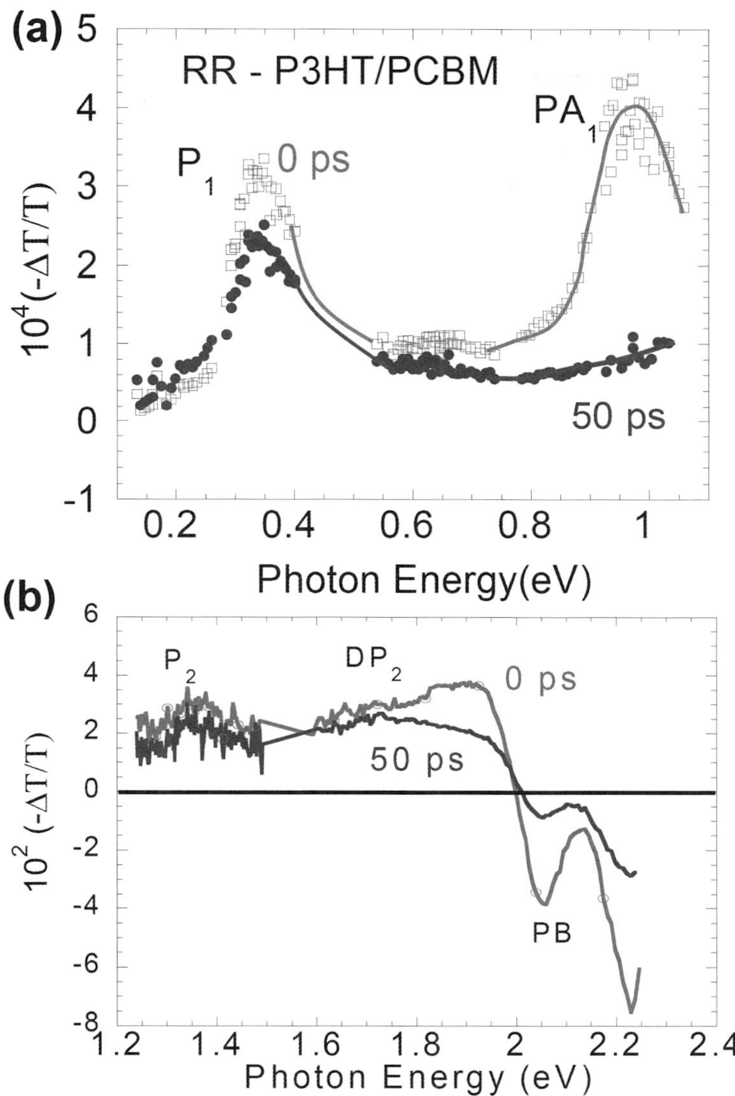

Figure 4. The transient PM spectrum of (1:1.2) RR-P3HT/PCBM blend in the mid-IR range (a) and near-IR/visible range (b) at t=0 and 50 ps, respectively. The bands PA$_1$, P$_1$ and DP$_1$, P$_2$ and, DP$_2$ PB are assigned.

exciton diffusion to the donor-acceptor (D-A) interfaces. In contrast, P_1 dynamics (Fig. 5(b)) reveals that these polarons are created instantaneously along with the excitons [14]; and in addition P_1 does not grow at the expense of PA_1 decay. Also P_1 remains constant after about 150 ps (Fig. 5(a)). Moreover PA_1 and P_1 dynamics do not change with the excitation intensities (not shown). We thus conclude that the photogenerated LP (P_1) and exciton (PA_1) species are uncorrelated. The branching ratio, η of photogenerated polarons/excitons has been estimated from the relative intensity ratio P_1/PA_1 [i.e. integrated area under each respective PA band]. For this we used the low intensity laser system in order to avoid the complex recombination processes typical of high intensity laser systems [32]; we thus estimated $\eta{\sim}80\%$ at t = 0 ps. This indicates that the initial LP population is relatively higher in the blend compared to the pristine RR-P3HT film (η=30% [11]), which could be explained as due to larger disorder and impurity density in the blend. On the other hand DP_1 at 0.25 eV grows exponentially, with a time constant τ=19 ps (Fig. 2(b)), which is very close to the τ_2 time constant of PA_1 decay. This shows that DP_1 rises on the expense of PA_1 decay, and this is compelling evidence that the delocalized polarons are in fact created once the excitons decay when arriving to the D-A interfaces. Nevertheless, we should point out that for both P_1 and DP_1, there is a slightly rising process at first 200 fs [Fig. 5(b)], which we contribute to the thermal dissociation of localized polaron pair at the P3HT/PCBM interfaces because of the extra pump photon energy [2, 33].

In addition the PM spectrum in the visible/near-IR range (Fig. 5(b) inset) shows a rise of DP_2 band in the ps range. We note that its dynamics was probed at 1.55 eV (within the DP_2 spectral range) using the low intensity laser system to avoid bimolecular recombination. We can clearly see that the 'rise dynamics' of DP_2 and DP_1 measured using the same laser system are identical (Fig. 5(b) inset), which gives another evidence for the generation of DP excitations on the expense of the excitons decay [17, 33]. The PM also shows a spectral feature associated with photoinduced electro-absorption modulation with phonon sidebands at ~1.9, 2.05, and 2.23 eV, respectively [11]; which indicates that most photoexcitations in the blends are in fact charged, and hence form photoinduced internal electric fields.

A striking consequence of the DP photogeneration in the ps time domain is the quantum interference AR between photoinduced infrared vibrations and DP_1 band, as shown in Fig. 4(a). The inset of Fig. 4(a) shows the cw PM spectrum of

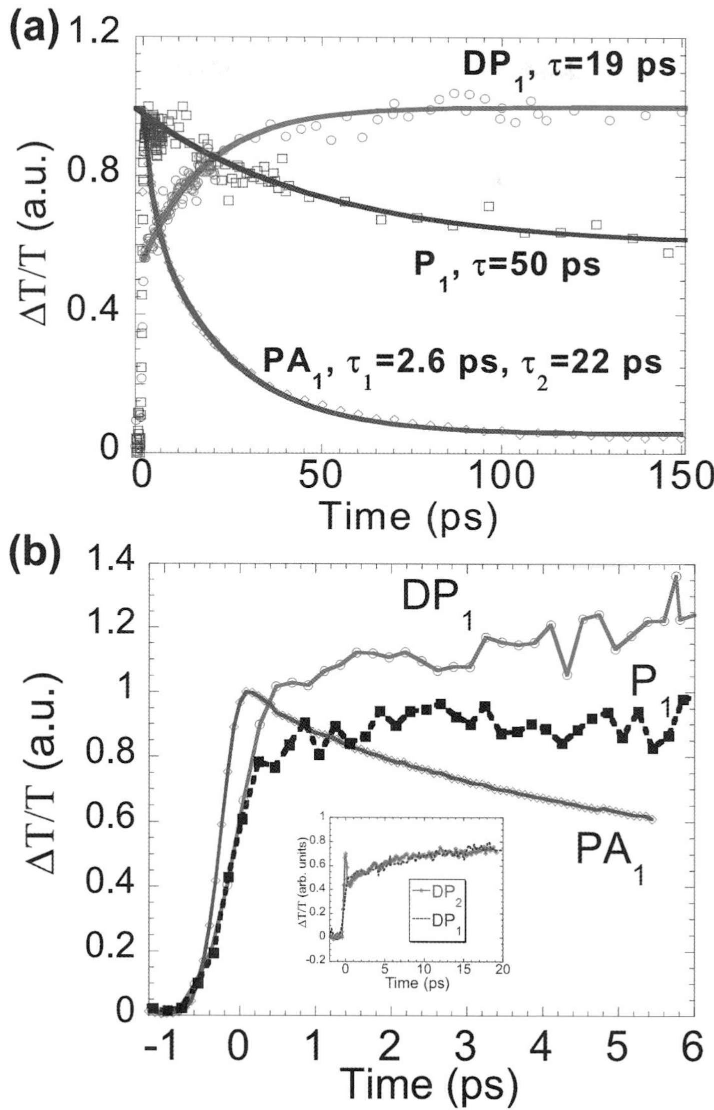

Figure 5. (a) and (b): Transient dynamics at various PA bands up to 150 ps and 5 ps, respectively. The lines through the data points in (a) are fitting, whereas the lines in (b) are to guide the eye. The inset in (b) compares the transient dynamics of DP_1 and DP_2 up to 20 ps.

the blend film for comparison. The dips at ~0.18 eV and ~0.16 eV in the transient PM spectrum clearly appear after ~200 ps with similar spectrum as that obtained in cw conditions. The AR dips are caused by the overlap between the discrete vibrational lines and DP_1 continuum band [34]. The complex AR structure in the cw PM spectrum was calculated before using the non-adiabatic version of the amplitude mode (AM) model coupled to charge density wave of a conductivity continuum band [34]. In the calculation, both vibronic and electronic excitations, as well as their quantum interference have to be taken into account to evaluate the conductivity $\sigma(\omega)$ (hence the absorption spectrum since imaginary $\sigma(\omega) \approx \alpha(\omega)$).

The conductivity spectrum $\sigma(\omega)$ consists of two parts [35, 36]; one is determined by the most strongly coupled phonons in the non-adiabatic limit; the other is related to the system response in the absence of phonons. In the charge-density wave approximation, the sharp features in $\sigma(\omega)$ are given by the relation [34]:

$$\sigma(\omega) \approx \frac{1 + D_0(\omega)[1 - \alpha_p]}{1 + D_0(\omega)[1 + C - \alpha_p]} \tag{1}$$

where C is a constant that presents a smooth electronic response; α_p is defined as polaron-vibrational 'pinning parameter' for the trapped polaron excitation; and $D_0(\omega)$ is the 'bare' phonon propagator. The latter is given [27] by the relation: $D_0(\omega) = \Sigma_n d_{0,n}(\omega)$, and $d_{0,n}(\omega) = \lambda_n/\lambda\{(\omega_n^0)^2/[\omega^2-(\omega_n^0)^2-i\delta_n]\}$, where ω_n^0, δ_n and λ_n are the 'bare' phonon frequencies, their natural linewidth (inverse lifetime) and electron-phonon (e-p) coupling constant, respectively; and $\Sigma\lambda_n = \lambda$, which is the total e-p coupling.

The poles of Eq. (1), which can be found from the relation: $D_0(\omega) = -(1-\alpha_p +C)^{-1}$ produce peaks (or IRAV's) in $\sigma(\omega)$. We have previously used the IRAV's, which appear as positive absorption lines to identify the charge state of photoexcitations in the ps PM spectra of less ordered PCPs' films [37]. In contrast the zeros in Eq. (1), which can be found by the relation: $D_0(\omega) = -(1-\alpha_p)^{-1}$ produce indentations (or ARs) in $\sigma(\omega)$. Eq. (1) was used before to successfully fit the cw AR features indicating the existence of a continuum band in RR-P3HT/PCBM blend films [34]. Therefore the obtained transient AR features in the ps transient PM spectrum provides direct evidence that the photogenerated DP species in the blend are *free charges*, since they form a continuum absorption band that is coupled to IRAVs.

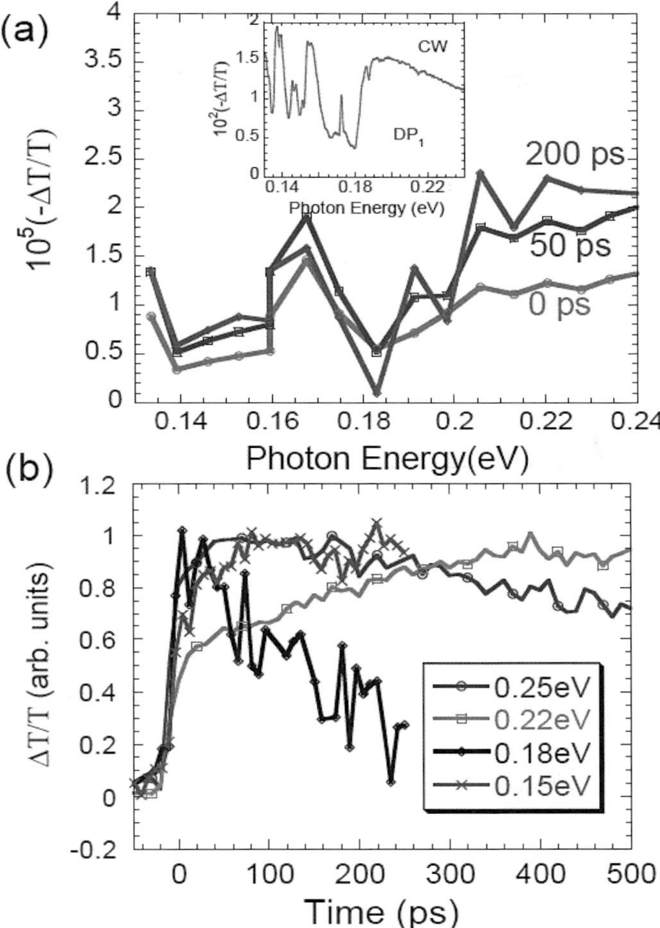

Figure 6. (a) Transient PA spectrum of RR-P3HT/PCBM blends from 0.13 eV to 0.24 eV measured at t=0 ps, 50 ps, and 200 ps, respectively. The inset in (a) shows the cw PA spectrum in the same spectral range. (b) Transient dynamics at various probe energies.

The transient dynamics of the PM spectrum near the AR features are shown in more detail in Fig. 4(b); dramatic different behavior for close-by energies is clearly seen. It is interesting to note that the PA dynamics at 0.18 eV and 0.16 eV are very different, although they both belong to the AR features. Apparently the anti-resonance effect is stronger at 0.18 eV, which is probably caused by the stronger e-p coupling of this mode (related to the C=C stretching vibration) [38].

4. Conclusions

In summary, for self-organized nanocrystalline P3HT/PCBM blends, there are three types of primordial photoexcitations in the fs-ps time domain. These are localized polarons; delocalized polarons; and singlet excitons. We found that localized polarons and singlet excitons are instantaneously photogenerated, the respective quantum yield depends on the way the films are prepared. In contrast, delocalized polarons which effectively contribute to the photocurrent are generated at the expense of excitons within ~20 ps, but do not correlate well with the generated localized polarons in fs-ps time domain. The high degree of phase separation in the blend results in the diffusion of excitons generated in the polymer domains towards the D-A interfaces, thereby forming an intermediate charge transfer exciton. We also identified the quantum interference anti-resonances between photoinduced infrared-active vibrations and delocalized polaron band in the ps time domain, indicating that the delocalized polarons form a continuum band in the blend film that indicates their ability to participate in charge transport.

In contrast, in amorphous C_{60}-doped MEH-PPV film, the intrachain excitons separate into charge transfer complex within 200 fs because of low degree of phase separation. On the other hand, the polarons are basically localized which is ascribed to the reason of poor solar cell power conversion efficiency.

Our findings indicate that the film morphology plays a crucial role in carrier photogeneration in donor-acceptor blends. More thorough exploration of the charge transfer processes and their relation with free carriers may improve the power conversion efficiency of organic solar cell and drive the development of novel photoactive materials.

Acknowledgments

We thank Drs. M. Tong and C. Yang for their help with the measurements. The work was supported in part by the DOE Grant No. DE-FG02-04ER46109.; NSF of China No. 61006014; the Fundamental Research Funds for the Central Universities (Grant No. 30920130111008).

References

1. T. M. Clarke and J. R. Durrant, *Chem. Rev.* **110**, 6736 (2010)
2. G. Grancini, et al., *Nature Materials* **12**, 29 (2013)
3. X. Yang et al., *Organic Electronics* **14**, 2058 (2013)

4. A.A. Bakulin, et al., *Science* **335**, 340 (2012)
5. R. Österbacka, et al., *Science* **287**, 839 (2000)
6. T. Drori, et al., *Phys. Rev. Lett.* **101,** 037401 (2008).
7. X. Jiang, et al., *Adv. Funct. Mater.* **12**, 587 (2002)
8. T. Kobayashi, et al., *Phys. Rev.* **B83**, 035305 (2011)
9. T. Drori, J. Holt, and Z.V. Vardeny, *Phys. Rev.* **B82** 075207 (2010)
10. F. Paquin, et al., *Phys. Rev. Lett.* **106**, 197401 (2011)
11. C.-X. Sheng, et al., *Phys. Rev.* **B75**, 085206 (2007)
12. J. Piris, et al., *J. Phys. Chem.* **C113**,14500 (2009)
13. D. Veldman, et al., *Adv. Func. Mater.* **19**, 1939 (2009)
14. J. Cabanillas-Gonzalez, et al., *Adv. Mater.* **23**, 5468 (2011)
15. C.-X. Sheng, et al., *Organ. Elec.* **13**, 1031 (2012)
16. S. Singh, et al., *Phys. Rev.* **B85**, 205206 (2012)
17. J. Guo, et al., *J. Am. Chem. Soc.* **132**, 9631 (2010)
18. T. J. Savenjie et al., *J. Phys. Chem. Lett.* **12**, 1368 (2011)
19. K. Aryanpour, et al, *Phys. Rev.* **B83**, 155124 (2011)
20. M. Tong, M. et al., *Phys. Rev.* **B75**, 125207 (2007).
21. N. Banerji, et al., *J. Phys. Chem.* **C115**, 9726 (2011).
22. I.A. Howard, *J. Am. Chem. Soc.* **132**, 14866 (2010).
23. A.L. Ayzner, *J. Phys. Chem.* **C113**, 20050 (2009)
24. L.G. Kaake, D. Moses, and A. Heeger, *J. Phys. Chem. Lett.* **4**, 2264 (2013)
25. G. Li, et al., *Nat. Mater.* **4**, 864 (2005).
26. Z.V. Vardeny et al., *Phys. Rev. Lett.*, **51,** 2326 (1983).
27. B. Horovitz, *Solid State Commun.* **41**, 729 (1982).
28. E. Ehrenfreund et al., *Phys. Rev.* **B36,** 1535 (1987).
29. A. Kohler, et al., *Nature* **392**, 903 (1998).
30. N.S. Sariciftci et al., *Science* **258**, 1474 (1992)
31. M. Hallermann, et al., *Adv. Funct. Mater.* **19**, 3662 (2009)
32. C. Silva, et al., *Phys. Rev.* **B64**, 125211 (2001)
33. J. Guo et al., *J. Am. Chem. Soc.* **132**, 6154 (2010)
34. R. Österbacka et al., *Phys. Rev. Lett.* **88**, 226401 (2002)
35. M.J. Rice, *Phys. Rev. Lett.* **37**, 36 (1976)
36. B. Horovitz, H. Gutfreund, and M. Weder, *Phys. Rev.* **B17**, 2796 (1978)
37. P.B. Miranda, D. Moese, and A.J. Heeger, *Phys. Rev.* **B64**, 081201 (2001)
38. E. Ehrenfreund, et al., *Phys. Rev.* **B36**, 1535 (1987)

ULTRAFAST OPTICAL PROBING OF CARRIER MOTION IN CONJUGATED POLYMERS AND BLENDS FOR SOLAR CELLS[*]

VIDMANTAS GULBINAS, ANDRIUS DEVIŽIS and DOMANTAS PECKUS[†]

Center for Physical Sciences and Technology, Savanoriu 231, LT-02300 Vilnius, Lithuania.

DIRK HERTEL

Department of Chemistry, Physical Chemistry, University of Cologne, Luxemburgerstr. 116, 50939 Cologne, Germany

We discuss application of time-resolved carrier mobility measurement methods for investigation of the carrier motion in organic semiconductors. We demonstrate that ultrafast optical probing of the electric field dynamics by means of the field-induced second harmonic generation together with conventional electrical methods enable monitoring of the charge carrier motion starting from their generation until extraction from the samples. We shortly discuss the electric field dynamics in neat conjugated polymer and in blend of organic molecules with fullerene.

1. Introduction

Organic optoelectronics has undergone tremendous development over the recent decades. There are already many commercial products based on organic technology. Manufacturing technologies of organic optoelectronic devices based on solution and printing promise low processing costs, environment-friendly fabrication and disposal. Large scale OLED (organic light emitting diode) displays have fundamental advantages over other existing technologies in terms of the energy consumption efficiency. Mechanical flexibility is another attractive and unique feature of organic devices. However, there is a lot of space for improvement, and deeper fundamental knowledge about electronic processes in organic materials is required.

Charge carrier mobility is one of the crucial parameters that determine applicability of organic semiconductors. Despite record braking bulk mobilities

---

[†] Work partially funded by the European Social Fund under the Global Grant measure.

above 10 cm^2/Vs in single cystalline organic materials [1], polymers have characteristic mobilities of the order of $10^{-2} - 10^{-6}$ cm^2/Vs [2]. A common approach is that charge transport in disordered organic materials occurs via temperature-activated hopping and is relatively slow [3]. Based on calculations of the charge transfer integrals in organic solids [4] and effective-mass calculations of charge carriers in conjugated polymers [5] a charge mobility as high as 1000 cm^2/Vs can be predicted under ideal conditions. The calculated effective carrier mass agrees with experimental results obtained in single crystalline polydiacetylenes [6]. Thus, it appears that high mobilities in organic solids might be feasible. This is further substantiated by recent studies of optical coherence in these materials [7]. Microwave conductivity measurements in a ladder-type polyphenylene (MeLPPP) have shown that the one-dimensional mobility is close to 600 cm^2/Vs [8]. On the contrary, the average bulk mobility in amorphous organic semiconductors is orders of magnitude lower. Apparently, the conventional mobility, which is commonly assumed as being constant in time, might not be sufficient to fully describe charge transport on the nano scale, such as in organic devices. Considering that the size of organic device structures ranges from almost molecular dimensions to micrometer length, it is a challenge to probe charge transport on vastly different time scales ranging from femtoseconds to microseconds. Insight into mobility dynamics could improve charge transport models. However, it was not possible to probe the time-resolved charge carrier drift in disordered organic materials on ultrafast time scales due to the lack of an appropriate experimental method. In conventional experiments the carrier mobility is averaged in time, and information about its dynamics is not accessible. Fast time and short length scales are important because active layers of the organic opto-electronic devices such as OLEDs or solar cells are of the order of tens of nanometers only. Consider, for example, photoexcitation of a polymer in a solar cell. On the scale of this dimension non-stationary effects can manifest themselves as reported in [9]. Juška *et al.* employed integral mode TOF [10, 11] to elucidate initial transport of the photogenerated charge carriers. They obtained the same fast initial transport distance for two different polymers RRa-PHT and PFB. It exceeded few tens of nanometers and was electric field-dependent. The initial transport on a picosecond time scale escaped determination due to limited time resolution of TOF measurements. Higher time resolution was demonstrated by Lee *et al.*, who implemented Auston strip line technique ensuring time resolution of tens of picoseconds [12]. Their investigations showed fast, temperature-independent photoconductivity during initial hundreds of picoseconds in PPV polymer. It was attributed to the interband carrier generation. Several attempts have been made to examine initial transport processes with better time resolution using pulse-radiolysis time-resolved microwave conductivity (TRMC) [13] or terahertz spectroscopy methods [14]. All of the above mentioned techniques have serious drawbacks that limit their application. With TRMC, free carriers are probed on the molecular dimensions only, often in a liquid solution. Highly sophisticated modeling has to be used to extract information on the mobility and the obtained values are time-averaged. Ultrafast terahertz spectroscopy allows time resolution of less than 1 ps [14] but the data interpretation is not straightforward. It was demonstrated that the Stark shift of absorption bands can be used as an ultrafast probe of charge drift dynamics in conjugated polymer MeLPPP [15]. Later, this method has been applied for the investigation of the initial carrier mobility in [6, 6]-Phenyl C_{61}-butyric acid methyl ester (PCBM) [16]. The time resolution of the Stark shift method is limited by the light pulse duration only. However, its application is limited to a relatively narrow class of materials with a clearly observed Stark shift, and its experimental realization is relatively difficult, particularly for materials without absorption bands in the visible spectral range. Moreover, treatment of the obtained experimental information is complicated because the absorbance dynamics caused by the Stark shift appears in the same spectral region and in the same time domain as the transient absorption related to excitons.

Here we describe an experimental method for the charge carrier mobility investigation on ultrafast time scales in organic films and devices based on time-resolved electric field-induced second harmonic generation (TREFISH). Electric field-induced second harmonic generation (EFISH) is commonly used for the determination of molecular hyperpolarizability and surface chemistry investigations [17, 18, 19]. A pump-probe version of the method has been applied to investigate the electric field dynamics on interfaces of inorganic semiconductors [20] and ultrafast electrical pulses propagating on thin-film microstrip lines [21]. Manaka et al. [22] also employed EFISH for direct observation of the carrier motion with low temporal resolution in OFET.

2. Experimental methods for the time-resolved mobility investigations

Mobility of charge carriers is the quantity used to describe the ability of the material to transport the electrical charge. Charge carriers that are present in a material manifest diffusive motion if there is no external electrical field. The diffusion coefficient D relates to the mobility μ *via* the Einstein's equation:

$$\frac{\mu}{D} = \frac{e}{kT}. \tag{1}$$

Once the external electric field is applied, carriers tend to drift along (holes) or against (electrons) the direction of the field. The average drift velocity is proportional to the strength of the electric field and drift mobility expresses this proportionality:

$$\langle \vec{v} \rangle = \pmb{\mu}_{\text{drift}} \cdot \vec{E} \,. \tag{2}$$

In general, mobility $\pmb{\mu}_{\text{drift}}$ is the tensor of the second rank with six independent components [23]. Equation (2) becomes scalar in fully amorphous materials: $\langle v \rangle = \mu_{\text{drift}} \cdot E$. In addition, the mobility in disordered materials is field-dependent, which makes definition (2) slightly ambiguous. Nevertheless mobility is still commonly used quantity, to describe electrical properties of these materials.

Time of flight (TOF). The idea of TOF measurement is very simple. The film made of the investigated material is sandwiched between two electrodes. One of the electrodes is transparent to light. Short exposure to light creates charge carriers near the transparent electrode and they start to drift towards the opposite electrode under the applied external electrical field. The photocurrent is measured with the oscilloscope. Time duration between the light pulse and the moment when the carriers reach the electrode (the current drops) is called the drift time t_{d}. Knowing the drift time the mobility can be easily calculated for given values of the film thickness and the applied voltage:

$$\mu = \frac{D^2}{V_0 \cdot t_{\text{d}}} \tag{3}$$

Mobility of holes and electrons is measured separately by changing bias polarity. Kepler [24] and LeBlanc [25] were the first who used this technique to measure carrier mobility in organic solids.

TOF also manifests a sharp peak of current at the moment of photoexcitation. This peak is caused by non-stationary transport as the photogenerated carriers initially have excess energy and rapidly move relaxing within DOS. This non-stationary phase takes place on a picosecond time scale and is not resolved by the TOF measurement because of the limited time resolution.

The applicability of TOF is limited to the class of materials where charge carriers can be generated by light. In opposite case the additional thin layer of photogeneration material helps to overcome this obstacle.

Pulsed-radiolysis time-resolved microwave conductivity (PR-TRMC). This technique was developed at the University of Delft (Netherlands) and firstly reported in [26]. It uses absorption of microwave as a probe for the carrier mobility. Originally a nanosecond pulse of electrons from the accelerator having the energy of several MeV was used to irradiate the sample and to create charge carriers. The following change of the microwave absorption reflects the mobility of created carriers: $\Delta\sigma = e \cdot n \cdot (\mu_e + \mu_h)$. Here n is the density of carriers, μ_e and μ_h are mobilities of the electrons and holes. Thus, obtaining the mobility requires information on the carrier density, which is estimated from the absorbed energy of the electron pulse, the energy needed to create the pair of carriers and their recombination probability. The electric field strength can be tuned by changing the power of microwave radiation. It should be noted that this technique provides information about the carrier mobility within short distances down to intrachain movement. Depending on the microwave power and frequency the probed spatial range falls into the range of several nanometers. Mobility obtained at high microwave frequencies may be treated as the upper limit for macroscopic mobility [4]. The advantage of PR-TRMC is that it can be applied to bulk material as well as isolated chains of polymers in solutions. It is contactless, thus, the result is not affected by any interfacial phenomena. However, mobility is averaged over electrons and holes. Therefore, special intentionally added traps for one sort of carriers have to be used to obtain the mobility of the other.

Terahertz spectroscopy. Usually femtosecond pulses of the Ti:sapphire laser are used to generate terahertz pulses in a photoconductive switch or in a non-linear crystal (ZnTe for instance). Detection of the terahertz pulses after passing the sample is done in a reverse manner used to generate terahertz pulses. The amplitude and the phase of transmitted terahertz pulse are detected. Therefore, it allows obtaining the refractive index and conductivity of the investigated material within a certain frequency range. If steady-state properties are investigated, the technique is known as terahertz time-domain spectroscopy. A sample can be excited with a femtosecond light pulse prior to the arrival of the terahertz pulse. This pump-probe experiment (time-resolved terahertz spectroscopy) enables observation of the dynamic changes of conductivity with a picosecond time resolution. In order to obtain mobility from the terahertz spectroscopy experiment one needs a complementary information concerning density of charge carriers, which has to be acquired by other methods. A detailed review on terahertz spectroscopy is given in [27].

3. Optical probing of electric field by EFISH

In sandwich-type organic electronics devices an investigated functional material fills the space between electrodes forming a plate capacitor-like sample. If this capacitor is charged and material exposed to the light pulse afterwards, the drifting photogenerated charge carriers discharge the capacitor and the electric field in the material declines. If this decay is recorded, estimations on the motion of photogenerated charge carriers can be made. One can observe the voltage (electric field) kinetics by electrical measurements. However, time resolution of electrical measurements is always limited by the time constant RC of the installation circuitry and does not exceed subnanoseconds in the best case. The investigated material itself can be used to measure the electric field strength optically if some of its optical properties depend on the applied field.

The second harmonics cannot be generated in amorphous and other centrosymmetrical materials. The centrosymmetry can be broken by the application of an external electric field. This effect is well-known electric field-induced second harmonic generation (EFISH) phenomenon, which is commonly used for the determination of the molecular hiperpolarizability [17, 18]. Any process changing the electric field distribution in the material will affect the measured temporal second harmonic (SH) signal; thus, the SH intensity can be considered being a good probe for changes of the electric field strength due to charge motion. The time-resolved EFISH (TREFISH) method is applicable to all π-conjugated organic materials, and its time resolution is limited only by the duration of the optical pulses used.

The second harmonic generation efficiency is proportional to the hiperpolarizability of the molecules [17, 18] and to the second power of the applied field. Thus, the electric field E inside the investigated material can be monitored by measuring the second harmonic intensity I_{2H}:

$$E \sim \sqrt{I_{2H}} \ . \tag{4}$$

If one applies voltage U_0 to the capacitor-like sample, it becomes charged and the amount of charge is:

$$q = CU_0 = \varepsilon\varepsilon_0 \frac{A}{D} U_0 ; \tag{5}$$

where C is the sample capacitance, A is the area of electrodes and D is the distance between electrodes, $i.e.$ thickness of the investigated film. The electric field, assuming homogeneous distribution inside the investigated material, equals to: $E_0 = U_0 / D$. As an example, Figure 1 shows the second harmonic intensity versus voltage applied to methyl substituted ladder-type poly(para-phenylene) (MeLPPP) sample.

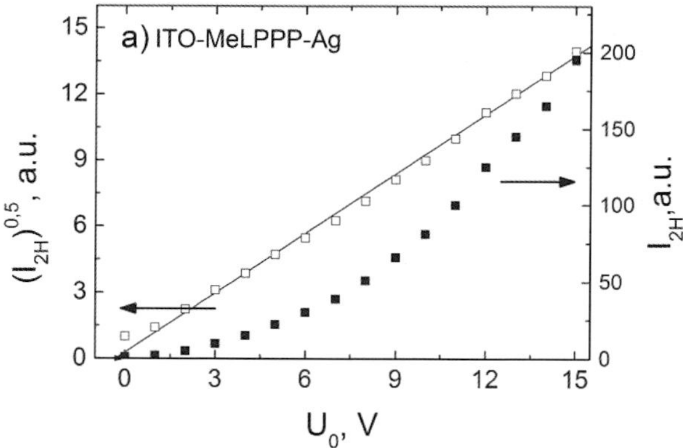

Fig. 1 Second harmonic intensity (■) and square root of second harmonic intensity (□) versus voltage applied to MeLPPP sample with silver electrode.

The TREFISH measurement is carried out in a pump-probe scheme. The pump pulse creates charge carriers, and the subsequent pulse probes the electric field strength in the polymer film. The electric field dynamics $\Delta E(t)$ may be obtained by changing time delay t between pump and probe pulses. In the case of relatively small field variations compared to E_0, the second harmonic kinetics intensity $I_{2H}(t)$ relates to $\Delta E(t)$ as:

$$\Delta E(t) \approx \left[\left(I_{2H}(t)/ I_{2H}^0 \right)^{1/2} - 1 \right] E_0 . \tag{6}$$

I_{2H}^0 is the second harmonic intensity without excitation. If ultrashort light pulses are used for the sample excitation and for probing of the second harmonic generation efficiency, the electric field, and thus the photocurrent dynamics, may be determined with the femtosecond time resolution. The time resolution is limited by the duration of the optical pulses only, however the time domain of such measurements is limited at the most to several ns by the length of the optical delay line. The photocurrent dynamics at longer times may be obtained by the current or voltage mode TOF measurements with a typical time resolution of tens of ns. Thus, by combining both methods the photocurrent kinetics from the carrier photogeneration until their extraction from the sample can be determined.

A drift of photogenerated charge carriers creates the photocurrent and the field-induced polarization of photogenerated neutral excitons creates the displacement current. Therefore, the change of the electric field can be expressed as the sum of the exciton $\Delta E(t)_{exc}$ and charge carrier $\Delta E(t)_{CC}$ contributions:

$$\Delta E(t) = \Delta E_{exc}(t) + \Delta E_{CC}(t); \qquad (7)$$

If the field change is small in comparison with the applied field, the exciton contribution may be approximately expressed as:

$$\Delta E_{exc}(t) \approx -n_{exc}(t)(\Delta\alpha)E_0 / 2\varepsilon\varepsilon_0; \qquad (8)$$

$n_{exc}(t)$ is the exciton density, $\Delta\alpha$ is the change of the polarizability of the excited chain or chain segment with respect to the ground state. Excitons cause a rapid drop of the field during the excitation pulse action and the field recovers when the excitons decay. Their time-integrated contribution to the transported charge is zero. Since $\Delta E_{exc}(t)$ is proportional to the exciton density, its kinetics may be obtained from time-resolved fluoresecence or transient absorption measurements. Proportionality of $\Delta E_{exc}(t)$ to the applied electric field helps to determine its absolute value by measuring the electric field kinetics at different applied voltages. The exciton contribution has to be subtracted from the electric field dynamics $\Delta E(t)$ in order to obtain the charge carrier contribution $\Delta E(t)_{CC}$.

The charge carrier contribution relates to photocurrent $j(t)$:

$$\Delta E(t)_{CC} = \Delta q(t)_{CC} / C \cdot D = \frac{1}{C \cdot D} \int_0^t j(t')dt', \qquad (9)$$

where $\Delta q(t)_{CC}$ is the amount of charge transported by the photocurrent. The $\Delta E_{CC}(t)$ can also be expressed by the averaged drift distance $\langle l(t)\rangle$ and the density $n_{CC}(t)$ of photogenerated charge carriers:

$$\Delta E(t)_{CC} = n_{CC}(t)e\langle l(t)\rangle / \varepsilon\varepsilon_0. \qquad (10)$$

If one seeks to obtain the carrier mobility, the current and the number of carriers have to be known. The current equals to a temporal derivative of equation (9). The charge carrier concentration $n(t)_{CC}$ is more complex to deal with. The concentration changes due to carrier photogeneration and due to extraction from the sample. For example, it has been shown that the majority of carriers are generated during the first picoseconds after excitation in MeLPPP [15]. The carrier concentration increase due to generation may be expressed as:

$$n_{CC}(t) = n_{CC}^0 \cdot g(t). \qquad (11)$$

Here n_{CC}^0 is the carrier concentration at the end of generation process and $g(t)$ is the function of the carrier generation. Finding $g(t)$ is a separate task. It can be obtained from the field-modulated transient absorption data [31] or electric field-assisted fluorescence quenching measurements. In case of a thin film with low optical density, charge carriers are generated homogeneously over the film thickness. Moreover, photogeneration of charges is much faster than their extraction at the electrodes. Thus, the concentration of photogenerated charge carriers in the volume of a polymer film may be approximately written as follows:

$$n(t)_{CC} = n_{CC}^0 \cdot g(t) \cdot \left[1 - \langle l(t)\rangle / D\right]. \tag{12}$$

The photocurrent is proportional to the carrier concentration and to the drift speed:

$$j(t) = A \cdot e \cdot n(t)_{CC} \cdot \frac{d\langle l(t)\rangle}{dt} = A \cdot e \cdot n_{CC}^0 \cdot g(t) \cdot \left[1 - \langle l(t)\rangle / D\right] \frac{d\langle l(t)\rangle}{dt} \tag{13}$$

The latter equation can be solved numerically with respect to $\langle l(t)\rangle$:

$$\frac{l_{k+1} - l_k}{\Delta t} = \frac{j_k}{Aen_{CC}^0 g_k} \cdot \frac{1}{1 + l_k / D},$$

$$l_{k+1} = \left[l_k + \frac{j_k}{Aen_{CC}^0 g_k} \frac{1}{1 + l_k / D}\right]\Delta t. \tag{14}$$

where k numerates successive values in the data array, or analytically for the prompt carrier generation:

$$\Delta q(t)_{CC} = \int_0^t j(t')dt' = Aen_{CC}^0 [\langle l(t)\rangle - \langle l(t)\rangle^2 / 2d]. \tag{15}$$

Averaged mobility is obtained by calculating a temporal derivative of $\langle l(t)\rangle$ according to definition:

$$\mu = \frac{1}{E_0} \cdot \frac{d\langle l(t)\rangle}{dt}. \tag{16}$$

If mobility on a short time scale is the subject of interest and there is no need to account for the carrier extraction it may be calculated more straightforwardly by taking a temporal derivative of $\Delta E_{CC}(t)$:

$$\mu = \frac{j(t)}{AeE_0 n(t)_{CC}} = \frac{\varepsilon\varepsilon_0}{eE_0 n(t)_{CC}} \cdot \frac{dE(t)_{CC}}{dt}. \tag{17}$$

Fig. 2 shows experimental setup for TREFISH and TOF measurements. It is based on an amplified femtosecond Ti:Sapphire laser generating 130 fs duration pulses at 810 nm at the 1 kHz repetition rate. For tunable wavelength excitation the parametric generator *TOPAS-C* is used. A probe beam of fundamental wavelength acts on the sample at an angle of incidence of approximately 45 degrees. It is adjusted to be *p*-polarized, because only the electric field of the *p*-polarized electromagnetic wave has a non-zero component in the direction of the applied DC electric field. The second harmonic intensity of the reflected probe beam is detected with a photo multiplying tube (PMT) blocking the fundamental and scattered excitation light by a bandpass filter. In order to extend lifetime of the sample the external voltage U_0 is applied to the sample in a pulsed mode by the square pulse generator synchronized to the laser. To avoid injection of the charge carriers, a positive bias is applied to the metal electrode (reversely biased device).

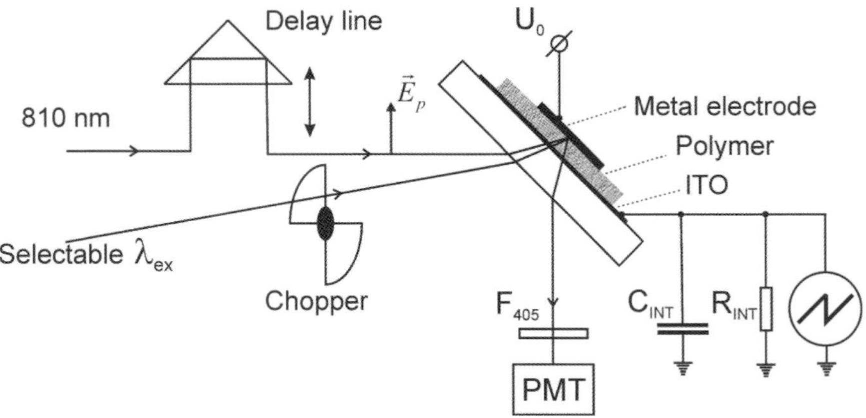

Fig. 2 Experimental setup for TREFISH and integral mode TOF measurement.

4. Charge carrier motion dynamics inneat MeLPPP conjugated polymer

Methyl substituted ladder-type poly(para-phenylene) (MeLPPP) is a widely investigated polymer [15, 28, 29, 30]. Ladder type structure of MeLPPP is the most prominent feature that prevents torsional defects and facilitates charge transport. We address charge the charge transport dynamics in MeLPPP by using TREFISH technique and conventional photocurrent investigations.

Fig. 3 shows the raw data obtained by the TREFISH technique for a 100 nm thick MeLPPP sample at different values of the applied external voltage. Curves represent the second harmonic intensity as a function of time delay between pump and probe pulses normalized to the second harmonic intensity without a pump pulse. Thus, the ratio equals to one at negative delays when the probe pulse arrives prior to the excitation pulse. The kinetics strongly depend on the applied voltage, which shows that several processes contribute to the second harmonic dynamics. At 4V there is an instantaneous (pulse width limited) drop of the second harmonic intensity during the sample excitation. After few hundreds of picoseconds the second harmonic intensity almost recovers close to its initial value. At higher voltages the recovery is not complete. Increase of the applied voltage causes decay of the second harmonic intensity on a nanosecond time scale. At 20V (close to the limit the sample can stand) the recovery vanishes.

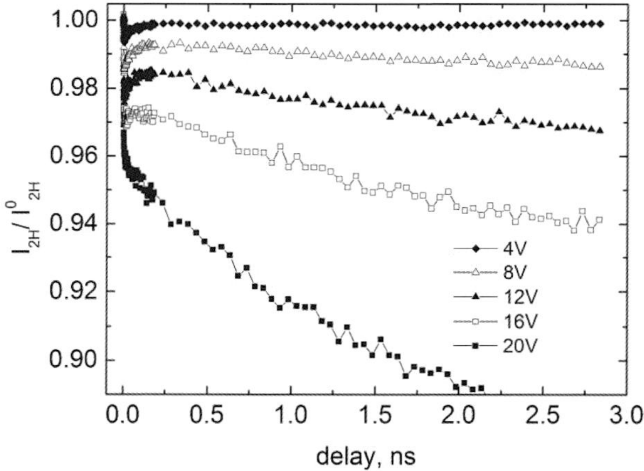

Fig. 3 Second harmonic kinetics in MeLPPP sample at different applied voltages. Excitation intensity is below $10\ \mu W/cm^2$, wavelength – 460 nm.

Fig. 4 presents the electric field dynamics calculated from the second harmonic data using eq. (6). Application of eq. (6) is square root transform, therefore the same features remain as in Fig. 3. At low applied voltages there is a pulse width limited drop of the electric field which partly recovers within a few hundreds of picoseconds.

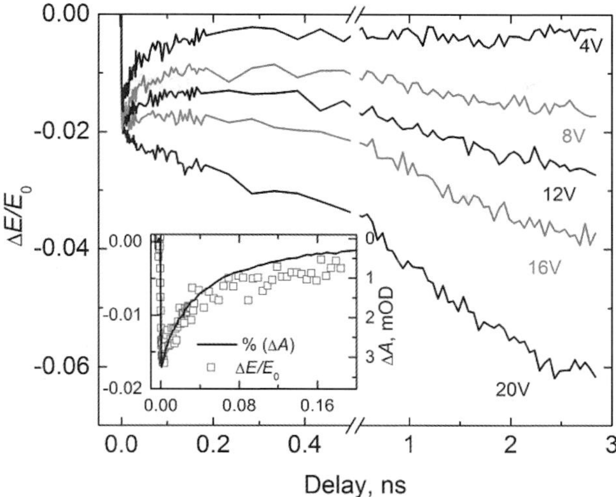

Fig. 4 Electric field dynamics obtained from the second harmonic kinetics presented in Fig. 3. Insert compares the electric filed and transient absorption kinetics at 810 nm (460 nm excitation) and 4V of the applied voltage.

Transient absorbtion kinetics at 810 nm, which was measured simultaneously with the 2H intensity, shows the pulse width limited rise of the sample absorption at zero delay time and a successive decay of this excitation induced absorption with a characteristic time constant of several tens of picoseconds. The transient absorption kinetics is similar to that obtained in previous investigations and attributed to the exciton density relaxation [31]. The kinetics shows weak dependence on the applied electric field and its shape is very similar to the 2H kinetics measured at low applied voltages (see insert in Fig. 4). It is obvious they are very akin. Indeed, transient absorption relaxing on a time scale of tens of ps is typical of the exciton dynamics in MeLPPP [29, 32]. Therefore, the field dynamics at low voltages should be assigned to the exciton contribution expressed by eq. (8). According to the eq. (8) the exciton contribution linearly depends on the applied field. Thus, at higher voltages the exciton contribution should scale linearly with the applied voltage, and the field kinetics can be decomposed into the exciton and charge carrier contributions. At weak field the exciton contribution dominates on a picosecond time scale. By increasing the applied field strength, the charge carrier contribution becomes dominating. It should be noted, that the transient absorption at 810 nm changes by less than 20 % by increasing applied voltages, and shows that the field-induced exciton quenching is not very strong.

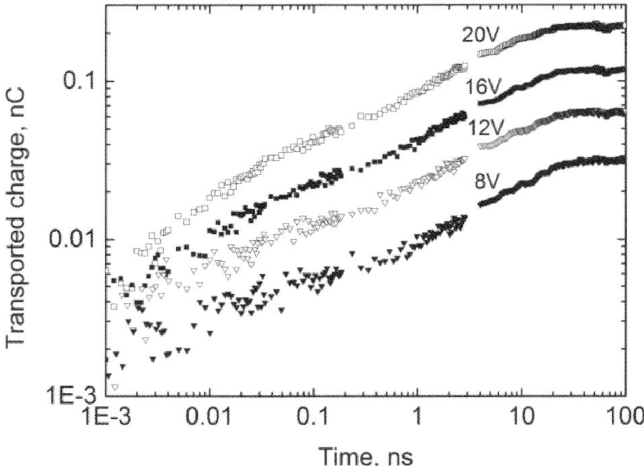

Fig. 5 Charge transported by photocurrent versus time-span after a pump pulse at different applied voltages. TREFISH data (from 0 to 3 ns) are combined with TOF data (from 4ns to 100 ns).

Charge transported by photocurrent $\Delta q(t)_{CC}$ can be calculated from the charge carrier contribution to the electric field (eq. (9)). The same information about the integrated current only on a longer time scale was also obtained from the TOF measurements. Fig. 5 represents experimental data obtained from both TREFISH and integral mode TOF measurements – transported charge (integrated photocurrent) as a function of time. TREFISH data end at 3 ns, which is the end of the optical delay stage. Curves reach plateau at tens of nanoseconds. Evidently, all photogenerated charge carriers are extracted up to this time. The value of the transported charge strongly depends on the applied voltage. This is because the carrier generation depends on the electric field. The electric field assists in the separation of geminate charge pairs and creation of free charge carriers [2]. Thus, more charge is extracted at the higher voltage.

5. Charge carrier motion dynamics in merocyanine/fullerene blend for solar cells

We combined ultrafast TREFISH and conventional time-resolved photocurrent measurements to get information on carrier motion in blends with different fullerene concentration. Fig. 6 shows the electric field kinetics in the investigated blends calculated from the time-resolved EFISH data (initial 3 ns) and from the voltage kinetics measured by the oscilloscope at longer times.

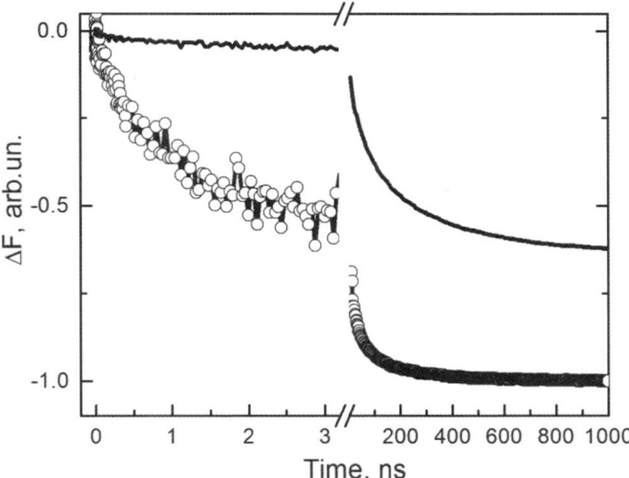

Fig.6. Electric field kinetics at 2 V applied voltage in the merociqanine/ fullerene blend with 50% (open circles) and 10% (line) fullerene concentration calculated from the TREFISH data and from voltage kinetics measured by oscilloscope.

The kinetics obtained for the 50% blend film shows the fast electric field decrease on the time domain of initial 3 ns. Approximately half of the total field decrease takes place during this time, while the remaining field decay is much slower, taking place on tens and hundreds of ns time scale. The electric field kinetics in 10% blend film is significantly different. It shows only a minor field decrease measured with TREFISH method, while the dominating field decrease takes place on tens and hundreds of ns time scale. The total field decrease is by about 30% smaller in film with 10% fullerene. Different electric field kinetics obtained in different samples should be related to different electron and hole motion and extraction dynamics. Electron mobilities are expected to be very different in samples with different fullerene content. The electron mobility is expected be low in samples with 10% fullerene concentration, which is below or close to the percolation threshold, being, e.g. at about 18% for spherical particles. Therefore we attribute the fast decrease phase to the electron motion in percolating fullerene domains. Field decrease by about 50% of the total decrease value during the initial 3 ns in film with 50% fullerene and its clear saturation suggest that all electrons are extracted from this sample during 3 ns, while hole extraction is consequently much slower, lasting for hundreds of ns. Slow electron extraction from samples with 10% fullerene is a natural consequence of low electron mobility due to weak percolation of PCBM domains. The hole mobility is expected to be less sensitive to the fullerene content because merocyanine concentration is sufficiently high in all the samples to reach its percolation threshold.

6. Conclusions

Time resolved optical probing of the electric field dynamics by means of EFISH generation is a useful tool for the investigation of the charge carrier generation and motion in organic semiconductors. It enables direct measurement of the carrier drift dynamics with subpicosecond time resolution.

References

1. H.Klauk (Ed.), *Organic electronics, Materials, Manufacturing and Applications*, Wiley–VHC VerlagGmbh& Co. KGaA, Weinheim (2006).
2. D. Hertel, and H. Bässler, *Chem. Phys. Chem*, **9**, 666 (2008).
3. H. Bässler, *Phys. Stat. Sol. B*, **175**, 15 (1993).
4. V. Coropceanu, J. Cornil, D. A. da Silva Filho, Y. Olivier, R. Silbey, and J.-L. Bredas, *Chem. Rev.*, **107**, 926 (2007).
5. J. W. van der Horst, P. A. Bobbert, M. A. J. Michels, and H. Bässler, *J. Chem. Phys.*, **114**, 6950 (2001).
6. A. Horvath, G. Weiser, C. Lapersonne-Meyer, M. Schott, and S. Spagnoli, *Phys. Rev. B*, **53**, 13507 (1996).
7. F. Dubin, R. Melet, T. Barisien, R. Grousson, L. Legrand, M. Schott, and V. Voliotis, *Nature Phys.*, **2**, 32 (2005).
8. P. Prins, F. C. Grozema, J. M. Schins, S. Patil, U. Scherf, and L. D. A. Siebbeles, *Phys. Rev. Lett.*, **96**, 146601 (2006).
9. D. Moses, J. Wang, G. Yu, and A. J. Heeger, *Phys. Rev. Lett.*, **80**, 2685 (1998).
10. G. Juška, K. Genevičius, R. Österbacka, K. Arlauskas, T. Kreouzis, D.D.C. Bradley, and H. Stubb, *Phys. Rev. B*, **67**, 081201 (2003).
11. R. Österbacka, K. Genevičius, A. Pivrikas, G. Juška, K. Arlauskas, T. Kreouzis, D.D.C. Bradley, H. Stubb, *Synthetic Metals*, **139**, 811 (2003).
12. C. H. Lee, G. Yu, D. Moses and A. J. Heeger, *Phys. Rev. B*, **49**, 2396 (1994).
13. R. J. O. M. Hoofman, M. P. de Haas, L. D. A. Siebbeles, and J. M. Warman, *Nature*, **392**, 54 (1998).
14. E. Hendry, M. Koeberg, J. M. Schins, H. K. Nienhuys, V. Sundström, L. D. A. Siebbeles, A. Bonn, *Phys. Rev. B*, **71**, 125201 (2005).
15. Gulbinas, V. Y. Zaushitsyn, V. Sundström, D. Hertel, H. Bässler, and A. Yartsev, *Phys. Rev. Lett.*, **89**, 107401 (2002).
16. J. Cabanillas-Gonzalez, J. T. Virgili, A. Gambetta, G. Lanzani, T. D. Anthopoulos, D. M. de Leeuw, *Phys. Rev. Lett.*, **96**, 106601 (2006).

17. Ch. Bosshard, G. Knöpfle, P. Pretre, and P. Günter, *J. Appl. Phys.*, **71**, 1594 (1992).
18. J.W. Perry, in *Nonlinear Optical-Properties of Molecules and Materials,* in *Materials for Nonlinear Optics* (edited by S. R. Marder, E. J. Sohn, and G. D. Stucky), American Chemical Society, Washington, D.C., p. 67 (1991).
19. R. M. Corn, and D. A. Higgins, *Chem. Rev.*, **94**, 107 (1994).
20. Y. D. Glinka, T.V. Shahbazyan, I. E. Perakis, N. H. Tolk, X. Liu, Y. Sasaki, and J. K. Furdyna, *Appl. Phys. Lett.*, **81**, 3717 (2002).
21. M. Nagel, C. Meyer, H.-M. Heiliger, T. Dekorsy, H. Kurz, R. Hey, and K. Ploog, *Appl. Phys. Lett.*, **72**, 1018 (1998).
22. T. Manaka, E. Lim, R. Tamura, and M. Iwamoto, *Nature Photon.*, **1**, 581 (2007).
23. E.A.Silinsh, and V.Čapek, Organic Molecular Crystals. Interaction, Localization and Transport Phenomena, AIP Press, New York (1994).
24. R. G. Kepler, *Phys. Rev.*, **119**, 1226 (1960).
25. O. H. LeBlanc, *J. Chem. Phys.*, **33**, 626(1960).
26. J. M. Warman, P. de Haas, and A. Hummel., *Chem. Phys. Lett.*, **22**, 480 (1973).
27. G. Lanzani (Ed.), *Photophysics of Molecular Materials*, Wiley-VCH Verlag GmbH & Co. KGaA, Weinheim, (2006).
28. D. Hertel, H. Bässler, U. Scherf, H. H. Hörhold, *J. Chem. Phys.*, **110**, 9214 (1999).
29. W. Graupner, G. Cerullo, G. Lanzani, M. Nisoli, E. J. W. List, G. Leising, S. De Silvestri, *Phys. ReV. Lett.*, **81**, 3259 (1998).
30. A. Haugeneder, M. Neges, C. Kallinger, W. Spirkl, U. Lemmer, J. Feldmann, *J. Appl. Phys.*, **85**, 1124 (1999).
31. G. Cerulo, S. Stagira, M. Nisoli, S. De Silvestri, G. Lanzani, G. Kranzelbinder, W. Groupner, and G. Leising, *Phys. Rev. B*, **57**, 1280, (1998).
32. V. Gulbinas,Yu. Zaushitsyn, H. Bässler, A. Yartsev, and V. Sundström, *Phys. Rev. B*, **70**, 035215 (2004).

SINGLET FISSION IN ORGANIC CRYSTALS

LIN MA

Division of Physics and Applied Physics, School of Physical and Mathematical Sciences, Nanyang Technological University, 637371, Singapore

CHRISTIAN KLOC

Division of Materials Science, School of Materials Science and Engineering, Nanyang Technological University, 639798, Singapore

CESARE SOCI

Division of Physics and Applied Physics, School of Physical and Mathematical Sciences, Nanyang Technological University, 637371, Singapore

MARIA E. MICHEL-BEYERLE

Division of Physics and Applied Physics, School of Physical and Mathematical Sciences, Nanyang Technological University, 637371, Singapore

GAGIK G. GURZADYAN

Division of Physics and Applied Physics, School of Physical and Mathematical Sciences, Nanyang Technological University, 637371, Singapore

Singlet fission attracts an extensive attention in recent years due to its potential to improve the efficiency of organic solar cells. Ultrafast time-resolved measurements provide a powerful tool for direct determination of singlet fission pathways and rates. In this chapter we present and discuss the experimental data on one- and two-photon, as well as consecutive two-quantum induced singlet fission in organic crystals rubrene and perylene. Temperature and excitation wavelength dependent rates of the singlet fission are discussed in terms of the activation energy. Ultrafast singlet fission was demonstrated to proceed directly from upper vibrational states of S_1, from upper excited states S_N, bypassing relaxed S_1 state, as well as from two-photon excited states.

1. Introduction

1.1. *Singlet fission*

Singlet fission is a process in which a singlet excited molecule, usually in a densely packed organic solid, shares its energy with a neighboring molecule in its electronic ground state, both molecules forming a pair of triplet states in a spin allowed process.[1] Singlet fission was first proposed in 1965 to explain the delayed fluorescence in anthracene crystal.[2] And it is proven by the magnetic field effect study on tetracene crystal by Geacintov[3] and Merrifield[4] in 1969. Last decade, with the boom and development in solar energy as the clean energy source, increase of the efficiency of the solar cells becomes a hot issue.

The thermodynamic limit of energy conversion efficiency in single cell photovoltaic devices is determined by the spectral mismatch between the solar radiation and the absorption of the active semiconductor. Two factors affect this limit, hence the maximum efficiency attainable in single-junction cells: 1. solar photons with energy lower than the optical bandgap of the semiconductor are not absorbed; 2. photons with energy higher than the bandgap generally produce a single electron-hole pair, which means the energy in excess of the bandgap energy is lost via thermalization processes. Once the bandgap of the active semiconductor has been optimized to match the solar spectrum, one way to increase the thermodynamic efficiency in single cell photovoltaics is the generation of multiple electron-hole pairs per absorbed photon with energy in excess of the bandgap.[5,6] This process, which in bulk semiconductors is known as impact-ionization and in low-dimensional systems is often referred to as Multiple-Exciton Generation (MEG),[7,8] can be highly enhanced in quantum dots thanks to the large electron-electron interaction resulting from quantum confinement. MEG was also reported in the case of semiconducting single-walled carbon nanotubes (SWNTs), where absorption of single photons with energies corresponding to three times the energy gap of (6,5) SWNTs resulted in an exciton generation efficiency of 130% per single photon absorbed.[9]

Singlet fission is an analogue of the MEG in the organic materials; the difference is that in this case, triplet excitons are generated from singlet exciton with the theoretical quantum yield up to 200%. The recent development and advances on singlet fission are reviewed in detail by Smith and Michl.[10,11] Singlet fission in semiconducting polymer thienylenevinylene was recently reported in Ref. 12.

Here, we present results of our studies on singlet fission in organic crystals rubrene and perylene. Singlet fission from upper excited electronic states as well as singlet fission induced via two-photon or consecutive two-quantum (two-step) absorption are discussed in detail.

2. Materials and Methods

Rubrene and perylene single crystals were grown by physical vapor transport (PVT) technique.[13] The reddish source powder of rubrene (Sigma-Aldrich, >99% purity) was purified by sublimation for at least 2 times before the crystal growth. The carrier gases used in PVT technique are Argon (Ar), Nitrogen (N_2), Hydrogen (H_2) and forming gas which is a mixture of H_2 and N_2 with varied mole fraction. Semi-transparent two-zone furnace is used to grow the crystal. Rubrene source powder is placed at the zone of 300°C and the other zone is set to a slightly lower temperature at 260°C. The carrier gas flows from high temperature zone to the lower, thus the sublimated rubrene molecules are condensed at the wall of the inner quartz tube at the lower temperature zone. The harvested as-grown rubrene single crystals are mostly flakes or needles shape, the size can reach 10-15 mm. They are selected under polarized microscope, only those who with smooth surface and large size are chosen for steady-state and transient absorption and fluorescence measurements. The thickness of rubrene single crystal used in the experiment is about 10 μm, characterized by SEM and steady state absorption spectrum.

Steady state absorption spectra were obtained by a UV-Vis spectrophotometer (Cary 100Bio, Varian) at 1.0 nm resolution, and the steady state fluorescence emission and excitation spectra were recorded by the spectrofluorometer (Fluorolog-3, HORIBA Jobin Yvon).

The transient absorption (TA) spectra were measured by the optical femtosecond pump probe spectroscopy; the output of titanium-sapphire (Legend Elite, Coherent) regenerative amplifier seeded by the oscillator (Micra, Coherent) was used as a pulse laser source. The fundamental wavelength was 800 nm, pulse width 65 fs, pulse repetition rate 1 kHz, and average power 3.5 W. 90% of the fundamental beam was converted into the visible and UV range by use of optical parametric oscillator (Topas, Light Conversion) and second- and sum-frequency mixing nonlinear processes. The remaining 10% was used to generate white light continuum in CaF_2 plate, i.e. probe beam. The details on the pump probe setup and measurement method are described in Ref. 14.

Time-resolved photoluminescence (TRPL) measurements were carried out at room temperature by time-correlated single photon counting (TCSPC) technique with resolution of 10 ps (PicoQuant PicoHarp 300). The output of Titanium sapphire laser (Chameleon, Coherent Inc. 100 fs, 80 MHz) with second- and third harmonic generation was used for exciting the fluorescence. The temperature dependent time-resolved fluorescence was measured by combining the TCSPC technique with an exchange gas continuous flow cryostat

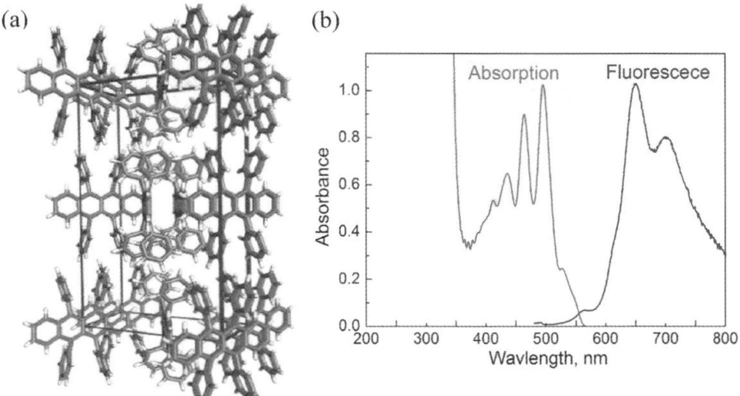

Figure 1. Rubrene crystal structure[15] (a) and steady state absorption and fluorescence spectra[16] (b).

(OptistatCF, Oxford Instruments).

The fluorescence up-conversion measurements were carried out by a commercial system FOG 100 (CDP Systems Corp., Russia). The second harmonic of a Titanium sapphire laser (Chameleon, Coherent Inc.) at 400 nm (100 fs, 80 MHz) was used as the excitation source. The fluorescence (540-700 nm) was collected by parabolic mirror and focused into a 0.5 mm BBO crystal (cut angle 38°, ooe interaction) together with the fundamental radiation (800 nm) to generate the sum-frequency radiation. The resulting radiation (322-373 nm) after passing through a double monochromator (CDP2022D) was detected by a photomultiplier based photon counting electronics.

3. Singlet Fission in Organic Crystals

Singlet fission in rubrene and perylene single crystals was studied under various excitation conditions, i.e. one- and two-photon excitation, consecutive two-quantum excitation. Singlet fission from upper excited electronic states bypassing the lowest excited S_1 state is reported and discussed in detail. Fission rates were found to depend on the excitation wavelength and temperature.

3.1. *Singlet fission: one quantum excitation*

3.1.1. *Singlet fission from the lowest excited singlet state S_1*

The rubrene crystal structure and the steady state spectra are presented in Fig. 1 (a) and (b). Transient absorption spectra in rubrene single crystal thin film were measured under 500 nm excitation (Fig. 2a). There are two positive bands with

114

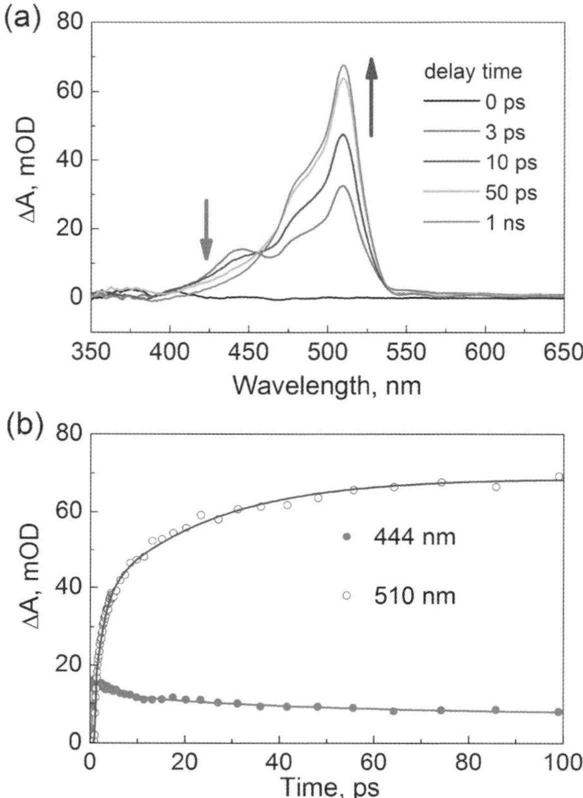

Figure 2. (a) Transient absorption spectra for the rubrene thin single crystal under 500 nm excitation. (b) Transient kinetics at 444 and 510 nm.[16]

maxima at 444 and 510 nm. Transient band at 444 nm decays biexponentially with $\tau_1 = 5.2$ ps, $\tau_2 = 51$ ps (Fig. 2b), which correlated well with the fluorescence lifetime TCSPC measurements: 40 ps[16]. In consideration of the transient absorption spectra of the rubrene in solution, the 444 nm band of single crystal is assigned to the singlet-singlet absorption $(S_1 \rightarrow S_N)$.[16]

Transient band at 510 nm develops with two rise components $\tau_1 = 2.3$ ps, $\tau_2 = 23$ ps and lives long: doesn't show decay in 1 ns time window. For the 510 nm band, compared with the reported transient absorption results of the rubrene solution studied by flash-photolysis[17], 510 nm band position is in agreement with the triplet-triplet absorption $T_1 \rightarrow T_N$ (10 nm red shift). The decay of singlet state is accompanied by the formation of triplet states. An isosbestic point at 455 nm is an indication of a single transformation process, in which singlet states are

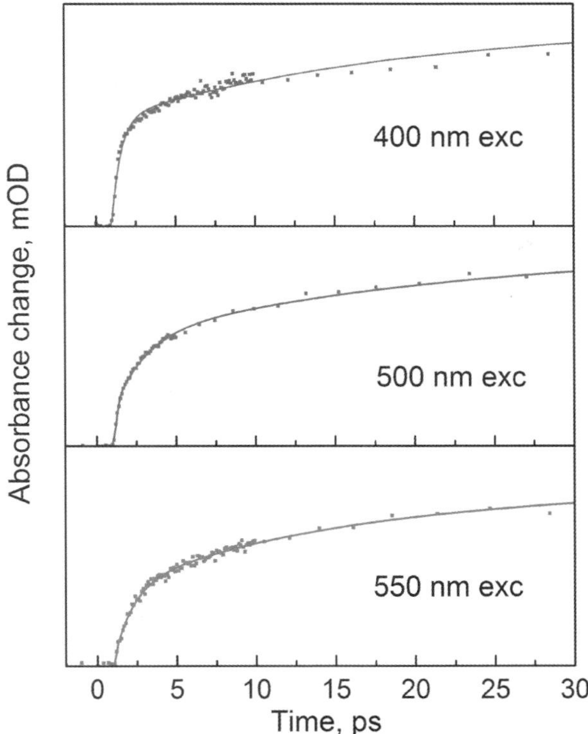

Figure 3. Transient absorption kinetics at 510 nm (T_1 state) under one-photon excitation (400-550 nm).

converted into triplet states ($S_1 \rightarrow T_1$). Transient absorption spectrum was also measured for rubrene in toluene solution, however, only the TA band of S_1 state was observed. Absence of triplet signal in solution is due to the low intersystem crossing efficiency (0.05).[17] Therefore, the triplet formation in crystal is attributed to the singlet fission process ($S_1 \rightarrow 2T_1$). The triplet state forms with two time constants $\tau_1 = 2.3$ ps and $\tau_2 = 23$ ps, which corresponds to the direct fission from the upper vibrational states of S_1 and thermally activated fission from the lowest excited state S_1, respectively.

3.1.2. *Singlet fission from upper vibrational states of S_1*

We have applied various excitation wavelengths, 400, 500 and 550 nm, in order to excite different vibrational states of the first excited singlet state S_1 in rubrene crystal. Comparison of the triplets' rise time under different excitation wavelengths is shown in Fig. 3, which indicates the rise time of triplet states, i.e.,

116

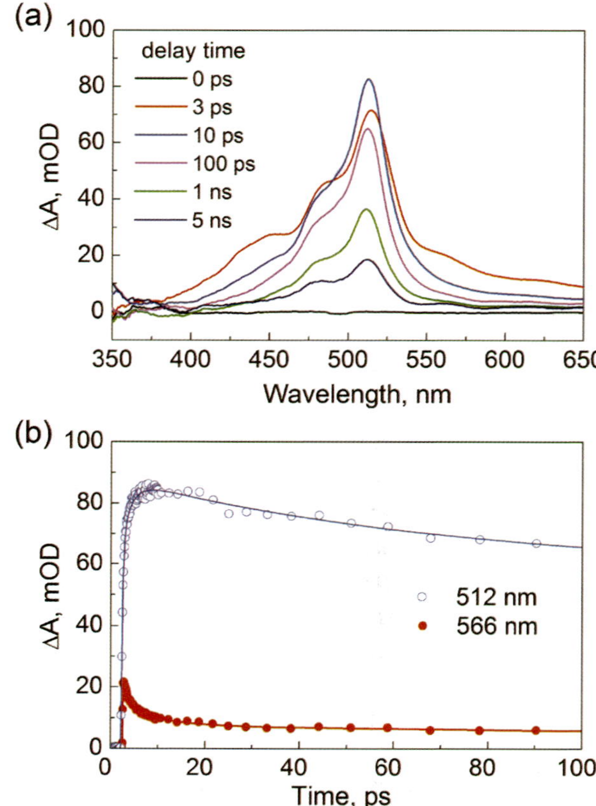

Figure 4. (a) Transient absorption spectra for the rubrene thin single crystal under 250 nm excitation. (b) Transient kinetics at 512 and 566 nm.[16]

singlet fission rate, is pump wavelength dependent.[18] The triplet states form faster with decreasing pump wavelength (i.e., increasing pump photon energy). It points out that singlet fission can proceed directly from the upper vibrational states of S_1 (400-550 nm). Singlet fission is faster from the upper excited states since it is exoergic. Detailed explanation of this effect is described below (section 3.14 and 3.2).

3.1.3. *Singlet fission from upper excited singlet states S_N*

Transient absorption under 250 nm excitation is shown in Fig. 4. The main feature still comes from the two absorption bands of S_1 and T_1. Compared with the spectra under 500 nm excitation, the TA spectra become much broader. We assign this broadening to the formation of polaron.[16] The isosbestic point at

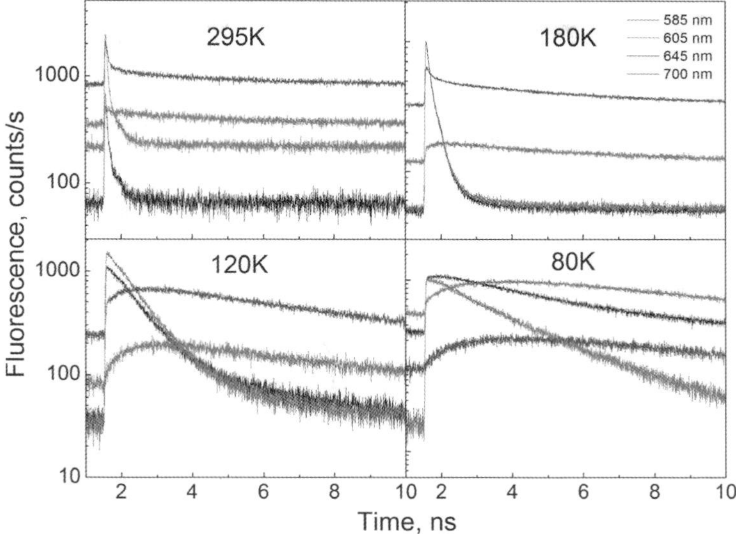

Figure 5. Temperature dependent TCSPC fluorescence kinetics in rubrene under 400 nm excitation.[19]

455 nm observed in the case of 500 nm excitation here has disappeared, which is indicative that there is more than one transformation processes. Moreover, the formation of triplet state becomes 10 times faster: $\tau_1 = 200$ fs and $\tau_2 = 2$ ps. The 200 fs ultrafast rise is due to direct fission from the upper excited singlet states S_N, i.e. $S_N \rightarrow 2T_1$. It also indicates that the internal conversion from S_N to S_1 is relatively slow: $\tau_{ic} > 200$ fs.[20] Singlet fission which was observed to occur directly from upper excited singlet states S_N can be considered as violation of Vavilov law, which states that the quantum yield of fluorescence is independent of the excitation wavelength i.e. all photophysical processes necessarily proceed from the lowest excited states S_1 or T_1. Singlet fission from upper excited electronic state $1B_u$ bypassing lower state $2A_g$ was recently reported in the semiconducting polymer thienylenevinylene.[12]

3.1.4. *Singlet fission: temperature dependence*

Apart from the transient absorption spectroscopy, time resolved fluorescence also provides a direct way for studying the singlet fission. Since the fluorescence is quenched mainly due to the singlet fission, the fluorescence lifetime reflects the rate of singlet fission. The temperature dependent time-resolved fluorescence was studied in rubrene crystal in order to characterize the thermally activated singlet fission. Time-resolved single photon counting (TCSPC) technique was

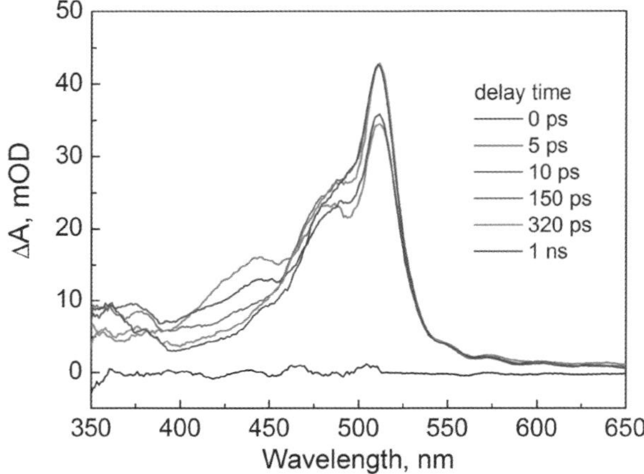

Figure 6. Transient absorption spectra under two-photon excitation at 750 nm.

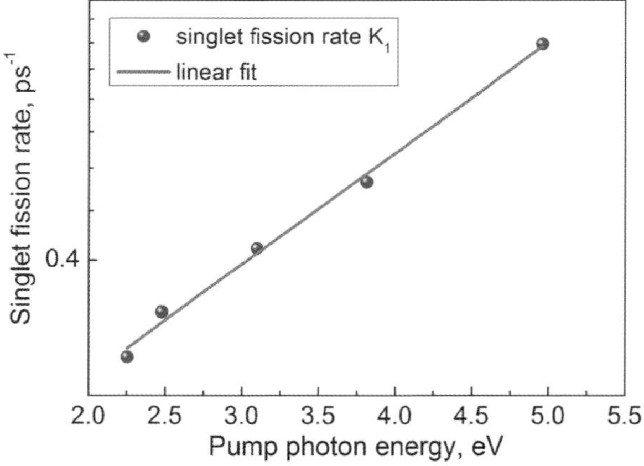

Figure 7. Arrhenius plot of singlet fission rate versus pump photon energy.[18]

applied for this purpose. Fluorescence kinetics at different emission wavelengths under various temperatures are shown in Fig. 5. At room temperature, all fluorescence kinetics are well fitted by two exponential components $\tau_1 < 100$ ps and $\tau_2 \sim 3$ ns. For the blue wing of the fluorescence emission, i.e. 585 and 605 nm, the short component τ_1 becomes longer with decreasing temperature due to the suppression of singlet fission. Analyzing the temperature dependent fluorescence lifetimes, we found that the singlet fission rate follows Arrhenius law:

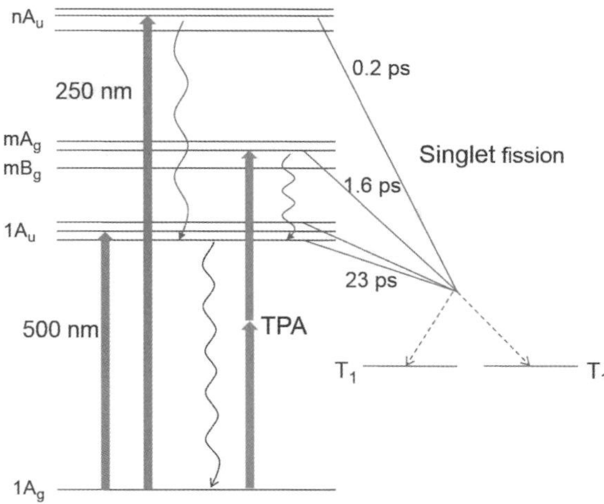

Figure 8. Schematic of the singlet fission induced by one- and two-photon absorption (TPA, 750 nm) in rubrene crystal.

$$k_{fiss} = A\exp(-\frac{E_a}{RT}) \tag{1}$$

where A is the pre-exponential factor, E_a is the activation energy, and R is the molar gas constant. The obtained activation energy E_a is 0.05 eV,[19] which corresponds to the energy difference $E(S_1) - 2E(T_1) = -0.05$ eV.[16]

3.2. *Singlet fission: two-photon excitation*

In order to study the two-photon excited state relaxation processes in rubrene single crystal, we applied the femtosecond pump probe technique with $\lambda_{exc} = 750$ nm, i.e. no linear absorption. Although the two-photon excited states have different symmetries relative to the one-photon excited states, the singlet fission is still observed (Fig. 6). Transient band with the maximum at 510 nm corresponds to triplet-triplet absorption. The relation between the singlet fission rate k_{fiss} versus the pump photon energy is shown in Fig. 7: the $\ln(k_{fiss})$ undergoes a linear increase with excitation photon energy, i.e. follows the Arrhenius law (Eq. 1).

3.3. *Singlet fission in rubrene: conclusive remarks*

Based on the above, we have demonstrated that singlet fission can proceed directly from the upper vibrational levels of S_1, upper excited singlet states S_N,

120

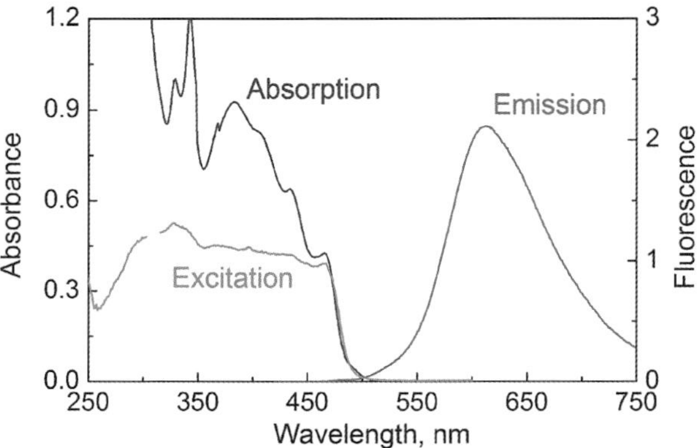

Figure 9. Steady state absorption, fluorescence and excitation spectra of α-perylene crystal.[21] For fluorescence emission spectrum λ_{exc} = 400 nm; for fluorescence excitation spectrum λ_{em} = 620 nm.

and two-photon excited states bypassing the relaxed S_1 state. The relaxation process of rubrene single crystal under various excitation conditions is shown in Fig. 8. Singlet fission from the S_1 state proceeds within 2 and 20 ps due to direct fission from upper and lowest vibrational levels of S_1, respectively. From the S_N states an ultrafast singlet fission occurs within 200 fs. Direct singlet fission from two-photon excited states takes place within 1.6 ps, and it competes with the fast internal conversion. 40% of the two-photon excited states relax to the lowest one-photon excited state S_1, where triplet states continue to be formed, even though 15 times slower via thermally activated fission.

3.4. *Singlet fission: consecutive two-quantum (two-step) excitation*

Singlet fission can proceed also after excitation of the high-lying excited electronic states via consecutive singlet-singlet absorption: $S_0 \rightarrow S_1 \rightarrow S_N$. In this case much lower radiation intensities are required compared with the above described two-photon excitation. This was demonstrated for the organic single crystal α-perylene.[21] Perylene and its derivatives are widely used in organic field effect transistors[22], light emitting diodes[23] and photovoltaics[24]. Perylene belongs to monoclinic crystal system and exists in two forms: α and β.[25] β-perylene is monomeric, with two molecules per unit cell; α-perylene is dimeric, with four molecules in a unit cell.

3.4.1. *Steady-state absorption and fluorescence of perylene*

Steady state absorption, fluorescence emission and excitation spectra of α-perylene are shown in Fig. 9. Due to the high absorption cross-section all optical

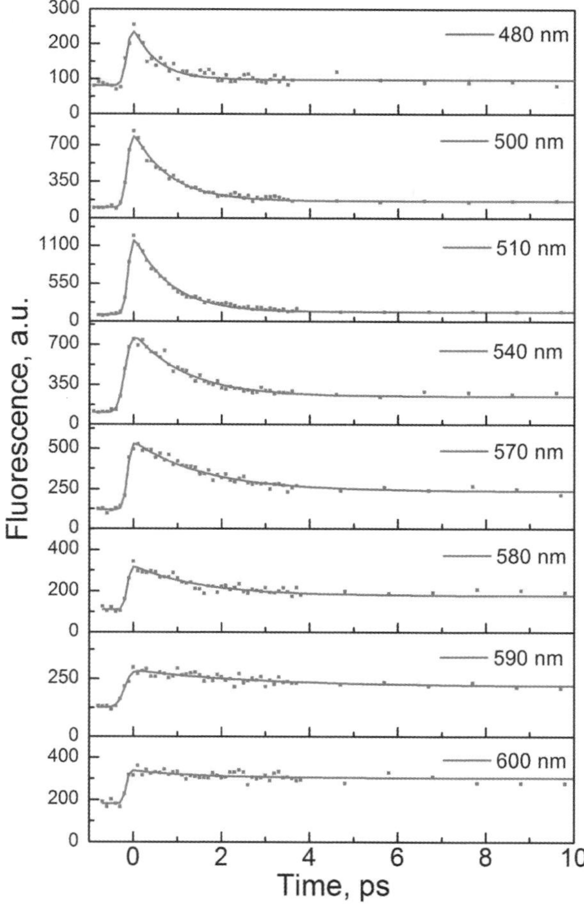

Figure 10. The fluorescence kinetics in α-perylene crystal, measured by fluorescence upconversion under 400 nm excitation at different emission wavelengths.

measurements were performed with thin samples. The fluorescence centered at 615 nm corresponds to the well-known excimer (self-trapped exciton) emission in α-perylene crystal.[26] The fluorescence spectrum is invariant under different excitation wavelengths. A discrepancy between the absorption and fluorescence excitation spectra was found. It indicates that below 465 nm quenching takes place. Comparison of the absorption and fluorescence excitation spectra shows that they correspond well only above 465 nm. However, below 465 nm there is a drop of the excitation spectrum indicating an efficient quenching process. This is due to dimer cation formation from ion pair (IP): the threshold according to Ref. 27 is 2.65 eV. At wavelengths below 350 nm there is a drastic drop of

122

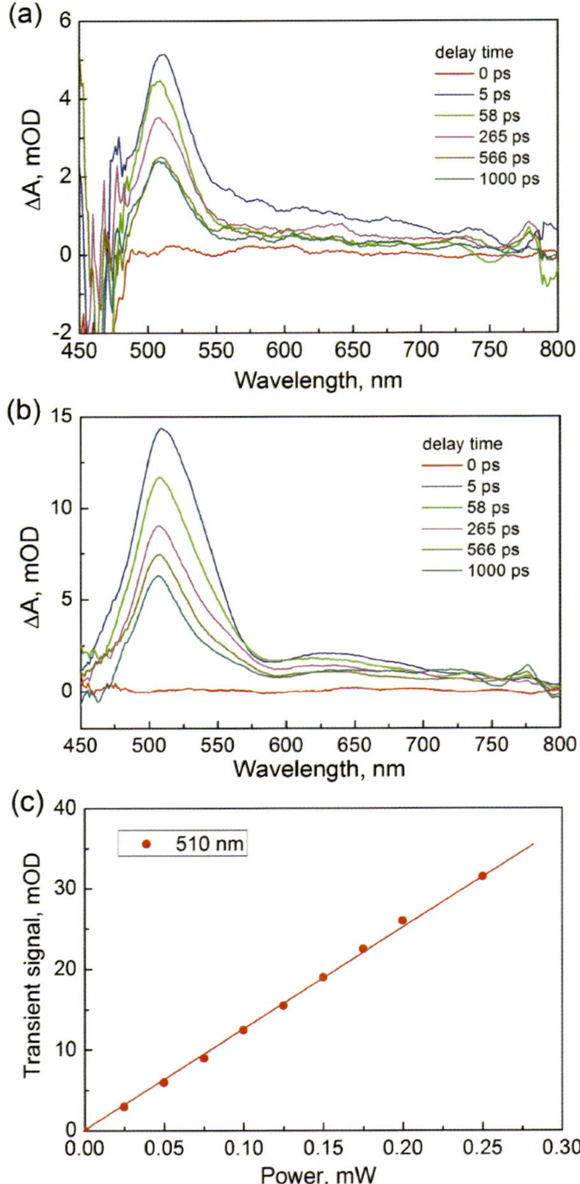

Figure 11. Transient absorption spectra for α-perylene crystal thin film at different delay times; λ_{exc} = 250 nm (a) and λ_{exc} = 340 nm (b). The power dependence of 510 nm TA signal measured at λ_{exc} = 340 nm (c).[21]

the fluorescence although the absorption becomes much larger. It is indicative of an efficient singlet fission which proceeds directly from the upper vibrational levels of S_1. The threshold of this process is 3.51 eV.[28,29]

3.4.2. *Fluorescence decay*

The fluorescence up-conversion is applied in order to study the ultrafast dynamics of α-perylene.[21] The results are shown in Fig. 10. The fluorescence lifetime becomes longer with increasing emission wavelength. At 460 nm, 92% of the fluorescence decays with a time constant $\tau_1 = 500$ fs. While at 630 nm, fluorescence decays mainly with $\tau_2 > 1$ ns. The wavelength dependent fluorescence lifetime indicates that the Y state has a fine structure and includes many sublevels, with lifetimes ranging from 500 fs to 3 ps. We should note that no rise-time was detected for all the fluorescence kinetics due to the strong spectral overlap between different emitting states. Based on our time-resolved fluorescence measurements, we conclude that the excited monomeric state relaxes to the "hot" excimer state (Y state) within 100 fs. Then these "hot" excimers, i.e. different sublevels of Y state, continue relaxing to the fully relaxed excimer state (E state) within 3 ps.

3.4.3. *Pump-probe measurements*

3.4.3.1. *One-quantum excitation: $\lambda = 250$ and 340 nm*

The TA spectra under 250 nm pump are shown in Fig. 11a. After the baseline correction (subtract the signal at negative delay times), the dominant TA peak locates at 505 nm, accompanied by a broad absorption band ranging from 600 to 800 nm. Similar spectra were also obtained at $\lambda_{exc} = 340$ nm (Fig. 11b).[21] The power dependence of the transient signal at 510 nm is linear (Fig. 11c). Considering the excimer absorption in α-perylene locating at 620 nm,[30] the broad TA band is assigned to the excimer absorption, i.e. including Y and E states. Based on the discussion on the fluorescence upconversion measurements, the "hot" excimer state Y consists of various sublevels covering the range of 4500 cm^{-1}. Therefore, the excimer TA band is broad. The 505 nm TA band is due to dimer cation; in perylene solution it is located at 503 nm.[31] In α-perylene crystal, the ion pair (IP) state, the precursor for the charge carrier, locates at 2.65 eV (468 nm).[27] It should be mentioned that at the negative delay times a pronounced peak located at 480 nm was observed,[21] which corresponds to the triplet state absorption. In diluted solution, only excited singlet state absorption at 700 nm was observed. No triplet state was found due to the low intersystem crossing ($\phi_{isc} = 0.01$).[32] Only in the highly concentrated solution we observed a

124

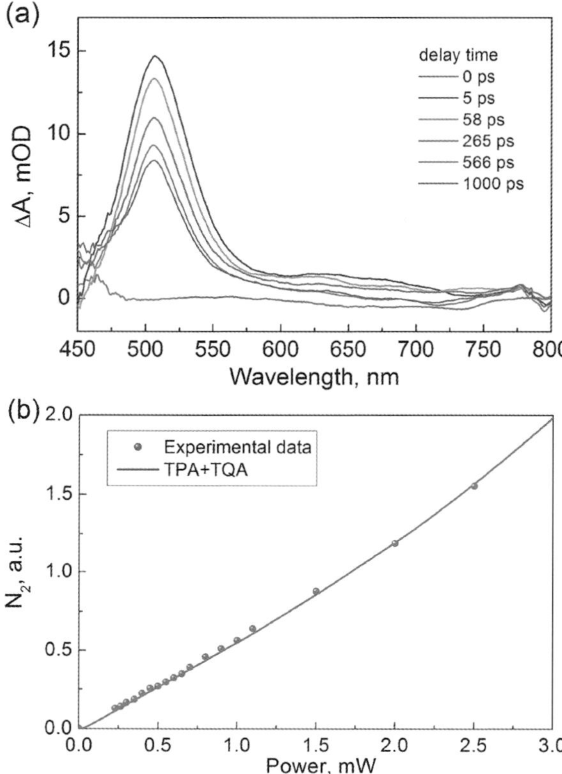

Figure 12. (a) Transient absorption spectra for α-perylene crystal thin film at different delay times; λ_{exc} = 475 nm, (b) the power dependence of 510 nm TA signal.[21]

weak positive TA band at about 490 nm at 5 ns delay which corresponds to the triplet absorption. Therefore, the triplet states observed in α-perylene crystal (Figs. 11a, b) are *not* generated by intersystem crossing. Hence triplet formation in the crystal is due to singlet fission. The only information in literature on singlet fission in α-perylene crystal is related to the magnetic field effect under the excitation with xenon lamp.[28,29] The threshold for singlet fission was estimated to be 3.51 eV (353 nm). The competing reaction was considered to be the fast excimer formation.

3.4.3.2. *Two-quantum excitation: λ = 475 nm*

Figure 12a shows the TA spectra under excitation at 475 nm, i.e. below the threshold for both fission and cation formation.[21] These TA spectra are similar to that for 250 or 340 nm excitation. The TA signals from dimer cation and excimer still govern the whole spectra. Moreover, at the negative delay times, the triplet absorption is still the dominant. However, as mentioned above, neither cation generation nor singlet fission under one-photon absorption at 475 nm should be expected. The power dependence of the transient signal at 510 nm is nonlinear (Fig 12b). This power dependence is explained in terms of two-photon absorption (TPA, i.e., simultaneous absorption of two photons) and consecutive two-quantum absorption (TQA, i.e., two-step absorption of two photons via intermediate excimer state).[21] Contribution of both TPA and TQA processes was discussed in detail in Refs. 33,34 for the nonlinear crystal KDP.

The rate equations for TQA process are:[21,33-35]

$$\frac{dN_0}{dt} = -\sigma_1 I N_0 + \frac{N_1}{\tau} \tag{2}$$

$$\frac{dN_1}{dt} = \sigma_1 I N_0 - \phi_2 \sigma_2 I N_1 - \frac{N_1}{\tau} \tag{3}$$

$$\frac{dN(TQA)}{dt} = \phi_2 \sigma_2 I N_1 \tag{4}$$

Where N_0, N_1 and N(TQA) are the populations of the ground state (S_0), intermediate state (S_1) and the two quantum excited state (S_N). σ_1 and σ_2 are the absorption cross sections of $S_0 \rightarrow S_1$ and $S_1 \rightarrow S_N$ transitions, respectively ($\sigma_1 = 7.6*10^{-17}$cm^2 at 475 nm[36]). ϕ_2 is the yield of the final product from the two quantum excited state S_N. Assuming rectangular pulse shape and based on different intermediate state lifetimes and the ratio of $\phi_2\sigma_2$ and σ_1, we have obtained the numerical solution of the rate equations (2)-(4). Taking also into consideration TPA process, a term N(TPA) = A*I^2 is added to N(TQA). The best fit is obtained for $\phi_2\sigma_2 = 50\sigma_1$ (Fig. 12b). The lifetime of S_1 is taken t = 2 ps which corresponds to the "hot" excimer Y state. Based on above considerations, we suggest that both singlet fission and dimer cation formation under 475 nm are due to the coexistence of TPA and TQA.

Acknowledgments

We thank Keke Zhang, Ke Jie Tan and Hui Jiang for providing high quality rubrene and perylene crystals.

References

1. C. E. Swenberg and N. E. Geacintov, *Organic Molecular Photophysics.* (Wiley & Sons Ltd., Chichester, Sussex, 1973), vol. 18.

2. S. Singh, W. J. Jones, W. Siebrand, B. P. Stoicheff, and W. G. Schneider, *J. Chem. Phys.* **42**, 330 (1965).

3. N. Geacintov, M. Pope, and F. Vogel, *Phys. Rev. Lett.* **22**, 593 (1969).

4. R. E. Merrifield, P. Avakian, and R. P. Groff, *Chem. Phys. Lett.* **3**, 386 (1969).

5. M. C. Hanna and A. J. Nozik, *J. Appl. Phys.* **100**, 074510 (2006).

6. M. C. Beard and R. J. Ellingson, *Laser & Photon. Rev.* **2**, 377 (2008).

7. R. J. Cava, F. J. DiSalvo, L. E. Brus *et al.*, *Prog. Solid State Chem.* **30**, 1 (2002).

8. R. D. Schaller, J. M. Pietryga, and V. I. Klimov, *Nano Lett.* **7**, 3469 (2007).

9. S. Wang, M. Khafizov, X. Tu, M. Zheng, and T. D. Krauss, *Nano Lett.* **10**, 2381 (2010).

10. M. B. Smith and J. Michl, *Chem. Rev.* **110**, 6891 (2010).

11. M. B. Smith and J. Michl, *Annu. Rev. Phys. Chem.* **64**, 361 (2013).

12. A. J. Musser, M. Al-Hashimi, M. Maiuri, D. Brida, M. Heeney, G. Cerullo, R. H. Friend, and J. Clark, *J. Am. Chem. Soc.* **135**, 12747 (2013).

13. C. Kloc, P. G. Simpkins, T. Siegrist, and R. A. Laudise, *J. Cryst. Growth* **182**, 416 (1997).

14. J. Shang, T. Yu, J. Lin, and G. G. Gurzadyan, *ACS Nano* **5**, 3278 (2011).

15. O. D. Jurchescu, A. Meetsma, and T. T. M. Palstra, *Acta Crystallogr. Sect. B* **62**, 330 (2006).

16. L. Ma, K. Zhang, C. Kloc, H.Sun, M. E. Michel-Beyerle, and G. G. Gurzadyan, *Phys. Chem. Chem. Phys.*, **14**, 8307 (2012).

17. A. Yildiz, P. T. Kissinger, and C. N. Reilley, *J. Chem. Phys.* **49**, 1403 (1968).

18. L. Ma, G. Galstyan, K. Zhang, C. Kloc, H. Sun, C. Soci, M. E. Michel-Beyerle, and G. G. Gurzadyan, *J. Chem. Phys.* **138**, 184508 (2013).

19. L. Ma, K. Zhang, C. Kloc, H. Sun, C. Soci, M. E. Michel-Beyerle, and G. G. Gurzadyan, *Phys. Rev. B* **87**, 201203 (2013).

20. W. G. Albrecht, H. Coufal, R. Haberkorn, and M. E. Michel-Beyerle, *Phys. Stat. Sol. B* **89**, 261 (1978).

21. L. Ma, K. J. Tan, H. Jiang, C. Kloc, *M. E. Michel-Beyerl*, and G. G. Gurzadyan, (2013), in press.

22. B. A. Jones, M. J. Ahrens, M.-H. Yoon, A. Facchetti, T. J. Marks, and M. R. Wasielewski, *Angew. Chem. Int. Ed.* **43**, 6363 (2004).

23. M. A. Angadi, D. Gosztola, and M. R. Wasielewski, *Mater. Sci. Eng. B* **63**, 191 (1999).

24. L. Schmidt-Mende, A. Fechtenkötter, K. Müllen, E. Moons, R. H. Friend, and J. D. MacKenzie, *Science* **293**, 1119 (2001).

25. J. Tanaka, *Bull. Chem. Soc. Jpn.* **36**, 1237 (1963).

26. T. Y. H. Nishimura, K. Mizuno, M. Iemura and A. Matsui, *J. Phys. Soc. Jpn.* **53**, 3999 (1984).

27. A. Furube, M. Murai, Y. Tamaki, S. Watanabe, and R. Katoh, *J. Phys. Chem. A* **110**, 6465 (2006).

28. W. G. Albrecht, M. E. Michel-Beyerle, and V. Yakhot, *J. Lumin.* **20**, 147 (1979).

29. W. G. Albrecht, M. E. Michel-Beyerle, and V. Yakhot, *Chem. Phys.* **35**, 193 (1978).

30. Z. Ludmer, L. Zeiri, and S. Starobinets, *Phys. Rev. Lett.* **48**, 341 (1982).

31. K. Kimura, T. Yamazaki, and S. Katsumata, *J. Phys. Chem.* **75**, 1768 (1971).

32. C. A. Parker and T. A. Joyce, *Chem. Commun. (London)* **0**, 108b (1966).

33. G. G. Gurzadyan and R. K. Ispiryan, *Appl. Phys. Lett.* **59**, 630 (1991).

34. G. G. Gurzadyan and R. K. Ispiryan, *J. Nonlinear Optic. Phys. Mat.* **01**, 533 (1992).

35. Z. Shaquiri, E. Keskinova, A. Spassky, and D. Angelov, *Photochem. Photobiol.* **65**, 517 (1997).

36. R. M. Hochstrasser, *J. Chem. Phys.* **40**, 2559 (1964).

128

MAPPING CARRIER DIFFUSION IN SINGLE SILICON CORE-SHELL NANOWIRES WITH ULTRAFAST OPTICAL MICROSCOPY

MINAH SEO

Center for Integrated Nanotechnologies, Los Alamos National Laboratory, Los Alamos, New Mexico 87545, USA

JINKYOUNG YOO

Center for Integrated Nanotechnologies, Los Alamos National Laboratory, Los Alamos, New Mexico 87545, USA

SHADI DAYEH

Center for Integrated Nanotechnologies, Los Alamos National Laboratory, Los Alamos, New Mexico 87545, USA

JULIO MARTINEZ

Center for Integrated Nanotechnologies, Sandia National Laboratories, Albuquerque, NM, 87185, USA

BRIAN SWARTZENTRUBER

Center for Integrated Nanotechnologies, Sandia National Laboratories, Albuquerque, NM, 87185, USA

SAMUEL PICRAUX

Center for Integrated Nanotechnologies, Los Alamos National Laboratory, Los Alamos, New Mexico 87545, USA

ANTOINETTE TAYLOR

Center for Integrated Nanotechnologies, Los Alamos National Laboratory, Los Alamos, New Mexico 87545, USA

ROHIT PRASANKUMAR[†]

Center for Integrated Nanotechnologies, Los Alamos National Laboratory, Los Alamos, New Mexico 87545

[†]To whom correspondence should be addressed. rpprasan@lanl.gov.

There has been an explosion in research on semiconductor nanowires (NWs) in recent years, primarily due to their variety of potential electronic and optoelectronic applications, including photodetectors, electrically-driven lasers, nanoscale transistors, and solar cells. Furthermore, recent success in the fabrication of NW heterostructures, composed of one or more layers with different properties, has enabled greater control of device operation for these applications. However, further progress is hindered by the limited knowledge of their properties. In particular, since interfaces between different layers in heterostructured NWs strongly influence properties such as the nature and speed of charge transport across the interface, it is essential to understand carrier dynamics in these nanosystems. Here, we use ultrafast optical microscopy to directly examine carrier dynamics and diffusion in single silicon core and Si/SiO_2 core/shell NWs with high temporal and spatial resolution in a non-contact manner. Spatially-resolved femtosecond optical pump-probe spectroscopy demonstrates the influence of surface-mediated mechanisms on carrier dynamics in a single NW, while polarization-resolved pump-probe spectroscopy reveals a clear anisotropy in carrier lifetimes measured parallel and perpendicular to the NW axis, due to density-dependent Auger recombination. Furthermore, separating the pump and probe spots along the NW axis enabled us to track space and time dependent carrier diffusion in radial NW heterostructures. These results enable us to reveal the influence of radial interfaces on carrier dynamics and charge transport in these quasi-one-dimensional nanosystems, which can then be used to tailor carrier relaxation in a single NW heterostructure for a given application.

1. Ultrafast carrier dynamics in single quasi-one-dimensional Si nanowires

In recent years, much interest has been shown in the optical properties of silicon nanowires (Si NWs) due to their wide range of potential applications, particularly in enabling nanoscale optoelectronic circuitry, transistors, light emitters, photodetectors, and Si NW-based solar cells [1-3]. These quasi-one-dimensional (1D) nanosystems have unique properties that depend on their size, shape, and alignment, which are often studied by photoluminescence (PL) spectroscopy. Recent success in the fabrication of axial and radial core-shell heterostructures, composed of one or more layers with different properties, on semiconductor NWs has enabled greater control of NW-based device operation for various applications [4-6]. However, further progress towards significant performance enhancements in a given application is hindered by the limited knowledge of carrier interactions in these structures. In particular, the strong influence of interfaces between different layers in NWs on transport makes it especially important to understand carrier dynamics in these quasi-one dimensional (1D) systems.

While time-integrated and time-resolved PL experiments have given some insight into the influence of the NW diameter [7], defect states [8], and incident light polarization relative to the long axis of the NW [9, 10], these measurements have not had sufficient time resolution to resolve several important carrier relaxation processes in these systems, including electron-phonon coupling and inter/intravalley scattering. Therefore, there remains a lack of basic understanding regarding how light interacts with individual NWs on an ultrashort time scale. In contrast, femtosecond optical pump-probe spectroscopy does have sufficient time resolution (<100 femtoseconds (fs)) to resolve fundamental dynamical processes in NWs. To date, nearly all ultrafast optical experiments on semiconductor NWs have been done on NW ensembles [11-18], in which the broad distribution of NW sizes and orientations can make it difficult to extract their fundamental physical properties, limiting the insight obtained from these studies.

These considerations have motivated the current interest in exploring carrier dynamics in individual NWs. However, due to the difficulty in isolating and finding the exact position of an individual NW on a substrate, as well as the small signal levels inherent to ultrafast optical experiments on single NWs, there have only been a few studies of femtosecond carrier dynamics in single NWs [19, 20], and until the work described here [21-23], there were none in single Si NWs, arguably the most important NW material of all. It is thus especially worth investigating the size- and polarization-dependence of carrier dynamics in individual nanowires to understand their fundamental properties without the influence of inhomogeneous broadening, which will then enable researchers to utilize the unique directional optical properties available from these quasi-1D nanostructures [24]. Since the surface to volume ratio varies directly with diameter and is relatively large in such systems, many properties of NWs are governed by their size, including carrier relaxation, surface recombination velocity, and diffusion length. The polarization of the incident light can also play an important role in NW properties and applications. For example, although the vast majority of solar cell modules are based on crystalline Si due to its high performance and relative ease in processing, there have been several attempts to make Si NW solar cells for enhanced charge transport [5, 25]. Because these devices have p-n junctions in either the radial [26] or axial [6, 27] directions, it is essential to understand the different physical processes that can influence the lifetimes of photoexcited carriers collected in a direction parallel or perpendicular to light absorption [28].

Here, we describe the first measurements of ultrafast carrier dynamics in single Si NWs, obtained by using non-degenerate pump-probe spectroscopy to excite and probe carriers above the indirect band gap of Si (E_g = 1.12 eV). In particular, we observed a strong position- and polarization-dependence of the transient carrier dynamics in single Si NWs, which gives deep insight into light-matter interactions in these quasi-1D-systems since the complications resulting from the broad NW size and alignment distribution in ensembles are avoided. We were able to spatially resolve diameter-dependent carrier dynamics along the axis of a tapered NW by varying the position of the focused beams, revealing the influence of surface trapping and recombination on carrier relaxation. This enabled us to quantify the impact of NW size on the carrier lifetime, surface recombination velocity, and diffusion length. We also found that the strong dependence of carrier dynamics in single Si NWs on the light polarization and NW diameter observed here is intimately related to density-dependent Auger recombination. Furthermore, we used ultrafast optical microscopy [21] to directly examine carrier relaxation and diffusion in single silicon core-only and Si/SiO$_2$ core-shell NWs with high temporal and spatial resolution in a non-contact manner. This enabled us to reveal strong acoustic phonon oscillations and experimentally map electron and hole diffusion currents in individual semiconductor NWs for the first time.

These new observations cannot be obtained using conventional contact-based methods for studying minority carrier transport in semiconductor systems such as scanning photocurrent microscopy (SPCM) or electron-beam-induced current (EBIC) analysis [28, 29], and as such these are the first spatiotemporal measurements of diffusion currents and phonon oscillations in single Si NWs. This novel approach to elucidating the optical and transport properties of Si NWs will have significant implications for semiconductor NW-based applications, ranging from photovoltaic devices to solar cells.

1.1. *Transient carrier dynamics measurement in ensemble and single Si NWs*

We used a non-degenerate femtosecond pump-probe system to measure polarization-dependent transient carrier dynamics in Si NWs. The output of a 30 femtosecond (fs) Ti:sapphire laser oscillator, operating at 80 MHz at a center wavelength of 840 nm, is divided into pump and probe beams, with the probe power <10% of the pump power (Figure 1(a)). The pump beam is then frequency doubled in a BBO crystal to generate 420 nm pulses. The initial magnitude of the normalized photoinduced change in probe transmission, $\Delta T/T$, with 420 nm pump and 840 nm probe pulses was ~10^{-5}. Using a 10X microscope objective lens, the back side of the sample is imaged onto a CCD camera to carefully overlap the pump (~25 μm diameter) and probe (~20 μm diameter) onto a certain position on a given SiNW. Placing broadband linear polarizers and wave plates in the path of both pump and probe beams enables us to measure $\Delta T/T$ signals as a function of the light polarization relative to the NW axis. All of our experiments were done at room temperature.

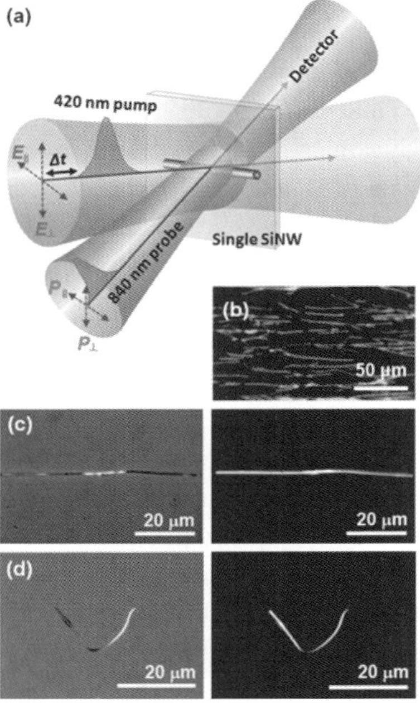

Figure 1. Schematic of non-degenerate pump-probe measurement in single Si NWs. (a) A conceptual illustration of a non-degenerate pump-probe measurement on a Si single nanowire in transmission. Optical microscope images are shown for (b) 50 nm average diameter ensemble Si NWs, (c) a straight single Si NW, and (d) a bent single Si NW. The left images were taken in bright field mode and the right images were taken in dark field mode for (c) and (d), respectively [21].

The Si NWs used for the experiments described in this section were grown by the vapor-liquid-solid (VLS) synthesis method using chemical vapor deposition with 50% silane in hydrogen and catalytic Au growth seeds [30]. The NWs were dry transferred onto a sapphire substrate with an areal density of $\sim 5.5 \times 10^6$ NWs/cm^2 and average diameters of $d \sim 50$ nm and lengths of $l \sim 80$ µm. These well aligned NW arrays are used in our measurements on ensembles (Figure 1(b)). Individual Si NWs were picked up from SiNW ensembles using a nanomanipulator developed at our center that consists of a field-emission SEM (JEOL 6701F) modified to include 2 probe tips. These NWs were then placed onto a gold patterned sapphire substrate, which enables us to easily and repeatably find their positions. Two types of NWs, straight (Figure 1(c)) and bent (Figure 1(d)), are examined in our experiments.

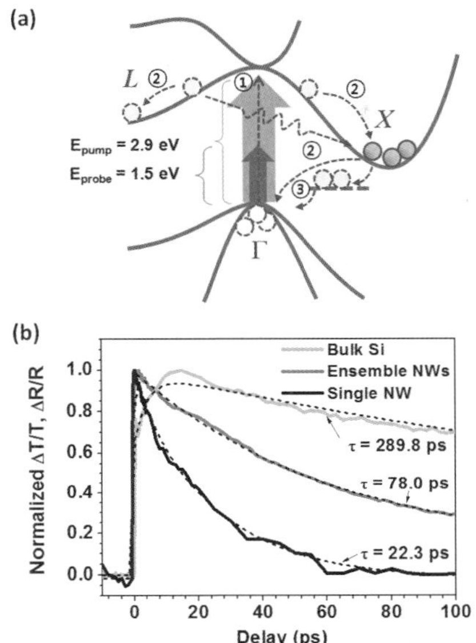

Figure 2. Band diagram and transient carrier dynamics spectrum of Si. (a) Upon photoexcitation, carriers undergo several processes: (1) photoexcitation, (2) relaxation to conduction minima (intravalley scattering) and/or intervalley scattering, and (3) recombination and/or surface/defect state trapping. (b) Comparison of transient carrier dynamics between ensemble- and single-Si NWs and bulk silicon, measured for $E_\parallel$ and $P_\parallel$.

Ultrafast optical pump-probe experiments were performed in reflection on bulk silicon and in transmission on SiNW ensembles and a single SiNW sample (Figure 2). In general, the photoinduced change in the probe reflection in bulk Si is sequentially induced by electron-hole pair excitation, intraband carrier energy relaxation, and trapping/recombination processes (Figure 2(a)) [31]. More specifically, in our experimental configuration, the 420 nm pump pulse excites electron-hole pairs at the Γ point, which rapidly scatter to states well above the conduction band minima in the X- and L- valleys. The carriers initially excited in the near-surface region of bulk Si cause a state-filling induced change in the indirect absorption of the 840 nm probe beam into the X valley, as indicated by the positive sign of the measured signal (Figure 2(b)). The non-equilibrium carriers then rapidly relax to lower energy states in the X- and L-valleys (Figure 2(b)), recombining or getting trapped at surface or defect states on long time scales. In contrast to bulk Si, curve fits to our data reveal that the $\Delta T/T$ signal decays in several tens of picoseconds for the ensemble NW sample and the single NW sample. The faster decay process in the NW ensemble as compared to bulk Si can be explained by the increased influence of surface traps and

134

recombination centers as the degree of spatial confinement increases. Carrier relaxation is even faster in the single NW, although this depends on the specific size and morphology of a given NW.

1.2. *Diameter and polarization dependence of carrier dynamics in single Si NWs*

The single NW that we focus on in this section has a tapered shape, as determined through SEM measurements (Figure 3(a)). Due to this tapered geometry, the measured photoinduced transmission change at different positions on the same NW directly reveals the diameter dependence of the carrier dynamics (Figure 3). The normalized $\Delta T/T$ signals at four different positions (P_1, P_2, P_3, and P_4) are shown here, which were obtained with 12.0 μJ/cm^2 pump fluence and both pump and probe polarizations parallel to the NW axis. As the NW diameter decreases while approaching its tip, carriers relax much more rapidly. This tendency is explicitly shown by plotting the time constant, τ, obtained from exponential curve fits to the measured $\Delta T/T$ data, versus the NW diameter (Figure 3(b)). This is consistent with previous measurements on NW ensembles, which demonstrated that surface-mediated mechanisms dominate their carrier dynamics, as the large surface-to-volume ratio of these one-

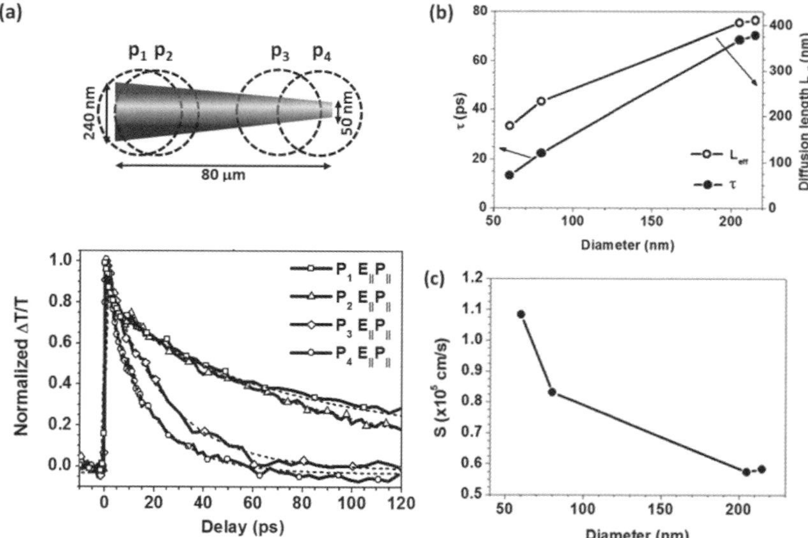

Figure 3. Position- and diameter-dependence of carrier dynamics. (a) Position-dependent transient transmission on a tapered single SiNW for E$_\parallel$ and P$_\parallel$. (b) Relaxation time, τ, and diffusion length, L_{eff}, vs. diameter, d, for the tapered single SiNW. (c) Diameter dependence of the surface recombination velocity, S [21].

dimensional systems makes them much more sensitive to surface effects than the corresponding bulk material [12, 15, 32-35].

Knowledge of the relaxation time constant also allows us to quantify transport properties, such as the carrier diffusion length. The diffusion length can be described by $L_{eff} \propto \sqrt{D_{eff}\tau}$ [29], where L_{eff} is the effective diffusion length, D_{eff} is the diffusion constant, and τ is the time constant described above. Our NWs are not expected to show any quantum confinement effects due to their relatively large diameter (d>10 nm); therefore, the parameters of undoped bulk Si can be used to interpret our results. Since we excite and observe both electron and hole dynamics in our experiments, we use D_{eff}=24 cm^2/s, defined here as an average diffusion constant of bulk silicon for electrons and holes [36]. The carrier diffusion length displays a strong dependence on nanowire diameter (Figure 3(b)), clearly showing that the diffusion length increases as the NW diameter increases (to a value of L_{eff}=410.5 nm on the large diameter (d=215 nm) side of the NW). We find that the extracted diffusion lengths in our nanowires have somewhat larger values than those measured using different techniques [37]. This may be due to our use of a non-contact optical method for directly measuring the carrier lifetime, which eliminates contact-induced measurement effects. Surface recombination velocities, S, can also be estimated from our results with a simple model [29, 38]; we find the extracted values to compare reasonably well with others [31] on average, with a strong diameter dependence as observed in ZnO NWs [39] (Figure 3(c)). The very low S values (<6x10^4 cm/s) obtained at large diameters (d>150~200 nm) are expected as surface states become less important in influencing recombination processes [31]. It is worth noting that the obtained S value approaches the value of 3x10^4 cm/s for bulk Si, which has been shown elsewhere [31].

The time-dependent photoinduced transmission change in single NWs can also be affected by the light polarization, as revealed through polarization-dependent experiments on single and ensemble NW samples. Figure 4a compares the measured transmission changes in SiNW ensembles for four different pump and probe polarization combinations; here, we use $E_{\parallel}$ and $E\perp$ to indicate pump polarizations parallel and perpendicular to the NW axis, and $P_{\parallel}$ and $P\perp$ to indicate probe polarizations parallel and perpendicular to the NW axis, respectively. The initial $\Delta T/T$ signal is maximum when both pump and probe are polarized parallel to the NW axis and minimum when both beams are polarized perpendicular to the NW axis. This is expected since light absorption is maximum for light polarized parallel to the NW axis [9, 10]; however, there was no significant anisotropy in the measured relaxation times (Figure 4(a) inset),

136

consistent with our previous experiments, which were conducted at much higher pump fluences and with different pump and probe wavelengths [14].

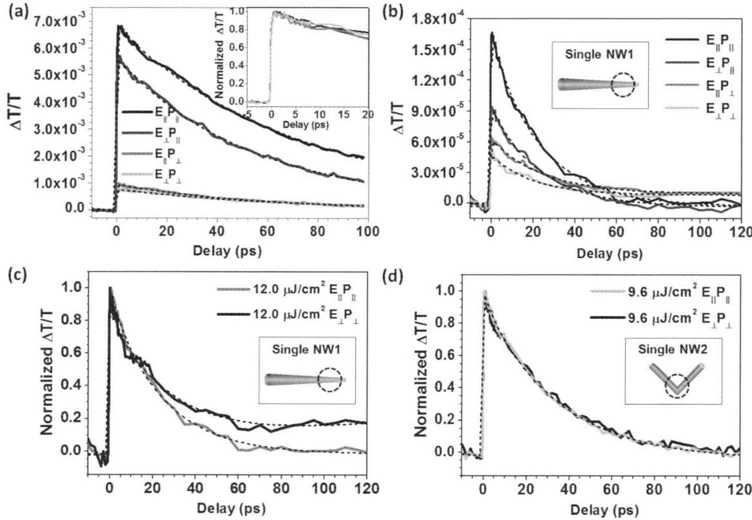

Figure 4. Polarization dependence of carrier dynamics. (a) Polarization-dependent ultrafast transmission measurements on ensemble Si NWs. The inset shows the normalized transient transmission for different pump and probe polarization combinations. (b) Photoinduced transient transmission measured on a single SiNW for different pump and probe polarizations. The normalized transient transmission for $E_{||}$, $E_\perp$, and $P_{||}$, $P_\perp$ is shown (c) at the small end of the straight NW and (d) at the middle of the bent NW [21].

Similarly, it is also observed that the $\Delta T/T$ signal is maximum for $E_{||}$ and $P_{||}$ and minimum for $E_\perp$ and $P_\perp$ from a measurement near the tip of the single SiNW (Figure 4(b)). In stark contrast to the measurements on the NW ensemble, however, there is a clear anisotropy in the relaxation times measured for light polarized parallel and perpendicular to the long axis of a single SiNW, especially for smaller NW diameters. Figure 4c shows the normalized $\Delta T/T$ signals extracted from Figure 4(b) for $E_{||}$ and $P_{||}$ (red line) and $E_\perp$ and $P_\perp$ (blue line), respectively. Overall, a faster decay was observed for $E_{||}$ and $P_{||}$ than $E_\perp$ and $P_\perp$. This anisotropy in the measured dynamics decreases as the NW becomes larger and its properties become more bulk-like, which has also been observed in Ge NWs[38] and in core/shell GaAs/AlGaAs NWs [40]. This is due to decreases in the dielectric contrast between the NW and the surrounding material for optically thick NWs with a radius $>\lambda/n_{nanowire}$ ($d{\sim}170$ nm in our case) [16]. This can also help explain the absence of polarization dependent-dynamics in the NW ensemble: even in well aligned NW ensembles (Figure 1(b)), the tapered shape of these NWs necessitates that there is a distribution of various diameters at any given point. The $\Delta T/T$ signals will be dominated by the larger NWs, which may explain why we measured minimal polarization anisotropy in the carrier dynamics for ensembles, both here and in our previous experiments. In contrast, a bent single NW with symmetric geometry has exactly the same normalized $\Delta T/T$ signal, irrespective of the light polarization (Figure 4(d)), which supports the fact that the absorption is the same for both polarizations and any anisotropy in the measured dynamics is therefore minimal.

In the single straight SiNW, more of the pump is absorbed for the parallel polarization, leading to a higher initial carrier density in this configuration. Density-dependent effects such as Auger recombination, which typically cause decreasing carrier lifetimes with increasing carrier density, can thus lead to the observed polarization dependence of the carrier dynamics [12, 17, 18]. In ref. [21], we performed density-dependent measurements on the single straight Si NW and fit the resulting data with an Auger recombination model for 1D structures. This revealed that the physical origin of the density dependence is Auger recombination, and that this can explain the observed polarization anisotropy. Our experiments thus demonstrated that carrier relaxation in single Si NWs is primarily governed by surface-mediated mechanisms and Auger recombination, with their relative importance depending on the NW size and morphology as well as the polarization and intensity of the incident light [21].

2. Mapping carrier diffusion in single silicon core-shell nanowires

Our next set of experiments focused on experimentally mapping electron and hole diffusion currents in single bare and core/shell Si NWs, using ultrafast optical microscopy (UOM) to track charge carrier transport through space and time [21]. While earlier UOM experiments by our group, as well as other groups, studied carrier dynamics at fixed positions along a semiconductor NW (as described in section 1) [20, 21], these experiments probed carrier concentration gradients dynamically along the length of the NW, which enabled us to observe a striking difference in carrier relaxation and diffusion for Si NWs with or without a high quality SiO_2 shell layer. We could then construct separate diffusion current maps for electrons and holes, described in more detail below. Our results thus demonstrate how shell passivation influences carrier concentration gradients and improves the efficacy of electron and hole diffusion currents in nanoscale semiconductor channels.

2.1. *Ultrafast optical microscopy on single core-shell Si NWs*

Our UOM system and a conceptual illustration of UOM with spatially separated pump and probe beams are shown in Figures 5(a) and (b), respectively. The same laser system and pump/probe wavelengths as described in section 1 were used for these experiments; the primary difference is that a 20X objective was used to focus both beams, and a 50X objective was used to collect the transmitted light, making the spatial resolution ~2 μm. The Si NWs used in these studies were fabricated using a top-down process, with a combination of e-beam lithography and Si deep reactive ion etching followed by thermal oxidation and stripping steps to form pristine NW surfaces (Figure 5(c)), after which they were transferred onto a transparent sapphire substrate, as shown in Figure 5(d), taken in our UOM system.

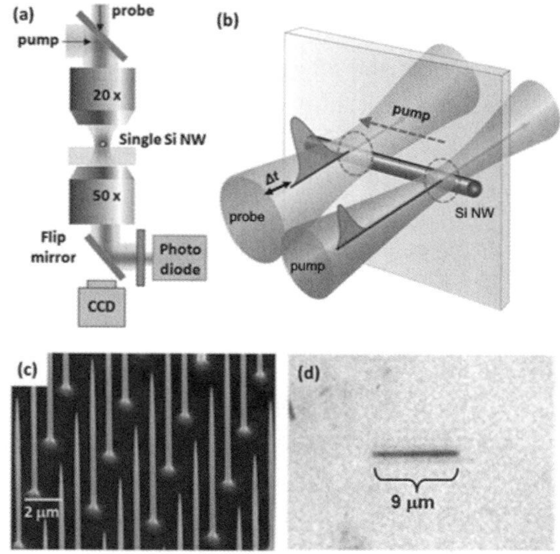

Figure 5. Ultrafast optical microscopy system. (a) Experimental setup for ultrafast optical microscopy on single nanowires. (b) A conceptual illustration of UOM with spatially separated pump and probe beams. (c) SEM image of pillar-type Si NWs on the silicon substrate. (d) Optical microscope image of a single Si NW on a sapphire substrate, taken in our UOM system [22].

2.2. *Ultrafast optical microscopy on single core-shell Si NWs*

UOM experiments were conducted on a single Si NW (255 nm) and a single Si/SiO$_2$ NW (255 nm core and 75 nm thick shell) with the probe beam fixed near one end of the NW and the spatial separation between the pump and probe beams, l, varied to a maximum of 5 µm along the NW axis. It is important to note that these single Si NWs are perfectly crystalline and have very small diameter variation along their axis (<10 nm) throughout the bulk of the NW where our measurements were performed (unlike the tapered NWs described in section 1), making them excellent quasi-1D systems to investigate carrier diffusion along the NW axis. Figures 6 (a) and (b) summarize spatiotemporal $\Delta T/T$ signals measured on the core-only and core/shell NWs. The rise time increases with l for both the bare Si NW and the Si/SiO$_2$ NW, revealing strong acoustic phonon oscillations in both cases. In ref. [22], we demonstrated that these were primarily attributed to Brillouin oscillations.

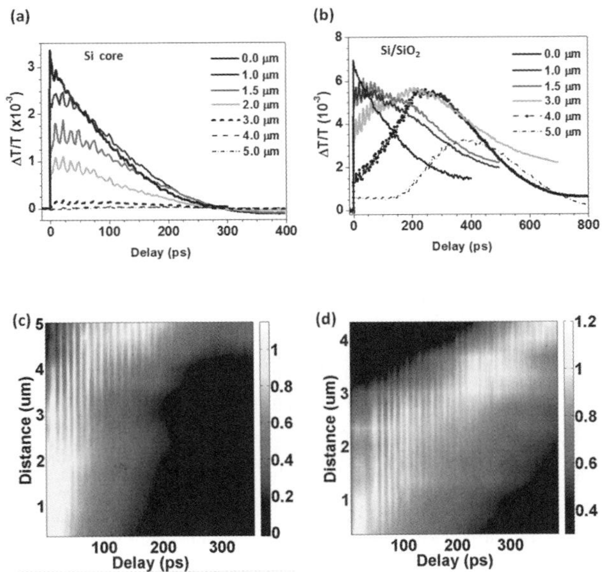

Figure 6. Mapping carrier dynamics in 255 nm diameter Si NWs. Photoinduced transmission changes for various pump-probe separations for (a) the bare Si NW and (b) the 150 nm SiO$_2$ shell encapsulated Si NW. Two-dimensional maps of the normalized $\Delta T/T$ signals as functions of separation and time delay are also shown for both NWs in (c) and (d) [22].

Interestingly, the maximum $\Delta T/T$ signal for the Si/SiO$_2$ NW stays nearly constant up to 4 µm, but rapidly decreases as a function of l for the bare Si NW. This drastic decrease in $\Delta T/T$ for the bare Si NW is caused by carrier trapping and recombination occurring at the surface of the NW [12, 15, 32, 34, 41]. In contrast, the SiO$_2$ layer in core/shell NWs reduces surface trapping and

140

recombination as described above, enabling us to directly track carriers as they diffuse from one end to the other in a single semiconductor NW.

The dramatic differences in time- and space-dependent carrier dynamics between the two NWs can be visualized in Figures 6(c) and (d)), which show contour plots of the normalized $\Delta T/T$ signal as a function of pump-probe separation and time delay. The diffusion velocity is defined as $v_{diff}=l/t_r$, where t_r is the time delay where the $\Delta T/T$ signal is maximum. For the bare Si NW, $v_{diff} \sim 5 \times 10^6$ cm/s, while $v_{diff} \sim 2 \times 10^6$ cm/s for the Si/SiO$_2$ core-shell NW. This can be inferred intuitively, since for bare Si NWs, surface recombination induces axial carrier concentration gradients larger than in core/shell Si/SiO$_2$ NWs, resulting in effectively higher diffusion velocities for the bare Si NWs.

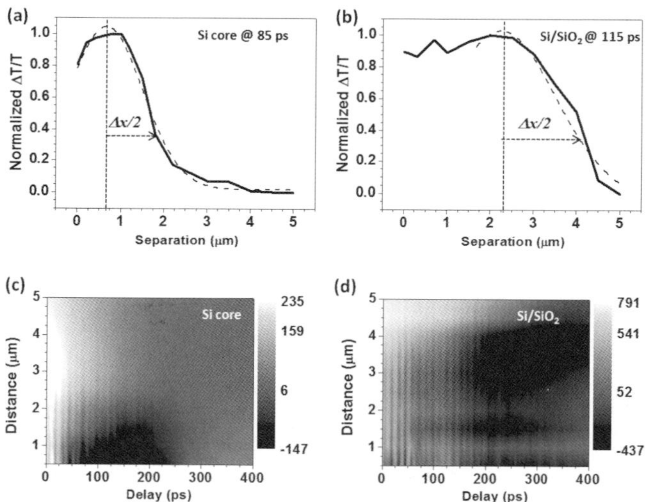

Figure 7. Mapping diffusion currents in Si NWs. Position-dependent carrier distribution along the NWs at specific times for (a) a Si core-only NW and (b) a Si/SiO$_2$ NW. Two dimensional diffusion current maps constructed from Figures 2(c) and (d), respectively, are also shown for (c) the core-only NW and (d) the core-shell NW. White denotes positive currents, black denotes negative currents [22].

To construct spatiotemporal diffusion current maps for our NWs, we first extract position-dependent carrier distributions for the Si core-only NW (Figure 7(a)) and the Si/SiO$_2$ core-shell NW (Figure 7(b)) from the contour plots in Figures 6(c) and (d) before normalization. Then, from the shape of the carrier distribution as a function of position at a specific time, we can directly obtain the diffusion coefficients for both NWs using a method based on the Haynes-

Shockley experiment [42]. Specifically, we can calculate the average diffusion coefficient for electrons and holes, D_{avr}, using

$$D_{avr} = \frac{(\Delta x)^2}{16 t_d} \qquad (1)$$

where Δx is the width of the carrier distribution (calculated by fitting with a Gaussian function) at a certain time, t_d. This relation is only valid without drift and when neglecting recombination, so we measured Δx for $t_d < \tau$ for each NW (85 ps for the Si core-only NW and 115 ps for the Si/SiO$_2$ NW). The extracted diffusion coefficients are 35 cm^2/s for the Si core-only NW and 53 cm^2/s for the Si/SiO$_2$ NW, respectively, which are somewhat larger than but still comparable to the values for bulk Si [36].

In ref. [22], we used a simple model to demonstrate that the diffusion current, J, is positive for electrons and negative for holes; importantly, this allows us to separate out the contributions of electrons and holes to the total diffusion current. Spatiotemporal maps of the diffusion current derived from the $\Delta T/T$ contour plots using equation (3) are depicted for both the bare Si NW (Figure 7(c)) and the core-shell NW (Figure 7(d)). For the bare Si NW, we can notice that electron diffusion currents evolve and decay over a relatively short time scale and throughout the entire length of the NW. This is due to the strong concentration gradient of electrons (Figure 6(c)) resulting from surface recombination throughout the entire NW length. For the core/shell NW, where the electron concentration is sustained throughout most of the NW length (Figure 6(d)), making the concentration gradients smaller, we find that electron diffusion currents are appreciable toward the far end of the NW away from the pump location (Fig. 7(d)), and are as a consequence sustained for significantly longer times than for the bare Si NW. The relatively short-lived nature of the electron diffusion currents in both types of NWs corroborates with earlier studies showing that surface states in Si NWs are dominantly acceptor-type and trap electrons [43].

In contrast, the heavier effective masses of holes cause a delay in the evolution of their diffusion currents. In addition, due to their lower diffusion coefficients, the magnitude of the hole diffusion current is observed to be lower than that of electrons. Consequently, for bare Si NWs, hole diffusion currents appear only over short distances from the probe spot, while many of the diffused electrons get trapped by surface states far away from the pump location. For the core/shell NWs, in the absence of strong recombination at the surface, a larger number of holes will propagate down the NW to recombine with the electrons

142

that have traveled to the NW end away from the pump spot. Hole diffusion currents therefore will be sustained over longer distances in the core/shell NWs.

The experimental observation of both electron and hole diffusion currents in a semiconductor nanowire is uniquely attained by our measurement technique and captures some of the most basic principles of semiconductor physics in a simple and straightforward analysis; in fact, our results essentially correspond to a "textbook" case of carrier diffusion in a quasi-1D semiconductor after an impulse excitation [42]. Furthermore, our experiments enable us to extract several fundamental parameters in these NWs, including the surface recombination velocity, diffusion coefficients, and diffusion velocities, without the influence of contacts. The results presented here thus have implications for semiconductor devices in a wide variety of applications that use minority carriers as the main charge transport mechanism, such as in optoelectronics, bipolar transistors, and sensitive photodetectors [4, 44]. Finally, our work further demonstrates that ultrafast optical microscopy will be particularly useful for future investigations of radial and axial NW heterostructures, where charge transport across heterointerfaces is of particular interest for advanced device architectures.

References

1. X. Duan, Y. Huang, R. Agarwal, C. M. Lieber, *Nature* **421**, 241 (2003).
2. C. Thelander *et al.*, *Materials Today* **9**, 28 (2006).
3. P. J. Pauzauskie, P. Yang, *Materials Today* **9**, 36 (2006).
4. M. S. Gudiksen, L. J. Lauhon, J. Wang, D. C. Smith, C. M. Lieber, *Nature* **415**, 617 (2002).
5. E. C. Garnett, P. Yang, *Journal of the American Chemical Society* **130**, 9224 (2008).
6. B. Tian *et al.*, *Nature* **449**, 885 (2007).
7. D. D. D. Ma, C. S. Lee, F. C. K. Au, S. Y. Tong, S. T. Lee, *Science* **299**, 1874 (March 21, 2003, 2003).
8. T. V. Torchynska *et al.*, *Physical Review B* **65**, 115313 (2002).
9. D. D. D. Ma, S. T. Lee, J. Shinar, *Applied Physics Letters* **87**, 033107 (2005).
10. J. Wang, M. S. Gudiksen, X. Duan, Y. Cui, C. M. Lieber, *Science* **293**, 1455 (August 24, 2001, 2001).
11. P. Parkinson *et al.*, *Nano Letters* **9**, 3349 (2009).
12. R. P. Prasankumar, P. C. Upadhya, A. J. Taylor, *physica status solidi (b)* **246**, 1973 (2009).
13. A. Othonos, E. Lioudakis, U. Philipose, H. E. Ruda, *Applied Physics Letters* **91**, 241113 (2007).

14. A. Kar *et al.*, *IEEE Journal of Selected Topics in Quantum Electronics* **17**, 889 (2011).
15. R. P. Prasankumar, S. Choi, S. A. Trugman, S. T. Picraux, A. J. Taylor, *Nano Letters* **8**, 1619 (2008).
16. J. Giblin, V. Protasenko, M. Kuno, *ACS Nano* **3**, 1979 (2009).
17. I. Robel, B. A. Bunker, P. V. Kamat, M. Kuno, *Nano Letters* **6**, 1344 (2006).
18. H. Htoon, J. A. Hollingsworth, R. Dickerson, V. I. Klimov, *Physical Review Letters* **91**, 227401 (2003).
19. G. V. Hartland, *Chemical Science* **1**, 303 (2010).
20. C. R. Carey, Y. Yu, M. Kuno, G. V. Hartland, *Journal of Physical Chemistry C* **113**, 19077 (2009).
21. M. A. Seo *et al.*, *Applied Physics Letters* **100**, 071104 (2012).
22. M. A. Seo *et al.*, *Nano Letters* **12**, 6334 (2012/12/12, 2012).
23. M. Seo *et al.*, *Opt. Express* **21**, 8763 (2013).
24. J. C. Johnson *et al.*, *Nat Mater* **1**, 106 (2002).
25. L. Tsakalakos *et al.*, *Applied Physics Letters* **91**, 233117 (2007).
26. B. M. Kayes, H. A. Atwater, N. S. Lewis, *Journal of Applied Physics* **97**, 114302 (2005).
27. S. Hoffmann *et al.*, *Nano Letters* **9**, 1341 (2009).
28. M. D. Kelzenberg *et al.*, *Nano Letters* **8**, 710 (2008).
29. J. E. Allen *et al.*, *Nat Nano* **3**, 168 (2008).
30. S. Picraux, S. Dayeh, P. Manandhar, D. Perea, S. Choi, *Journal of the Minerals, Metals and Materials Society* **62**, 35 (2010).
31. A. J. Sabbah, D. M. Riffe, *Physical Review B* **66**, 165217 (2002).
32. T. Hanrath, B. A. Korgel, *The Journal of Physical Chemistry B* **109**, 5518 (2005).
33. S. A. Dayeh, C. Soci, P. K. L. Yu, E. T. Yu, D. Wang, *Applied Physics Letters* **90**, 162112 (2007).
34. R. Calarco *et al.*, *Nano Letters* **5**, 981 (2005).
35. S. A. Dayeh, E. T. Yu, D. Wang, *Small* **5**, 77 (2009).
36. D. F. Edwards, *Handbook of Optical Constants of Solids*. E. D. Palik, Ed., (Academic, Newyork, 1998).
37. Y. Hashimoto, Y. Murakami, S. Maruyama, J. Kono, *Physical Review B* **75**, 245408 (2007).
38. Y. Dan *et al.*, *Nano Letters* **11**, 2527 (2011).
39. A. Soudi, P. Dhakal, Y. Gu, *Applied Physics Letters* **96**, 253115 (2010).
40. A. Persano *et al.*, *Applied Physics Letters* **98**, 153106 (2011).
41. A. H. Chin *et al.*, *Nano Letters* **7**, 626 (2007).
42. B. G. Streetman, S. Banarjee, S. K. Banerjee, *Solid State Electronic Devices* (Prentice Hall, 1995).
43. J. Jie *et al.*, *Advanced Functional Materials* **18**, 3251 (2008).
44. X. Duan, Y. Huang, Y. Cui, J. Wang, C. M. Lieber, *Nature* **409**, 66 (2001).

EXCITON DYNAMICS AND ITS REGULATION ABILITY IN PHOTOSYNTHESIS

V. BALEVICIUS, Jr. and L. VALKUNAS*

*Faculty of Physics, Vilnius University,
Vilnius, Sauletekio al. 9-III, LT-10222, Lithuania
Institute of Physics, Center for Physical Sciences and Technology,
Vilnius, Savanoriu Avenue 231, LT-02300, Lithuania * E-mail: leonas.valkunas@ff.vu.lt*

D. ABRAMAVICIUS

*Faculty of Physics, Vilnius University,
Vilnius, Sauletekio al. 9-III, LT-10222, Lithuania*

The possible role of quantum coherence in photosynthesis is the subject of extensive discussions. Therefore the overview of the theoretical models used to describe the exciton dynamics and decay in molecular aggregates, is presented. The main aspects, revealed by several theoretical approaches, are demonstrated on the simplest molecular aggregate — a molecular dimer. The possible role of the quantum coherence in photosynthesis from the broader perspective of energy transfer and its non-photochemical quenching is also discussed.

Keywords: Coherent dynamics; Energy transfer; Non-photochemical quenching.

1. Introduction

To efficiently harvest solar light, photosynthetic organisms are equipped with pigment–protein antenna complexes. These complexes are involved in the initial stage of photosynthesis, starting with the absorption of the solar light by the pigment molecules and followed by the transfer of the accumulated energy to the reaction center (RC), where this energy is stabilized as a chemical potential.[1,2] As a result of the structural arrangement and spectral composition of the pigment molecules, at low excitation intensities the efficiency of this process is close to unity.

Spectral variability of photosynthetic light-harvesting pigment–protein complexes is usually attributed to either excitonic interactions between pigment molecules or to their interactions with the protein surrounding.[1,2]

Such attribution is strongly supported by stationary and time-resolved spectroscopic data of various photosynthetic pigment–protein complexes. The variation of the protein environment of an individual pigment molecule introduces differences in transition energies and determines the time-scale of their changes.[3] The interaction with the local protein environment is easily described in two limiting cases, corresponding to the static and dynamic disorder of the transition energies of the pigment molecules. The static disorder corresponds to the slow protein movement with respect to the measurement time, while the dynamic disorder reflects the opposite limiting case of fast vibrations of the environment. Hence, the interpretation essentially depends on the relevant time-scale of the experiment or the process. The electronic molecular excitations have the dephasing time of a few hundred femtoseconds. The "slow" or "fast" should be interpreted with respect to this time-scale. The fast fluctuations are averaged and result in the exciton transport and dephasing.[4] The entire set of such vibrations might be considered as the bath, while the interaction of these vibrations with molecular electronic excitations is treated perturbatively within the framework of the density matrix theory.[5,6] The slow fluctuations are treated as the static energy displacements from the mean values. According to such theoretical scheme, the exciton dynamics can contain both coherent and incoherent, dephasing and decoherence components. These effects originate from the disruption of the phase relationship between excitonic wavefunctions of the molecules due to the interaction of the electronic excitations with the intra- and inter-molecular vibrations.

Recent development of nonlinear spectroscopies, such as the two-dimensional photon echo (2D PE) spectroscopy,[4,7–9] is getting widely available for studies of exciton coherence in photosynthetic pigment-protein complexes.[10,11] 2D PE spectroscopy was a key tool demonstrating a complex pathway network of the energy transfer in LH2[12] and LH3,[13] the peripheral light-harvesting complexes from photosynthetic bacteria, and long-lasting coherence in Fenna–Matthews–Olson (FMO) complexes[8,14] and in LHCII, the major light-harvesting complex from PSII of plants.[15] Recently the 2D PE spectra were also recorded for the RC from photosynthetic bacteria[16] and for other molecular aggregates, such as polymers[17] or the cylindrical (bi-tubular) J-aggregates.[18] Apart from clear identification of exciton transfer between pigment molecules or their clusters, quantum coherence and population oscillations were also observed. All the new data gave rise to the currently active research area sometimes referred to as "quantum biology", which aims to evaluate the importance of coherence, entanglement and noise in the energy transport of biological molecular complexes.[19–25]

The capture of a solar photon in the form of the molecular excitation and its delivery into the RC, where further energy conversion takes place, is one of the main functions of the light-harvesting aggregates. In addition to this physical function, the aggregates have another — biological self-regulation function, which is a physiologically significant strategy evolved by plants.[26] Rapid excitation density control in photosystem II (PSII), termed as non-photochemical quenching (NPQ), ensures the robustness of plant photosynthesis under fluctuating light, even at very high intensities. The NPQ is in essence the mechanism controlling the dissipation of the excess excitation energy. For a reasonable effect to be achieved, NPQ should be competitive with the excitation trapping by an open RC. The NPQ phenomenon is usually attributed to some activated quenching species which allows the excitation to undergo a rapid non-radiative decay. The exact location of the quencher within the antenna and its precise nature are a matter of on-going debate, with both pigment molecules, i.e., chlorophylls (Chl) and carotenoids (Car), being put forward as essential components of the quenching mechanism. Usually the quenching mechanisms are attributed (in no order of preference) to: (i) a Chl–Chl dimer showing the charge transfer (CT) state character[27] or via the formation of Chl–Chl excimeric states;[28] (ii) a CT state, which appears in the Chl and Car (xanthophyll) dimer, resulting in generation of the cation radical state of the xanthophyll;[29,30] (iii) the excitonic coupling of Chl to a short-lived xanthophyll excited state,[31,32] and (iv) the direct energy transfer from the Chl pool to a particular Car (lutein).[33] Here we will consider the excitation dynamics and the possible role of coherence in determining the excitation transfer and trapping, and discuss the possible mechanism of NPQ.

2. Excitons

2.0.1. *Frenkel exciton model of molecular aggregates*

The quantum-mechanical formulation of molecules and their complexes, embedded in various environments, is the starting point in molecular aggregate theory. First, we assume that we can solve the Schrödinger equation of the isolated molecules. Since the optical excitation is resonant with one particular electronic transition of the molecules, we are interested only in the characteristics of two electronic states — the electronic ground and excited state wavefunctions of the j-th molecule, $\varphi_j^{(g)}$ and $\varphi_j^{(e)}$.

These functions are taken as the basis set for the problem formulation: the wavefunction of an aggregate is constructed as a direct product of the wavefunctions of the isolated molecules (the Heitler–London approximation):

$$\Phi^{(g)} = \prod_j^N \varphi_j^{(g)} \tag{1}$$

for the ground state and

$$\Phi_i^{(e)} = \varphi_i^{(e)} \prod_{j \neq i}^N \varphi_j^{(g)} \tag{2}$$

for the excited state.

The aggregates can have multiple excitations. While a molecule can be only in the ground or excited state, a set of N molecules has room for up to N excitations. However, only the double-exciton states, where two molecules are excited in the aggregate, are relevant for the third-order nonlinear spectroscopy. These can be denoted by

$$\Phi_{ij}^{(f)} = \varphi_i^{(e)} \varphi_j^{(e)} \prod_{l \neq i,j}^N \varphi_l^{(g)}. \tag{3}$$

This basis set is easily translated into an excitation creation and annihilation operator picture. The vacuum state $|g\rangle$ is the ground state of the aggregate having no excitations (its wavefunction is $\Phi^{(g)}$). The state, where the m-th molecule is excited (a single-exciton state) is represented by $|m\rangle \equiv \hat{B}_m^\dagger |g\rangle$, and for a pair of excited molecules (a double-exciton state) $|mn\rangle \equiv \hat{B}_m^\dagger \hat{B}_n^\dagger |g\rangle$. A molecule cannot be excited twice, which is guaranteed by choosing the Pauli commutation relations[4,34]

$$\left[\hat{B}_n, \hat{B}_m^\dagger\right] = \delta_{mn}\left(1 - 2\hat{B}_m^\dagger \hat{B}_m\right). \tag{4}$$

The aggregate Hamiltonian can be represented using these operators:

$$\hat{H} = \sum_i^N \varepsilon_i \hat{B}_i^\dagger \hat{B}_i + \sum_{i \neq j}^N J_{ij} \hat{B}_i^\dagger \hat{B}_j, \tag{5}$$

where ε_i is the transition energy of the molecular excitation in the presence of other molecules in their ground states, and J_{ij} is the resonance interaction.

According to the Heitler–London approximation, the lowest (ground) state is a single state, and thus, all other properties of the system are given

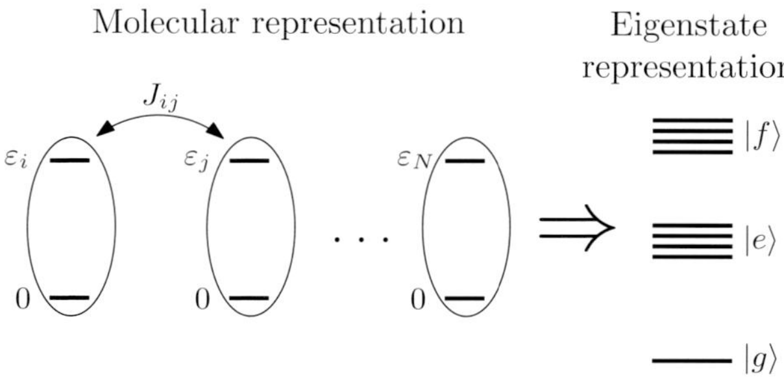

Fig. 1. Scheme of an excitonic aggregate represented by Eq. 5. Transformation of the molecular aggregate into the eigenstate (exciton) representation results in bands of states (shown on the right).

with respect to that state. An aggregate containing N chromophores has a single ground state, N singly-excited states and $N(N-1)/2$ doubly-excited states. This setup of states can be represented as shown in Fig. 1. The exciton eigenstate properties are obtained by diagonalizing the Hamiltonian defined by Eq. (5). Consequenty, the N single-excitons $|e\rangle$ are related to the molecular excitations $\hat{B}_m^\dagger|g\rangle$ by a unitary transformation matrix ψ_{me}, composed of the eigenfunctions

$$|e\rangle = \sum_m \psi_{me} \hat{B}_m^\dagger |g\rangle. \tag{6}$$

The exciton states emerge as a consequence of the optical excitations. Properties of the exciton wavefunctions ψ_{me} become central in determining the dynamical properties of the excitation.

2.0.2. *Excitations coupled to the bath and their correlation functions*

In the realistic molecular aggregate the electronic degrees of freedom are coupled to the nuclear degrees of freedom constituting the environment (the intra- and inter-molecular vibrations, e.g., the fluctuations of the protein scaffold in the case of light-harvesting complexes), which introduces damping and relaxation. To include this effect, we additionally couple the aggregate to the harmonic bath, represented in terms of bosonic operators $\hat{b}_k^\dagger$ (bath excitation creation) and $\hat{b}_k$ (conjugate annihilation). The corresponding Hamiltonian terms are

$$\hat{H}_{\mathrm{B}}+\hat{H}_{\mathrm{SB}} \equiv \sum_\alpha w_\alpha \left(\hat{b}_\alpha^\dagger \hat{b}_\alpha + \frac{1}{2} \right) + \sum_{\alpha j} c_{\alpha j} \hat{B}_j^\dagger \hat{B}_j \left(\hat{b}_\alpha^\dagger + \hat{b}_\alpha \right). \tag{7}$$

The first term describes the bath oscillators and the second couples them to the system excitations. Here, the coefficients $c_{\alpha j}$ are assumed to be real.

In terms of the degrees of freedom, the bath is usually considered to be much larger than the system, thus, it is described using thermodynamical or statistical concepts, and its thermodynamic state is not affected by the system. The main bath characteristic is the temperature ($k_{\mathrm{B}}T \equiv \hbar\beta^{-1}$, where k_{B} is the Boltzmann constant). The bath at a fixed temperature performs equilibrium fluctuations, which in turn induce the fluctuations of chromophore transition energies via the second term on the r.h.s. of Eq. (7). Such fluctuations can be characterized by the molecular transition energy correlation functions[4,35]

$$C_{mn}(t) = \mathrm{Tr}_{\mathrm{B}}\left(\langle m|e^{i\hat{H}_{\mathrm{B}}t}\hat{H}_{\mathrm{SB}}e^{-i\hat{H}_{\mathrm{B}}t}|m\rangle\langle n|\hat{H}_{\mathrm{SB}}|n\rangle\rho_B\right)$$
$$= \sum_{\alpha} c_{\alpha m}c_{\alpha n}\mathcal{Z}(\beta,\omega_\alpha,t), \tag{8}$$

where ρ_B is the equilibrium density matrix of the bath and we have introduced the phonon Green's function

$$\mathcal{Z}(\beta,\omega,t) \equiv \cos(\omega t)\coth(\beta\omega/2) - i\sin(\omega t). \tag{9}$$

Thanks to the fluctuation–dissipation theorem,[34,36] we can obtain a more convenient form, which allows us to isolate the temperature-independent part of the correlation function in the frequency representation. It is the odd part of the Fourier transform of the correlation function, denoted by the spectral density, which characterizes purely the spectral properties of the bath:

$$C''_{mn}(\omega) = \pi\sum_{\alpha} c_{\alpha m}c_{\alpha n}[\delta(\omega - \omega_\alpha) - \delta(\omega + \omega_\alpha)] \tag{10}$$

and the correlation function can then be given by

$$C_{mn}(t) = \frac{1}{\pi}\int_{-\infty}^{\infty} d\omega\,\frac{1}{1 - e^{-\beta\omega}}e^{-i\omega t}C''_{mn}(\omega). \tag{11}$$

A usual assumption is that fluctuations of different molecules are uncorrelated, which gives $C''_{mn}(\omega) = \delta_{mn}C''_n(\omega)$. Additionally, a popular model for the spectral density is that of an overdamped Brownian oscillator (or the Debye spectral density):[34,36]

$$C''_n(\omega) = 2\lambda_n\frac{\omega\Lambda_n}{\omega^2 + \Lambda_n^2}, \tag{12}$$

150

where Λ_n^{-1} is the decay time of the bath correlation of the n-th chromophore and λ_n, the reorganization energy, describes the system–bath coupling strength.

2.1. *Exciton dynamics and relaxation*

2.1.1. *Quantum master equation*

One of the most elaborated ways to treat the non-stationary dissipative dynamics of open quantum systems is the density operator theory. The dynamics of the total density operator, W, are governed by the quantum Liouville equation (we take $\hbar = 1$):

$$\frac{\mathrm{d}}{\mathrm{d}t}W = -\mathrm{i}\left[\hat{H}, W\right]. \tag{13}$$

In the total Hamiltonian we can distinguish three terms:

$$\hat{H} = \hat{H}_\mathrm{S} + \hat{H}_\mathrm{B}(\hat{p}, \hat{q}) + \hat{H}_\mathrm{SB}(\hat{q}). \tag{14}$$

The system (first term) is directly observable and should be attributed to the exciton Hamiltonian. Usually, the bath (second term) is considered to be in the thermal equilibrium, and the system–bath coupling, represented by $\hat{H}_\mathrm{SB}(\hat{q})$, can be defined as a product of the system and bath operators:

$$\hat{H}_\mathrm{SB}(\hat{q}) = \sum_n \hat{S}_n \hat{q}_n, \tag{15}$$

where $\hat{S}_n$ is the system operator (usually a projector) and $\hat{q}_n$ is the associated generalized coordinate of the bath.

We introduce the interaction picture of the density operator

$$W_\mathrm{I}(t) = U_0^\dagger(t) W(t) U_0(t), \tag{16}$$

where $U_0(t)$ denotes the evolution generated by the Hamiltonians $\hat{H}_\mathrm{S} + \hat{H}_\mathrm{B}(\hat{p}, \hat{q})$. The system–bath coupling Hamiltonian, Eq. (15), is likewise defined in the interaction picture and is denoted by $\hat{H}_\mathrm{I}(t)$.

Often only a few degrees of freedom constitute the observable system, while the others affect the system, but are not directly observed. We can derive an approximate closed equation for the *reduced density matrix* of the system ρ. Several approximations have to be used in order to arrive at this result. The first approximation is the factorization of the total density matrix W into the system and the bath components (Born approximation). This implies the absence of entanglement of the system and bath states. Moreover, considering that the bath effectively has an infinite number of degrees of freedom, the bath is taken to be in the equilibrium state, ρ_B, at all times. We thus write:

$$W_I(t) = \rho_I(t) \otimes \rho_B. \tag{17}$$

We can formally solve the quantum Liouville equation for the density operator in the interaction picture, and then plug the result back into the equation. Then, performing the trace operation over the equilibrium bath variables gives the Quantum Master Equation (QME) for the reduced density matrix in the interaction representation[4,6]

$$\frac{d}{dt}\rho_I(t) = - \int_{t_0}^{t} d\tau \, \mathrm{Tr}_B \left(\left[\hat{H}_I(t), [\hat{H}_I(\tau), \rho_I(\tau) \otimes \rho_B] \right] \right). \tag{18}$$

The Born approximation essentially allows us to isolate the reduced density matrix, and the trace operation brings-in the irreversibility into the dynamics. We have to understand Eq. (18) as follows: at time $t < t_0$ the system and the bath are uncoupled. Their dynamics are uncorrelated and the total density matrix is block-diagonal with respect to the system and the bath. At $t = t_0$ the interaction is switched on and the dynamics become correlated. Eq. (18) includes the system–bath correlations up to infinite order. The relaxation kernel thus carries the memory effects. If the bath is not in the equilibrium state at time t_0 or if it is correlated with the system, then the system has to be extended to include these correlation effects.

By performing the trace over the bath in Eq. (18) and taking the initial condition $t_0 \to -\infty$, the rate operator, governing the dissipative dynamics, is obtained as a function of the interaction delay times $t - \tau$. The evolution is still of infinite order in the system–bath interaction. It is convenient to introduce the delay time explicitly, which yields in the Schrödinger representation

$$\frac{d}{dt}\rho(t) = -i[\hat{H}_S, \rho(t)] - \int_{0}^{\infty} d\tau \, R(\tau)\rho(t - \tau). \tag{19}$$

2.1.2. *Redfield theory for exciton relaxation*

The obtained integro-differential form of the QME, Eq. (19), is rather complicated. It can be simplified by using the Redfield approximation, i.e., assuming the second-order approximation for the relaxation kernel.[4,37] Additionally, the Markovian approximation is invoked. To that end it is assumed, that the system-bath interaction is weak and, thus, the system density matrix in the interaction picture, $\rho_I(t)$, is a slowly-evolving function, as compared to the decay time of the relaxation tensor. Under these conditions it can be taken out of the integral and the time-local equation with time-independent rate matrix is obtained. In the Schrödinger picture the Redfield equation is obtained

$$\frac{\mathrm{d}}{\mathrm{d}t}\rho(t) = -\mathrm{i}[\hat{H}_{\mathrm{S}}, \rho(t)] - K\rho(t). \tag{20}$$

The Redfield relaxation superoperator, K, can be simplified considerably if the system operators are expanded in an arbitrary orthogonal basis $|a\rangle$. Then the Hamiltonian reads

$$\hat{H} = \sum_{ab}(h_{ab} + \tilde{h}_{ab}\hat{q}_{ab})|a\rangle\langle b| + \hat{H}_{\mathrm{B}}(\hat{p}, \hat{q}). \tag{21}$$

This form can be directly applied to the Frenkel exciton case considering the single exciton manifold, where the exciton relaxation and transport take place. We then have $h_{ab} = \delta_{ab}\varepsilon_a + J_{ab}$. The parameter $\tilde{h}_{ab}$ denotes the system–bath coupling amplitude and $\hat{q}_{ab}$ is the generalized coordinate of the bath, coupled to system Hamiltonian element ab. We obtain the relaxation matrix defined by the fluctuation correlation functions:

$$\begin{aligned}
K_{ab,a'b'} = \sum_{cd}\int_0^\infty \mathrm{d}\tau[&\delta_{bb'}\sum_e \tilde{h}_{ae}\tilde{h}_{dc}C_{ae,dc}(\tau)U_{ed}(\tau)U_{ca'}(-\tau)\\
&-\tilde{h}_{aa'}\tilde{h}_{cd}C_{cd,aa'}(-\tau)U_{b'c}(\tau)U_{db}(-\tau)\\
&-\tilde{h}_{dc}\tilde{h}_{b'b}C_{b'b,dc}(\tau)U_{ad}(\tau)U_{ca'}(-\tau)\\
&+\delta_{aa'}\sum_e \tilde{h}_{cd}\tilde{h}_{eb}C_{cd,eb}(-\tau)U_{b'c}(\tau)U_{de}(-\tau)].
\end{aligned} \tag{22}$$

The correlation function $C_{ab,cd}(\tau)$ describes the fluctuations of the Hamiltonian elements ab and cd.

A natural choice for the basis set for the Redfield relaxation superoperator is the eigenstate basis of the system Hamiltonian. This choice makes simulations much simpler and allows us to introduce the *secular approximation* and to define the requirements for the long-time limit. Let us assume that states $|a\rangle$ are eigenstates of the system Hamiltonian (the exciton states in the case of the exciton Hamiltonian). In that case, when $h_{ab} = \delta_{ab}\varepsilon_a$ is diagonal, the Redfield equation reduces to

$$\frac{\mathrm{d}}{\mathrm{d}t}\rho_{ab}(t) = -\mathrm{i}\omega_{ab}\rho_{ab}(t) - K_{ab,cd}\rho_{cd}(t), \tag{23}$$

where $\omega_{ab} = \varepsilon_a - \varepsilon_b$.

The secular Redfield relaxation equation can then be written in the form

$$\frac{\mathrm{d}}{\mathrm{d}t}\rho_{ab}(t) = -\mathrm{i}(\omega_{ab} - \mathrm{i}\gamma_{ab})\rho_{ab}(t) - \delta_{ab}\sum_b k_{ab}\rho_{bb}(t), \tag{24}$$

where γ_{ab} $(a \neq b)$ are complex numbers representing the dephasing rates of coherences, and k_{ab} are real numbers that represent the population transport rates. All these rates are given by one-sided Fourier transforms of the coordinate–coordinate correlation function.[4,6] The latter can be given in terms of the fluctuation spectral densities, defined by Eq. (11).

2.1.3. *Modified Redfield rates for exciton transfer*

The Redfield approach assumes that the bath is Markovian and certain types of fluctuations are independent, and all the states are in thermal equilibrium with the bath. However, it is important that the bath is often affected by the system, thus the bath equilibrium for different system states can be slightly shifted, and in short distances molecular fluctuations may be highly correlated. The modified Redfield theory includes these effects: it is non-perturbative with respect to diagonal fluctuations and it includes the correlations between the diagonal and off-diagonal fluctuations.[4,38]

The modified Redfield population transfer rate (from the state a to the state b) reads

$$
\begin{aligned}
k_{ba} = 2\mathrm{Re}\,|\tilde{h}_{ba}|^2 \int_0^\infty &\mathrm{d}\tau e^{i\omega_{ab}\tau} \{\ddot{g}_{ab,ba}(\tau) \\
&-[\dot{g}_{bb,ba}(\tau) - \dot{g}_{aa,ba}(\tau) + 2i\lambda_{ba,aa}][\dot{g}_{ab,aa}(\tau) - \dot{g}_{ab,bb}(\tau) + 2i\lambda_{ab,aa}]\} \\
&\times \exp[-g_{aa,aa}(\tau) - g_{bb,bb}(\tau) + g_{aa,bb}(\tau) \\
&+g_{bb,aa}(\tau) + 2i(\lambda_{aa,bb} - \lambda_{aa,aa})\tau],
\end{aligned}
\tag{25}
$$

where the so-called excitonic lineshape function, $g_{e_4 e_3, e_2 e_1}(t)$, has been introduced.[4,34] It is given by

$$
g_{e_4 e_3, e_2 e_1}(t) = \sum_m \psi_{m e_4} \psi^*_{m e_3} \psi^*_{m e_2} \psi_{m e_1} g_m(t),
\tag{26}
$$

where the auxiliary (monomeric) function $g_m(t)$ is defined as

$$
g_m(t) = \int_0^t \mathrm{d}\tau \int_0^\tau \mathrm{d}\tau' \int \frac{\mathrm{d}\omega}{2\pi} C''_m(\omega) \mathcal{Z}(\beta, \omega, \tau').
\tag{27}
$$

The dots and double dots in Eq. (25) denote the time derivatives, and $\lambda_{ab,cd}$ are the reorganization energies (Stokes shifts) given in the limit

$$
\lambda_{ab,cd} = -\lim_{t\to\infty} \dot{g}_{ab,cd}(t).
\tag{28}
$$

The modified Redfield rate expression is very suitable for excitons. It includes the full equilibration in the initial excited state. The way this expression interpolates between the Redfield and Förster resonance energy transfer (FRET) theories has been demonstrated.[4,39,40] It also includes correlations of the diagonal and off-diagonal fluctuations. The modified rate formula thus explicitly includes the Stokes shifts.

2.1.4. HEOM theory

Hierarchical equations of motion (HEOM) is a non-perturbative theory describing the exciton dynamics in the open quantum systems.[41,42] Because the usage of the full theory is computationally expensive here we present a modified HEOM theory, termed as hierarchical quantum master equation (HQME).[4,43] It is still non-perturbative, but is restricted to the following approximate form of the bath correlation function:

$$C_n(t) = \left(\frac{2\lambda_n}{\beta} - \frac{\beta\lambda_n\Lambda_n^2}{6} \right) e^{-\Lambda_n t} - i\lambda_n\Lambda_n e^{-\Lambda_n t} + \frac{\lambda_n\Lambda_n\beta}{3}\delta(t). \quad (29)$$

This form follows from Eq. (11) by using the overdamped Brownian oscillator spectral density, Eq. (12), and expanding the Bose–Einstein function up to $(\beta\omega)^1$ term:

$$\frac{1}{1 - e^{-\beta\omega}} \approx \frac{1}{\beta\omega} + \frac{1}{2} + \frac{\beta\omega}{12}. \quad (30)$$

Conventional high temperature approximation schemes use only the first two terms of this expansion. Here, all three terms are used and the rest of the correlation function is accounted as the Markovian-white-noise residue ansatz.[43]

The HQME for the correlation function given by Eq. (29) is written in the Liouville space as a hierarchy of coupled differential equations for auxiliary density operators (ADOs) denoted by $|\rho_{\mathbf{n}}(t)\rangle\rangle$:[43]

$$\frac{d}{dt}|\rho_{\mathbf{n}}(t)\rangle\rangle = -i\hat{\mathcal{L}}_e|\rho_{\mathbf{n}}(t)\rangle\rangle - \sum_{m=1}^{N}\left(\Lambda_m n_m + \delta\hat{\mathcal{R}}_m\right)|\rho_{\mathbf{n}}(t)\rangle\rangle \quad (31)$$

$$+ \sum_{m=1}^{N} n_m\hat{\mathcal{A}}_m|\rho_{\mathbf{n}_m^-}(t)\rangle\rangle + i\sum_{m=1}^{N}\hat{\mathcal{Q}}_m^\times|\rho_{\mathbf{n}_m^+}(t)\rangle\rangle \quad (32)$$

where the auxiliary superoperators

$$\delta\hat{\mathcal{R}}_m = \frac{\lambda_m\Lambda_m\beta}{3}\hat{\mathcal{Q}}_m^\times\hat{\mathcal{Q}}_m^\times, \quad (33)$$

$$\hat{\mathcal{A}}_m = i\left(\left(\frac{2\lambda_m}{\beta} - \frac{\beta\lambda_m\Lambda_m^2}{6}\right)\hat{\mathcal{Q}}_m^\times - i\lambda_m\Lambda_m\hat{\mathcal{Q}}_m^\circ\right) \quad (34)$$

are introduced. Here $\hat{\mathcal{Q}}_m^{\times} \bullet \Leftrightarrow \left[\hat{Q}_m, \bullet\right]$ denotes the commutator and $\hat{\mathcal{Q}}_m^{\circ} \bullet \Leftrightarrow \left\{\hat{Q}_m, \bullet\right\}$ is the anti-commutator. In Eq. (31) $|\rho_{\mathbf{0}}(t)\rangle\rangle$ corresponds to the physical reduced density operator, $\mathbf{n}$ is a vector of indices $\mathbf{n} \equiv (n_1, n_2, \ldots, n_N)$ and we use notation $\mathbf{n}_m^{\pm} \equiv (n_1, n_2, \ldots, n_m \pm 1, \ldots, n_N)$. All ADOs with a negative index are not physical and set to zero.

Formally the hierarchy in Eq. (31) is infinite, thus, the equations are not closed. Various truncations schemes could be used. The simplest one is based on the assumption, that all ADOs with their tier level $L = \sum_{m=1}^{N} n_m$ greater than truncation level L_{trunc} can be simply discarded. The truncation level is usually chosen to guarantee the convergence of the simulation results. Other truncation schemes are also possible.[44,45]

Since the HEOM theory is derived as an operator equation and makes no approximation for the bath, it is independent of the basis chosen for the problem solution. It can thus capture such effects as the polaron formation and exciton localization. Therefore the theory can describe much broader class of problems where the exciton concept and the corresponding exciton basis are not valid.

2.1.5. *Weak inter-chromophore coupling limit*

Energy transfer in molecular aggregates in the case of weak resonance coupling is usually considered within the framework of the FRET theory.[4,6] This is a widely employed method, working remarkably well even in situations, where the condition of weak chromophore–chromophore coupling might be questionable. However, for some application (such as 2D spectroscopy) FRET has several important closely related deficiencies. Namely, since FRET is an approach of the Fermi's golden rule type, it gives the population transfer rates, but no prescription for propagating the coherences. By the same token, FRET intrinsically assumes, that the excitations are localized on individual chromophores despite their mutual interaction. The latter deficiency can be overcome by using modified Redfield theory, but in this case the former still persists, which renders the description of coherent phenomena impossible.

It is possible, however, to formulate a dynamical description of the whole reduced density matrix in the weak resonance coupling limit.[46] The derivation is similar to that of the QME, except that this time the resonance coupling, $\hat{H}_{\mathrm{J}}$, instead of the system–bath interaction is treated as a perturbation. Such treatment leads to the equations of motion for the reduced density matrix $\bar{\rho}(t) = \mathrm{Tr}_{\mathrm{B}}(W_{\mathrm{I}}(t))$ in the basis of localized excitations:

$$\frac{\mathrm{d}}{\mathrm{d}t}\bar{\rho}_{ab}(t) = -i\sum_c J_{ac}(t)\bar{\rho}_{cb}(t) + i\sum \bar{\rho}_{ac}(t)J_{cb}(t)$$

$$-\sum_{cd}[R_{accd}(t)\bar{\rho}_{db}(t) - R^*_{cabd}(t)\bar{\rho}_{cd}(t) - R_{dbac}(t)\bar{\rho}_{cd}(t) + R^*_{bddc}(t)\bar{\rho}_{ac}(t)].$$

$$(35)$$

Here, the effective resonance coupling is defined as

$$J_{ab}(t) \equiv \langle a|\mathrm{Tr}_{\mathrm{B}}(\hat{H}_{\mathrm{J}}(t)\rho_{\mathrm{B}})|b\rangle = J_{ab}e^{i\omega_{ab}t - (1-\delta_{ab})(g_a^*(t) + g_b(t))}; \qquad (36)$$

and the relaxation tensor reads:

$$R_{abcd}(t) = \int_0^t \mathrm{d}\tau[J_{ab}J_{cd}M_{abcd}(t, t-\tau) - J_{ab}(t)J_{cd}(t-\tau)], \qquad (37)$$

where the auxiliary function is given by

$$M_{abcd}(t, \tau) = e^{F_{abcd}(t,\tau) + i\omega_{ab}t + i\omega_{cd}\tau}, \qquad (38)$$

and

$$F_{abcd}(t, \tau) = -g_a^*(t) - g_b(t) - g_c^*(\tau) - g_d(\tau)$$
$$- \delta_{ac}(g_a(t) - g_a(t-\tau) + g_a^*(\tau)) + \delta_{ad}(g_a(t) - g_a(t-\tau) + g_a^*(\tau))$$
$$+ \delta_{bc}(g_b(t) - g_b(t-\tau) + g_b^*(\tau)) - \delta_{bd}(g_b(t) - g_b(t-\tau) + g_b^*(\tau)). \qquad (39)$$

The connection between the reduced density matrix in the interaction picture $\bar{\rho}(t) = \mathrm{Tr}_{\mathrm{B}}(W_{\mathrm{I}}(t))$ and in the Schrödinger picture $\rho(t) = \mathrm{Tr}_{\mathrm{B}}(W(t))$ is given by the equation

$$\bar{\rho}_{ab}(t) = e^{i\omega_{ab}t + (1-\delta_{ab})[g_a(t) + g_b^*(t)]}\rho_{ab}(t). \qquad (40)$$

2.2. *Modelling of exciton dynamics in a dimer*

The simplest molecular aggregate — a molecular dimer — is a good model system already disclosing the effects caused by the excitonic quantum coherence.[25,47,48] Let us consider two resonantly coupled molecules, characterized by two site energies ε_1 and ε_2 and the inter-molecular coupling J. We denote $\Delta = \varepsilon_2 - \varepsilon_1$. The Hamiltonian of such system is solvable analytically by introducing the mixing angle θ so that the exciton eigenvectors are[4,6]

$$\begin{pmatrix} \cos(\theta) & -\sin(\theta) \\ \sin(\theta) & \cos(\theta) \end{pmatrix}, \tag{41}$$

where the mixing angle is given by the relation

$$\tan(2\theta) = \frac{2J}{\Delta}. \tag{42}$$

The eigenstate energies are

$$\varepsilon_\pm = \frac{\varepsilon_1 + \varepsilon_2}{2} \pm \frac{\Delta}{2}\sqrt{1 + \tan^2(2\theta)}. \tag{43}$$

Let us choose the following parameters, which are typical for the pigment molecules in the protein environment: $\varepsilon_1 = 10100\,\mathrm{cm}^{-1}$, $\varepsilon_2 = 10200\,\mathrm{cm}^{-1}$ and $J = 100\,\mathrm{cm}^{-1}$. The mixing angle is therefore $\theta \approx 2.6$ [rad], which leads to the eigen-energies of $10038.2\,\mathrm{cm}^{-1}$ and $10261.8\,\mathrm{cm}^{-1}$. To include the relaxation effects, we couple this dimer to the independent phonon baths, characterized by the overdamped Brownian oscillator spectral density, Eq. (12). The relaxation time-scale parameter for each monomer is $\Lambda = 0.53J$ ($\sim 100\,\mathrm{fs}$).

We will consider the exciton dynamics in such a system. We assume that the molecular dimer is excited by an ultrashort laser pulse. In that case the pulse bandwidth covers both exciton eigenstates so that they both are being excited simultaneously. Such laser excitation prepares the system in a highly non-equilibrium excited configuration, described by the density matrix

$$\rho(t=0) = \begin{pmatrix} 0.5 & 0.5 \\ 0.5 & 0.5 \end{pmatrix}. \tag{44}$$

We will follow the exciton dynamics in this system using various propagation methods with the same set of parameters, when this system is either weakly ($\lambda = 0.1J$) or strongly ($\lambda = J$) coupled to the bath.

2.2.1. Redfield dynamics

The full Redfield theory is a straightforward approach to tackle the non-equilibrium system dynamics. In this case all fluctuations are included up to the second order. In the weak system–bath coupling case (see Fig. 2(a)), the system remains highly coherent, as implied by oscillatory populations and density matrix coherences. After the initial coherent phase the populations cease to oscillate and approach equilibrium values monotonically. At

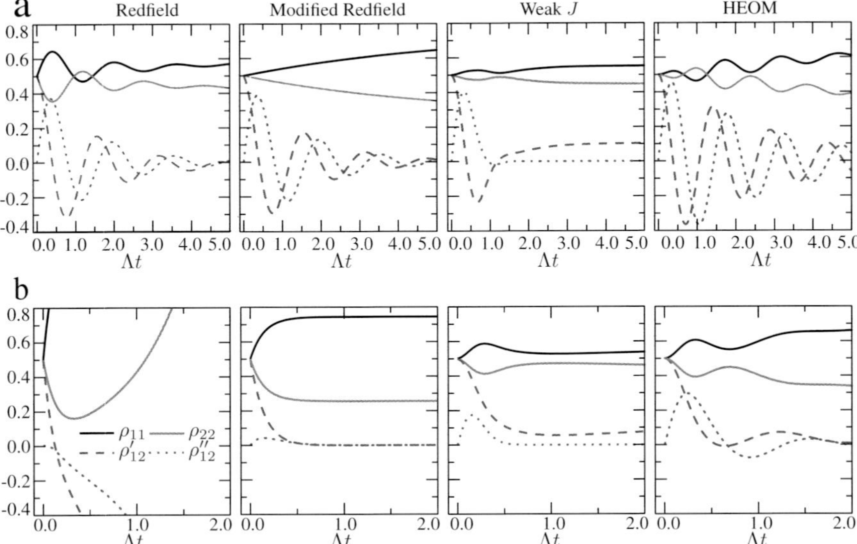

Fig. 2. The dynamics of the density matrix of the dimer with the initial condition set in Eq. 44 in the case of a weak (a; $\lambda = 0.1J$) and strong (b; $\lambda = J$) system–bath coupling. We have denoted $\rho'_{12} = \operatorname{Re} \rho_{12}$ and $\rho''_{12} = \operatorname{Im} \rho_{12}$.

this state the coherences are almost zero. A completely different picture is obtained in the strong system–bath coupling case, where the parameters of the Redfield equation make the solution diverge. This demonstrates that the Redfield equation has a limited parameter space of validity. It must be noted, that since the equation is up to the second order in fluctuations, the error of the solution accumulates in time. In the weak system-bath coupling case the result is physically feasible for $\Lambda t \sim 1$, however, it should be concluded, that it is not trustworthy without relating to a particular experiment. This is obvious in the case of strong system–bath coupling, which blows-up for $\Lambda t \sim 0.1$. The secular theories cure this problem by neglecting some rates, however, their results forexcitons are questionable when energy level spacings are of the order of the system-bath coupling strength.

2.2.2. *Modified Redfield*

The modified Redfield theory is intrinsically secular. In Fig. 2(a) the density matrix dynamics are presented in the weak coupling limit. As the dynamics of the populations and coherences are uncoupled, we can observe only exponential decay of non-equilibrium populations to the equilibrium. The coherences oscillate with their native frequencies, dictated by the energy gaps between the eigen-levels. In Fig. 2(b) the strong coupling case is shown. We find that the population dynamics are now confined in the proper range. In the strong system–bath coupling case we do not observe any considerable oscillations of the system coherences. The damping is thus larger than the excitonic coherence frequency and the coherences are overdamped.

2.2.3. *Weak resonance coupling regime*

The weak-J theory captures the other limit of strong system–bath coupling correctly. In this case, when the coupling with the bath is strong and comparable to the inter-chromophore coupling J, the theory correctly reflects the density matrix dynamics. This is in stark contrast to the Redfield theory, which diverges and fails. The long-time limit is, however, shifted from the correct thermal equilibrium, but the initial dynamics are properly recovered. The finite coherence elements at long times show, that the excitonic basis is no longer the eigenbasis of the solution and the site basis is more appropriate. Thus, in the long-time limit the system relaxes into the site basis equilibrium.

2.2.4. *HEOM dynamics*

The HEOM method is a way to compute the aggregate dynamics without any approximations. It thus correctly includes the polaron formation and the quantum transport effects as well as realistic coherence dephasing rates. As can be seen in Figs. 2(a) and (b), in the initial times it reflects the coherent dynamics, which transform into the dissipative ones. Only the short-time dynamics are shown in the figures. However, in the long-time limit it can be shown, that coherences in the strong system–bath coupling case do not relax to the equilibrium. This shows that the weak-J theory is more accurate in that limit than the Redfield or the modified Redfield and the site basis is more appropriate. The exciton states thus decay and the dynamics should be described in the site basis.

3. Discussion: relaxation of excitons in photosynthesis

3.1. *Applicability of various relaxation theories and approaches in light-harvesting complexes*

The essential step in developing a description of exciton relaxation in molecular aggregates is to make the right approximations. The most sophisticated

160

HEOM theory is an exact approach, applicable when the correlation function can be approximated by the exponential decay function. However, it scales very poorly with the system size. For instance, if the system consists of N chromophores and we assume that each site is coupled to an independent bath coordinate, we have to use an N-dimensional space for hierarchy of equations, thus the system becomes hardly tractable on a computer even for moderate system sizes (~ 10). However, using a multicore supercomputers, aggregates as large as ~ 50 have been processed.[49] Moreover, much progress has been made by employing graphics processing units (GPU) instead of conventional CPU clusters so the computing efficiency could be even more improved.[50] However, these are the technical aspects, which do not change the scaling law of the HEOM. Hence, the approximate methods are very important. Current applications are mostly limited to the overdamped model of the bath, while coloured bath theoretically is possible in HEOM.[51] Alternative exact methods, which include an arbitrary bath spectral density are available as well.[22,49,50,52–54]

As the HEOM are numerically exact, they are now taken as a reference when examining the approximate methods. The latter can still be used in specific parameter regions. For instance, in the case of weak system-bath coupling, the full Redfield theory is very suitable in short times. After the system reaches the dissipative incoherent relaxation phase, the secular theory can be used to reach the thermal equilibrium faster. In the case of strong system-bath couplings the weak-J theory gives correct coherent dynamics phase, and it encapsulates the formation of polarons. Also, the approximate methods have much better scaling with the system size. Hence they can be used for large systems.

As approximate theories introduce errors, the solution could diverge for specific parameters, as has been shown for the full Redfield theory. It has been shown by one of the authors that the Lindblad equation "fixes" the divergences.[55,56] The Lindblad equation of motion guarantees physically acceptable density matrix for all delay times, however, it does not provide a recipe to obtain the relaxation rates or to define the Lindblad parameters from the bath spectral densities.

The modified Redfield scheme is an upgrade to the full Redfield theory, which promises better exciton transfer rates. However, the theory uses the second-order approximation in the off-diagonal fluctuations, which limits the rates again to the weak coupling regime. This could be a proper approach when intermolecular couplings are small and we have only diagonal site-energy fluctuations. However, the weak-J theory should be better approximation in that case as it directly employs the proper small parameter.

The presented theories were demonstrated employing a simple dimer. The obtained results and conclusions can be extended to larger systems, since the excitonic dimer is the simplest system having all ingredients inherent in larger systems. It has to be emphasized, that the photosynthetic molecular aggregates are often disordered assemblies of molecules and the exciton delocalization length is of the order of a few molecules. Thus, the excitonic properties of a dimer must be quite close to the realistic properties of photosynthetic excitons. The other limit applies for the self-assembled regular J-aggregates, where tens or hundreds of molecules can contribute to a single exciton. In that case the conclusions drawn from a dimer should be used with caution.

The recent two-dimensional spectroscopy experiments, done on molecular aggregates, revealed a broad range of coherent processes that take place on the femtosecond time scale.[14,23,57,58] These experiments initiated the discussion on the importance of the quantum coherent effects in the biological functioning of the light harvesting in photosynthesis. Therefore the quantum relaxation theories with strong emphasis on the quantum coherence are recently being reviewed.[19,20,22,24] As was already pointed out, the full Redfield theory accounts for the damped evolution of the excitonic coherences. These coherences survive for a few picoseconds, as follows from the experiments, while they should decay in femtosecond range, as follows from the simulations. It has been suggested that the observed spectral beats in the 2D spectra reflect the impact of the molecular vibrations.[25,59–61] The latter is not surprising, knowing that the vibrational progression is very weakly expressed in the FMO absorption spectrum, yet the vibrations may influence the energy transfer rates.[62] It could also mean that the correlations between electronic and vibrational coherences may be important for the energy transfer in photosynthetic aggregates.

3.2. *Realistic photosynthetic conditions: exciton quenching mechanisms*

From a broader perspective, additional effects of the energy transport become essential for photosynthesis. Under the high light conditions the down-regulation of photosynthesis rather than the pure efficiency of energy transfer becomes essential for the survival of plants. The process of non-radiative decay of the excited states of Chls within the antenna complex is termed the non-photochemical quenching (NPQ). Although the exact location and mechanism of the NPQ are still a matter of on-going debate, it is rather obvious that the process must involve the main building blocks of the antennae — Chl and/or Car. Some of the proposed mechanisms involve excitonic coupling between Chl and Car or Chl and Chl–Car/Chl–Chl CT states. The possible involvement of Car is particularly appealing due to the short lifetime ($\sim 10\,\mathrm{ps}$) of the lowest excited state. Both the presence of CT states[29,63] and the Chl–Car interaction[31] under the NPQ conditions are experimentally observed, which supports the idea that an excitonic dimer could actually be the quenching center. The idea was also tested in experiments on artificial carotenoid–tetrapyrrole dyads.[64–66]

To study these ideas theoretically, an excitonic heterodimer could again be employed.[67,68] The main parameters of such a dimer are: site energies, resonance coupling, transition dipole moments, reorganization energies and excited state lifetimes. The parameters of the Chl molecules are well known,[1] while for the Car excited state (due to its dipole-forbidden transition to the first excited state S_1) or the CT state the site and reorganization energies are only inferred from indirect observations. Therefore the energy of the Car excited state ε_{Car} (as well as the CT state ε_{CT}) can be larger or smaller than that of the Chl. This also applies to the reorganization energies, accordingly λ_{Chl} and λ_{Car} (λ_{CT}). The excited state lifetimes of Chl and Car are accordingly $\tau_{Chl} = 2\,\mathrm{ns}$ and $\tau_{Car} = 5\,\mathrm{ps}$, and the lifetime of the CT state is similar to that of Chl $\tau_{CT} \approx \tau_{Chl}$.

Because of the absent transition dipole moment the resonance coupling J between Chl and Car is obviously not of the dipole–dipole type, however it can be as large as $\sim 100\,\mathrm{cm}^{-1}$.[69] Similar values can be also assumed for the resonance coupling with a CT state, which arises due to the overlap of the chromophore wavefunctions, and therefore it is also described beyond the dipole–dipole approximation.[70,71]

Non-radiative transitions to the ground state have been phenomenologically incorporated into the equation of motion for the reduced density matrix, defined by Eq. (19), by appending a term $\mathcal{K}\rho(t)$, where the $\mathcal{K}$ superoperator governing the relaxation to the ground state is given by:[67,68]

$$\mathcal{K}\bullet = -\sum_i \frac{\kappa_i}{2}\{Q_i, \bullet\}. \tag{45}$$

Here, κ_i denotes the experimentally known inverse lifetime of the population of the i-th site (in the site basis). According to such definition it contains only two types of non-zero elements: the population relaxation ($\mathcal{K}_{ii,ii}$) and the coherence decay due to corresponding population relaxation ($\mathcal{K}_{ij,ij}$).

The evolution of populations within the Chl–Car dimer can be calculated using the HEOM approach by including the superoperator Eq. (45). The initial conditions can be set by assuming an optical excitation of the dimer. Therefore, initially the density matrix, represented in the site basis, is just the population $\rho_{Chl\,Chl}(0) = 1$. It has been shown,[67] that independently of whether the $|Chl\rangle$ or $|Car\rangle$ is the lower state, the initial coherent oscillations are barely visible and can be neglected, and the overall process has a two-exponential character. The first exponent corresponds to the thermal redistribution of populations, and therefore this part is quite minor when $|Chl\rangle$ is the lower state. The second part (on the picosecond time scale) corresponds to the excitation relaxation to the ground state. The decrease takes place with the time constant of $\approx 12\,\mathrm{ps}$ ($\approx 8\,\mathrm{ps}$) when $|Chl\rangle$ is the lower (upper) state.

Estimating the excitation quenching efficiency within such a dimer, a few points are noteworthy. Firstly, the long-time decay rates in both cases are similar to the inverse lifetime of Car in comparison to the Chl lifetime. This shows that the excitonic mixing of lifetimes which takes place upon the transformation $|Chl\rangle \rightarrow |e_1\rangle$ is not essential to the global relaxation, contrary to some expectations.[31] Furthermore, the position of the Car state has only a minor effect on the net relaxation rate (cf. Ref. 65). But once below the Chl state, it can greatly reduce the population of $|e_1\rangle$ state on the sub-picosecond time scale due to thermalization.

The case of Chl–CT is different firstly because the excited state lifetimes of the constituent dimers are similar and a couple of orders larger than the thermal relaxation time. Therefore the quenching in this case would work mainly by removal of excitation from Chl due to thermal population redistribution. Hence it is important that $|CT\rangle$ state is the lower of the two.

3.3. *Whether electronic quantum coherence is relevant in photosynthesis*

One of the questions that has been under high scrutiny recently, is how quantum coherences contribute to the excitation energy transfer. It has been suggested that the FMO aggregate of green sulphur bacteria has developed a highly reliable balance between dephasing and coherence.[19] It necessarily plays a role in the energy transfer in an arbitrary molecular aggregate. The electronic coherence is the consequence of intermolecular interactions, which result in formation of excitons. Hence, the concept of excitons is obviously related to electronic coherences. If we disregard electronic coherences we must rule out the exciton delocalization phenomenon. The bath induced dephasing is also a necessary component in the energy transfer, since if the system was not coupled to the bath, the exciton dynamics would be reversible and no energy transfer would be possible. In the other limit, if the chromophores were only weakly coupled, the energy transfer would proceed only by the Förster mechanism. So in this respect, the inter-molecular couplings result in exciton delocalization and in electronic coherences.

From the different perspective, when we consider a single step of energy transfer, we should ask, whether the non-secular terms in the density matrix propagation are of any importance in energy transfer. This process is slightly different than the pure electronic coherence as this process affects the physically relevant energy transfer. The electronic coherences now become mixed with populations. As shown by our simulations, the initial quantum dynamics are always non-secular (according to HEOM) and that is the nature of the density matrix relaxation from a non-equilibrium initial preparation. It is thus a natural result due to the strong inter-molecular interactions and system–bath coupling.

However, if we consider the whole energy transport chain from the photon absorber to the RC, which proceeds in 10–100 ps, the question of coherences becomes less relevant, since the excitation lifetime in the absence of RCs is of the order of several nanoseconds. Hence, the ultrafast energy transport is not what the plants actually seek for. Instead, under real-life conditions, the photoprotection becomes very important. In the case of exciton quenching by coherently coupled molecules we have to consider mainly the long-time behaviour of the populations within the one-exciton manifold. Therefore the initial coherent dynamics of the wavepacket created by, e.g., optical excitation can be neglected.

Finally, it should be stressed that most of the conclusions obtained from the simulations based on specific models explain the properties of these models, which can be associated with the biological antennas *approximately*. The exciton models have large uncertainties and some parameters can hardly be determined. However, the parameters have been and are still being refined in the light of new experiments. Therefore the parameter sets should not be understood as the very "true" characteristics of the biological aggregates, but as the most probable values with respect to the set of experiments. It is possible that the future experiments may request a reconsideration of the present interpretations. In this respect the field of open quantum systems in photosynthesis is still open.

Acknowledgments

This research was funded by the European Social Fund under the Global Grant measure.

References

1. H. van Amerongen, L. Valkunas, and R. van Grondelle, *Photosynthetic Excitons* (World Scientific, Singapore, 2000).
2. R. E. Blankenship, *Molecular Mechanisms of Photosynthesis* (Wiley-Blackwell, 2002).
3. O. Rancova, J. Sulskus and D. Abramavicius, *J. Phys. Chem. B* **116**, 7803 (2012).
4. L. Valkunas, D. Abramavicius and T. Mancal, *Molecular Excitation Dynamics and Relaxation* (Wiley-VCH, Weinheim, 2013).
5. T. Renger, V. May and O. Kühn, *Phys. Rep.* **343**, 137 (2001).
6. V. May and O. Kuhn, *Charge and energy transfer in molecular systems* (Wiley-VCH, 2004).
7. M. Cho, H. M. Vaswani, T. Brixner, J. Stenger and G. R. Fleming, *The Journal of Physical Chemistry B* **109**, 10542 (2005).
8. T. Mančal, L. Valkunas, E. L. Read, G. S. Engel, T. R. Calhoun and G. R. Fleming, *Spectroscopy* **22**, 199 (2008).
9. M. Cho, *Two-Dimensional Optical Spectroscopy* (CRC Press, Boca Raton, 2009).
10. G. S. Schlau-Cohen, A. Ishizaki and G. R. Fleming, *Chemical Physics* **386**, 1 (2011).
11. G. D. Scholes, G. R. Fleming, A. Olaya-Castro and R. van Grondelle, *Nature Chem.* **3**, 763 (2011).
12. E. Harel and G. S. Engel, *Proc. Nat. Acad. Sci. USA* **109**, 706 (2012).
13. D. Zigmantas, E. L. Read, T. Mančal, T. Brixner, A. T. Gardiner, R. J. Cogdell and G. R. Fleming, *Proc. Nat. Acad. Sci. USA* **103**, 12672 (2006).
14. G. S. Engel, T. R. Calhoun, E. L. Read, T. K. Ahn, T. Mančal, Y. C. Cheng, R. E. Blankenship and G. R. Fleming, *Nature* **446**, p. 782 (2007).
15. G. S. Schlau-Cohen, T. R. Calhoun, N. S. Grinsberg, E. L. Read, M. Ballottari, R. Bassi, R. van Grondelle and G. R. Fleming, *J. Phys. Chem. B* **113**, 15352 (2009).
16. H. Lee, Y.-C. Cheng and G. R. Fleming, *Science* **316**, 1462 (2007).
17. E. Collini and G. D. Scholes, *Science* **323**, 369 (2009).
18. F. Milota, J. Sperling, A. Nemeth and H. Kauffmann, *Chemical Physics* **357**, 45 (2009).

19. F. Caruso, A. W. Chin, A. Datta, S. F. Huelga and M. B. Plenio, *J. Chem. Phys.* **131**, p. 105106 (2009).

20. P. Rebentrost, M. Mohseni and A. Aspuru-Guzik, *J. Phys. Chem. B* **113**, 9942 (2009).

21. F. Fassioli, A. Nazir and A. Olaya-Castro, *J. Phys. Chem. Lett.* **1**, 2139 (2010).

22. P. Nalbach, J. Eckel and M. Thorwart, *New. J. Phys.* **12**, p. 065043 (2010).

23. G. R. Fleming, S. F. Huelga and M. B. Plenio, *New. J. Phys.* **12**, p. 065002 (2010).

24. M. Sarovar, A. Ishizaki and G. R. Fleming, *Nat. Phys.* **6**, p. 462 (2010).

25. V. Butkus, D. Zigmantas, L. Valkunas and D. Abramavicius, *Chem. Phys. Lett.* **545**, 40 (2012).

26. A. V. Ruban, M. P. Johnson and C. D. Duffy, *Biochim. Biophys. Acta* **1817**, 167 (2012).

27. M. G. Müller, P. Lambrev, M. Reus, E. Wientjes, R. Croce and A. R. Holzwarth, *Chem. Phys. Chem.* **11**, 1289 (2010).

28. P. Horton, A. V. Ruban and R. G. Walters, *Annu. Rev. Plant Physiol. Plant Mol. Biol.* **47**, 655 (1996).

29. N. E. Holt, D. Zigmantas, L. Valkunas, X.-P. Li, K. K. Niyogi and G. R. Fleming, *Science* **307**, 433 (2005).

30. T. K. Ahn, T. J. Avenson, M. Ballottari, Y.-C. Cheng, K. K. Niyogi, R. Bassi and G. R. Fleming, *Science* **320**, 794 (2008).

31. S. Bode, C. C. Quentmeier, P.-N. Liao, N. Hafi, T. Barros, L. Wilk, F. Bittner and P. J. Walla, *Proc. Nat. Acad. Sci. USA* **106**, 12311 (2009).

32. P.-N. Liao, S. Bode, L. Wilk, N. Hafi and P. J. Walla, *Chem. Phys.* **373**, 50 (2010).

33. A. V. Ruban, R. Berera, C. Ilioaia, I. H. M. van Stokkum, J. T. M. Kennis, A. A. Pascal, H. van Amerongen, B. Robert, P. Horton and R. van Grondelle, *Nature* **450**, 575 (2007).

34. S. Mukamel, *Principles of Nonlinear Optical Spectroscopy* (Oxford University Press, New York, 1995).

35. V. Chernyak and S. Mukamel, *J. Chem. Phys.* **105**, 4565 (1996).

36. V. Butkus, L. Valkunas and D. Abramavicius, *J. Chem. Phys.* **137**, p. 044513 (2012).

37. A. G. Redfield, *IBM J. Res. Dev.* **1**, 19 (1957).

38. W. M. Zhang, T. Meier, V. Chernyak and S. Mukamel, *J. Chem. Phys.* **108**, 7763 (1998).

39. M. Yang and G. R. Fleming, *Chem. Phys.* **275**, 355 (2002).

40. D. Abramavicius and S. Mukamel, *J. Chem. Phys.* **133**, p. 184501 (2010).

41. A. Ishizaki and Y. Tanimura, *J. Phys. Soc. Jpn.* **74**, 3131 (2005).

42. Y. Tanimura, *J. Phys. Soc. Jpn.* **75**, p. 082001 (2006).

43. R. X. Xu, B. L. Tian, J. Xu, Q. Shi and Y. J. Yan, *J. Chem. Phys.* **131**, p. 214111 (2009).

44. R. X. Xu, P. Cui, X. Q. Li, Y. Mo and Y. J. Yan, *J. Chem. Phys.* **122**, p. 041103 (2005).
45. L. Chen, R. Zheng, Q. Shi and Y. J. Yan, *J. Chem. Phys.* **131**, p. 094502 (2009).
46. T. Mančal, V. Balevičius Jr. and L. Valkunas, *J. Phys. Chem. A* **115**, 3845 (2011).
47. A. Ishizaki and G. R. Fleming, *J. Chem. Phys.* **130**, p. 234111 (2009).
48. A. Gelzinis, D. Abramavicius and L. Valkunas, *Phys. Rev. B* **84**, p. 245430 (2011).
49. J. Strumpfer and K. Schulten, *J. Chem. Phys.* **131**, p. 225101 (2009).
50. C. Kreisbeck, T. Kramer, M. Rodriguez and B. Hein, *J. Chem. Theory Comput.* **7**, p. 2166 (2011).
51. R.-X. Xu and Y.-J. Yan, *Phys. Rev. E* **75**, p. 031107 (2007).
52. T. Guohua and W. H. Miller, *J. Phys. Chem. Lett.* **1**, p. 891 (2010).
53. G. Ritschel, J. Roden, W. T. Strunz and A. Eisfeld, *New J. Phys.* **13**, p. 113034 (2011).
54. J. Prior, A. W. Chin, S. F. Huelga and M. B. Plenio, *Phys. Rev. Lett.* **105** (2010).
55. B. Palmieri, D. Abramavicius and S. Mukamel, *J. Chem. Phys.* **130**, p. 204512 (2009).
56. D. Abramavicius and S. Mukamel, *J. Chem. Phys.* **133**, p. 064510 (2010).
57. E. Collini, C. Y. Wong, K. E. Wilk, P. M. G. Curmi, P. Brumer and G. D. Scholes, *Nature* **463**, 644 (2010).
58. S. Westenhoff, D. Paleček, P. Edlund, P. Smith and D. Zigmantas, *J. Am. Chem. Soc.* **134**, 16484 (2012).
59. D. Egorova, M. Gelin and W. Domcke, *J. Chem. Phys.* **126**, p. 074314 (2007).
60. N. Christensson, H. F. Kauffmann, T. Pullerits and T. Mancal, *J. Phys. Chem. B* **116**, 7449 (2012).
61. V. Tiwari, W. K. Peters and D. M. Jonas, *Proc. Natl. Acad. Sci. USA* **110**, 1203 (2013).
62. P. Nalbach, D. Braun and M. Thorwart, *Phys. Rev. E* **84**, p. 041926 (2011).
63. M. Wahadoszamen, R. Berera, A. M. Ara, E. Romero and R. van Grondelle, *Phys. Chem. Chem. Phys.* **14**, 759 (2012).
64. R. Berera, C. Herrero, I. H. M. van Stokkum, M. Vengris, G. Kodis, R. E. Palacios, H. van Amerogen, R. van Grondelle, D. Gust, T. A. Moore, A. L. Moore and J. T. M. Kennis, *Proc. Nat. Acad. Sci. USA* **103**, 5343 (2006).
65. M. Kloz, S. Pillai, G. Kodis, D. Gust, T. A. Moore, A. L. Moore, R. van Grondelle and J. T. M. Kennis, *J. Am. Chem. Soc.* **133**, 7007 (2011).
66. P.-N. Liao, S. Pillai, D. Gust, T. A. Moore, A. L. Moore and P. J. Walla, *J. Phys. Chem. A* **115**, 4082 (2011).
67. V. Balevičius Jr., A. Gelzinis, D. Abramavicius, T. Mančal and L. Valkunas, *Chem. Phys.* **404**, 94 (2012).
68. V. Balevičius Jr., A. Gelzinis, D. Abramavicius and L. Valkunas, *J. Phys. Chem. B* , doi:10.1021/jp3118083 (2013).

69. G. D. Scholes, R. D. Harcourt and G. R. Fleming, *J. Phys. Chem. B* **101**, 7302 (1997).
70. A. Warshel and W. W. Parson, *J. Am. Chem. Soc.* **109**, p. 6143 (1987).
71. T. Mančal, L. Valkunas and G. R. Fleming, *Chem. Phys. Lett.* **432**, 301 (2006).

ULTRAFAST INTRAMOLECULAR DYNAMICS IN NOVEL STAR-SHAPED MOLECULES FOR PHOTOVOLTAIC APPLICATIONS

OLEG V. KOZLOV

Zernike Institute for Advanced Materials, University of Groningen, Nijenborgh 4, 9747 AG Groningen, The Netherlands

International Laser Center and Faculty of Physics, M.V. Lomonosov Moscow State University, Leninskie Gory, 1, Moscow RU-119991, Russia

YURIY N. LUPONOSOV
SERGEI A. PONOMARENKO

Enikolopov Institute of Synthetic Polymeric Materials of the Russian Academy of Sciences, Profsoyuznaya St. 70, Moscow 117393, Russia

DMITRY YU. PARASCHUK

International Laser Center and Faculty of Physics, M.V. Lomonosov Moscow State University, Leninskie Gory, 1, Moscow RU-119991, Russia

NINA KAUSCH-BUSIES

Heraeus Precious Metals GmbH & Co. KG, Conductive Polymers Division (Clevios), Chempark Leverkusen Build. B202, D-51368 Leverkusen, Germany

MAXIM S. PSHENICHNIKOV

Zernike Institute for Advanced Materials, University of Groningen, Nijenborgh 4, 9747 AG Groningen, The Netherlands

In this paper, intramolecular charge dynamics in diluted solutions of novel star-shaped push-pull molecules is studied by ultrafast photoinduced absorption (PIA) spectroscopy. Three star-shaped donors with different conjugated chain lengths are used. From PIA dynamics, it is shown that in diluted solutions charge recombination is accompanied by structural reorganization of the solvent. Intramolecular recombination of photogenerated charges is estimated to occur in the 0.5-1.7 ns range, while solvent structural rearrangement takes place at a 10's of ps timescale.

1. Introduction

Nowadays, the most widely-used structure for the photovoltaic layer of organic solar cells (OSC) is the so-called bulk heterojunction (BHJ), which is a nanostructured mixture of donor and acceptor molecules. As donor materials, conjugated polymers are usually used, while the most successful acceptor materials are fullerene derivatives. With polymer-based organic solar cells, the important psychological threshold of 10% of power conversion efficiency (PCE) has just been surpassed [1].

Recently, it was realized that polymers could be substituted by small molecules (SM) in the OCS photoactive layer [2-9]. SM-based OSCs could potentially combine the advantages of conventional polymer solar cells with the benefits of small molecules such as good batch-to-batch reproducibility, well-defined molecular structure, easy mass-scale production, etc. Small molecules have already proved quite promising donor materials with PCE as high as 7% [10, 11].

In the family of SMs, one of the promising structures is the so-called star-shaped push-pull molecules [12, 13]. They have a number of advantages over more conventional linear molecules like better solubility, higher degree of purity and structural uniformity. Usually star-shaped molecules consist of a donor core unit and acceptor end groups connected through a π-conjugated bridge. As the donor core, triphenylamine (TPA) is typically used, and benzothiadiazole derivative [7, 14, 15] or dicyanovinyl (DCV) [16-19] are the most common acceptor groups. To achieve sufficient solubility in organic solvents, alkyl end chains may be attached to the end of the conjugated arms [19]. As a result, star-shaped molecules show broad absorptions in visible and near-UV ranges, high solubility, and good hole mobility. To date, the best OSCs based on the star-shaped SMs have demonstrated PCEs approaching 5% [20,21].

In essence, the ultimate goal of the solar cell's operation is photocurrent. To contribute to the photocurrent under light illumination free charges should be generated within the OSC's active layer that subsequently drift to the electrodes. The processes of free charges photogeneration take place at the sub-ns time scale, and naturally impact the overall efficiency of the OSC. In push-pull molecules, initial charge separation occurs within the molecule itself, producing a pair of spatially separated charges. In the absence of the external acceptor, these charges will eventually recombine to restore equilibrium. Therefore, further steps in the charge separation – when the electron is transferred from the donor to the (fullerene) acceptor – strongly depend on the efficiency of the initial charge splitting and on the time window provided by the intramolecular recombination. This highlights the importance of the initial step of charge separation.

In this paper we report on ultrafast dynamics in star-shaped molecules based on TPA donor core and DCV acceptor end groups. The length of the π-bridge was varied by including a different number (1-3) of conjugated thiophene units between the core and acceptors. Time resolved polarization-sensitive visible-pump IR-probe photoinduced absorption (PIA) spectroscopy was used to reveal the initial dynamics of charges and their back recombination. All experiments are performed in diluted solutions to eliminate intermolecular interactions. From the transients obtained, intramolecular charge recombination times for different molecules were estimated between 0.5 to 1.7 ns. The recombination time depends strongly on the chemical structure of the molecule, and decreases with increasing molecule size. Simultaneously, structural rearrangement of the (polar) solvent molecules occurs at faster timescales of about 10 ps for all samples.

2. Sample preparation and experimental

In this study, a family of three star-shaped push-pull molecules was used (Fig. 1). All molecules were based on triphenylamine as donor core with dicyanovinyl

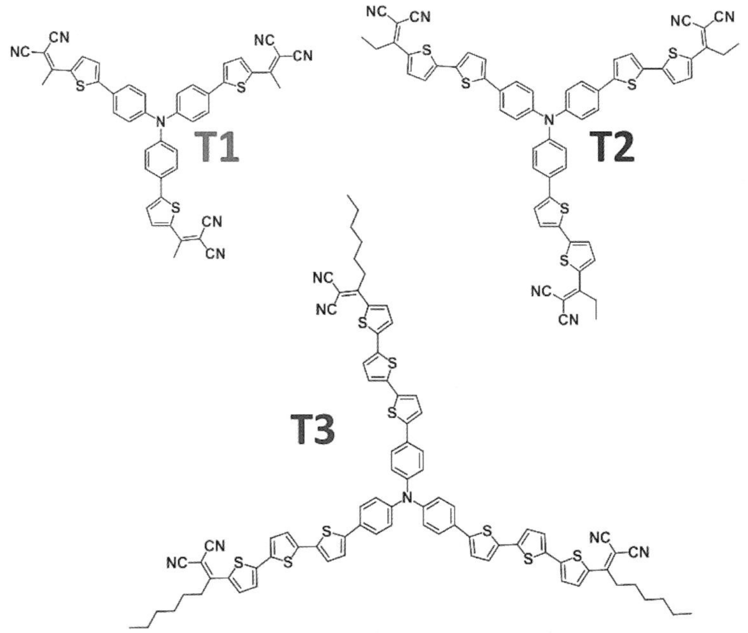

Fig. 1 Chemical structure of the investigated molecules.

172

acceptor groups attached to the core by a thiophene conjugated chain in a star-like manner. In such molecular design, the length of the conjugated chain is systematically varied which defines the bandgap of the molecule. Alkyl end groups were connected to the end of the molecule arms to achieve sufficient solubility in organic solvents. The molecules were synthesized by the method described in detail in Ref. [19, 21] resulting in one thiophene ring in the conjugated arm and one carbon atom in the alkyl chain (denoted as T1), two thiophene rings and two carbon atoms (T2), and three thiophene rings and six carbon atoms (T3).

For performing PIA experiments diluted solutions of star-shaped molecules were prepared aimed at elucidating the intramolecular dynamics only. Each type of molecules was dissolved separately in dichlorobenzene (DCB) at concentration of 0.025 g/l. Solutions were stirred by a magnetic stirrer for at least 12 hours at 50°C. Then the solutions were placed into 1 mm thick quartz cuvettes for PIA transient measurements or into a 1 mm thick barium fluoride cuvette for IR absorption and polaronic spectra measurements.

Absorption spectra of the samples were recorded with a Perkin-Elmer Lambda 900 spectrophotometer. Polarization-sensitive ultrafast PIA measurements were performed with home-build VIS-pump IR-probe setup with 100-fs time resolution described in details elsewhere [22]. The polarization of the probe pulse was rotated by 45° with respect to the polarization of the pump pulse, and after the sample component parallel or perpendicular to the pump polarization was selected by a polarizer. Wavelengths of the pump pulses were 540 nm for T1 compound and 560 nm for T2 and T3 molecules in accordance with absorption spectra of the solutions investigated. Due to the polaronic spectra, wavelength of the probe pulse was chosen of 3 μm.

PIA transients were measured in pump-probe delay time ranges from -1 to 1000 ps with a step size between 10 fs for short delays and 10 ps for long delays. Isotropic (population) part of the signal and photoinduced anisotropy were recalculated from the parallel and perpendicular parts of the photoinduced absorption transients using the following relations [23]:

$$\Delta T_{ISO}(t) = \frac{\Delta T_{\parallel}(t) + 2\Delta T_{\perp}(t)}{3} \qquad (1)$$

$$r(t) = \frac{\Delta T_{\parallel}(t) - \Delta T_{\perp}(t)}{3\Delta T_{ISO}(t)} \qquad (2)$$

where $\Delta T_{\parallel}$ and $\Delta T_{\perp}$ are relative transmission changes for the parallel and perpendicular components of the signal respectively.

Polaronic spectra were recorded using a similar pump-probe technique at another setup with a tunable wavelength of the IR pulses at the expanse of time resolution (~300 fs). In these measurements, wavelengths of the pump pulses were chosen similar to those in PIA measurements while the wavelength of the IR probe pulse was varied in a wide range from 3 to 10 μm.

3. Results

3.1. *Absorption spectra*

Absorption spectra of solutions of different star-shaped molecules are shown in Fig. 2. The spectra exhibit two absorption peaks at ~370-420 nm and ~500-540 nm which position shifts and separation decreases with increasing number of thiophene rings. The strongest peaks at 500-540 nm are typically assigned to intramolecular charge transfer [15, 17, 18], while the weaker peaks at 370-420 nm correspond to either the π-π* excitation of the conjugated backbone [15, 18] or also to the intramolecular charge transfer [17]. To reveal the nature of the high-energy peak, the absorption spectrum of a star-shaped molecule with two thiophene rings in the branch but without DCV acceptor groups (TPA-2T) is shown for comparison. This spectrum contains only of a 400-nm peak which is consistent with the charge-transfer origin of the low-energy peak and π-π* character of the high-energy one.

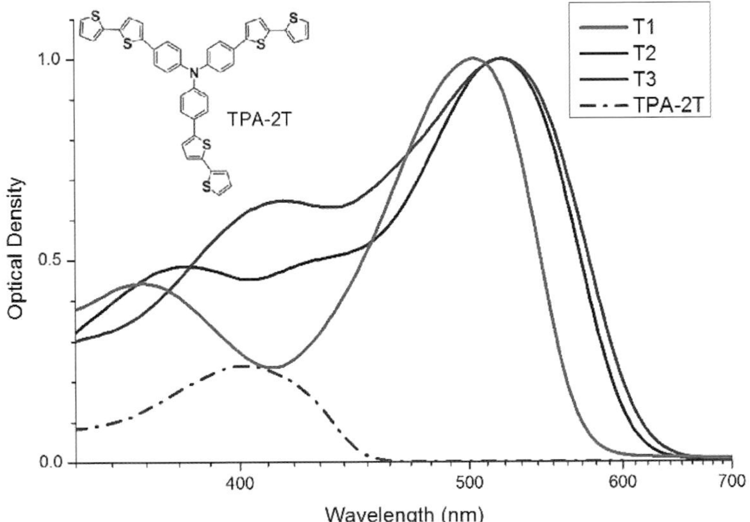

Fig. 2 Absorption spectra of all sample solutions. Absorption spectrum of a solution of the TPA-2T compound without the DCV acceptor groups is shown for comparison.

The wavelength of the pump pulse for PIA measurements was chosen at the red flank of the high-energy absorption peak (i.e. 540 nm for T1 and 560 nm for T2 and T3 compounds) to avoid excitation of hot vibrational states and interference with the low-energy absorption.

3.2. *Polaronic and infrared absorption*

To determine the concentration of photogenerated charges in the star-shaped molecules, IR polaronic absorption was used. It is well-known that polaronic spectra of conjugates systems appear as a response to charge separation along the conjugated backbone, and consist of several broad IR absorption bands [24, 25]. In particular, the high-energy (HE) band is located in the region of ~1 μm while the low-energy (LE) band is situated around 2-3 μm. In principle, both HE and LE bands are suitable for tracking dynamics of generated polarons; however, spectral fingerprints of excitonic excitations, stimulated emission and charge-transfer ground-state absorption are often located in the region of HE polaron absorption [22]. Consequently, the LE band is the most suitable one for monitoring charge dynamics since it provides spectral windows for a completely background-free detection of generated charges.

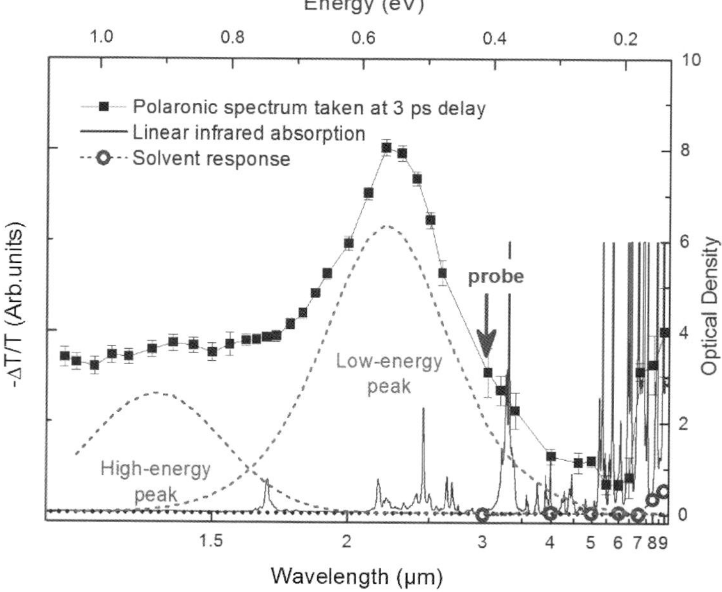

Fig. 3 (Color online) Polaronic (black curve) and IR absorption (blue curve) spectra for T3 solution in dichlorobenzene. The solvent (DCB) PIA response is shown for comparison by brown circles.

Polaronic and IR absorption spectra of the dichlorobenzene solutions of T3 molecules are shown in Fig. 3. The polaronic spectrum was recorded at a delay of 3 ps between the pump and probe pulses to show the distribution of initially generated photoinduced states. The PIA response consists of three peaks at 1.4 µm, 2.2 µm and 7-9 µm. The weak peak at 1.4 µm and the more intense 2.2-µm peak are assigned to intramolecular polarons while the 9-µm peak most probably is related to the solvent absorption (blue line in Fig. 3) with a subsequent temperature response of vibrational frequencies of the molecule under study. According to the polaronic spectrum, the wavelength of the probe pulse was chosen in the absorption-free region of the low-energy polaronic band at 3 µm.

3.3. *Photoinduced absorption dynamics*

Isotropic PIA transients for solutions of all molecular compounds are represented in Fig. 4. Signals are normalized to their maximum values. Since intermolecular interactions in diluted solutions are negligibly low, the transients represent intramolecular charge separation and recombination.

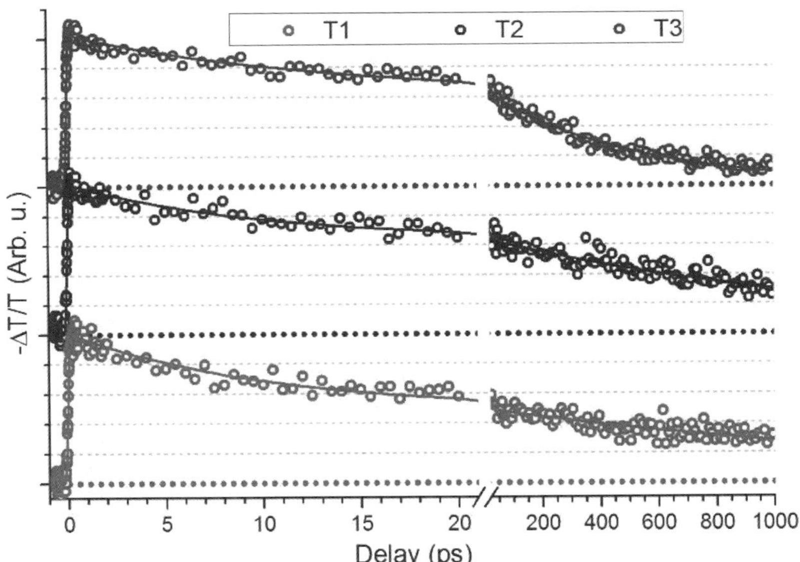

Fig. 4 PIA isotropic transients for solutions of star-shaped molecules. Transients are normalized to their maximum values.

The dynamics of all samples are quite similar: at zero delay, a very fast build-up of the pump-probe signal occurs within apparatus response function of the setup. Since the amplitude of the signal is supposed to be proportional to the

176

concentration of the generated charges (polarons), one can conclude that the intramolecular charge separation occurs instantaneously within the temporal resolution (~100 fs). After the instantaneous intramolecular separation of the charges, relatively slow charge recombination occurs. Signal dynamics occur at two different timescales. The initial timescale is similar for all molecules and takes place within several picoseconds. The second timescale comprises the interval in the ns range. This "slow" time depends on the chemical structure of the molecule, and decreases with elongation of the molecule's conjugated arm.

To estimate recombination times, the data were fitted using a biexponential function:

$$\Delta T_{ISO} = A_1 \exp(-t/\tau_1) + A_2 \exp(-t/\tau_2) \qquad (3)$$

convoluted with Gaussian apparatus function (σ~50 fs). Fit parameters are given in Table 1.

Table 1 Fit parameters of isotropic PIA for all samples

Molecule	A_1, %	τ_1, ps	A_2, %	τ_2, ns
T1	50±1	10±1	50±1	1.7±0.2
T2	37±1	10±1	64±1	1.3±0.1
T3	35±1	12±1	65±1	0.5±0.1

The first "fast" constant is almost identical for all three molecules and equal to ~10 ps. There might be two interpretations of this timescale. According to the first one, it represents the timescale of intramolecular recombination of polarons as it is similar to the time constant observed in other conjugated materials [26]. Contribution of this component decreases with increasing length of the conjugated arm. This is most probably related to the increase of the spatial scale of charge separation by adding extra thiophene rings in the star branch. Increasing the distance between the donor and acceptor units leads to stronger spatial charge separations which in turn leads to less efficient back recombination.

The second interpretation connects the fast time to a scale of polar solvent rearrangements as its reaction to the dipole moment induced by the pump at the star-shaped molecule. This rearrangement tends to minimize the total free energy of the system thereby changing the electronic structure of the star shaped molecule – the effect is closely related to solvation dynamics.

The "slow" part of the recombination dynamics depends strongly on the molecule's size. For instance, for the smallest T1 molecule this recombination time is equal to 1.7 ns while for T3 molecule it decreases to 0.5 ns. This recombination time most probably corresponds to relaxation of the excited state due to conformational reorganization of the molecule via interactions with the solvent. It is known that the efficiency of partial charge separation in push-pull molecules depends on the length of the spacer between the donor and acceptor units [27]. Thus, for T1 molecule partial charges on the acceptor units should be larger in magnitude that those for T2 and T3 molecule, which leads to more intensive molecule-solvent interaction and longer recombination time.

3.4. *Photoinduced anisotropy dynamics*

Anisotropy dynamics for solutions are shown in Fig. 5. For all samples, anisotropy decays from its initial value of about 0.25 to ~0.12 within 1 ps, and changes only insignificantly further on. Anisotropy decays were fitted with biexponential decay function with a zero baseline:

$$r(t) = r_1 \exp(-t/\tau_1) + r_2 \exp(-t/\tau_2) \tag{4}$$

and fit parameters are given in Table 2.

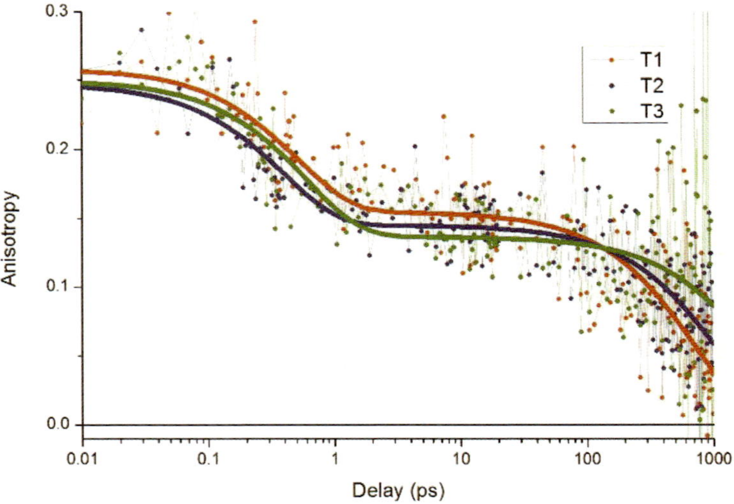

Fig. 5 Anisotropy decay for solutions of star-shaped molecules.

Table 2 Fit parameters of anisotropy decay

Molecule	r_1	τ_1, fs	r_2	τ_2, ns
T1	0.11±0.01	530±130	0.15±0.01	0.7±0.1
T2	0.13±0.01	400±60	0.14±0.01	1.1±0.1
T3	0.14±0.02	600±180	0.14±0.01	2.2±0.6

A long-time decay of the anisotropy is likely related to a rotational degree of freedom of the molecules in solution. However, in sub-ps region, anisotropy decay is too fast (within ~0.5 ps) to be assigned to any mechanical movements. Moreover, a short-time part of the decay looks quite similar for all three molecules, so one can conclude that in this case the anisotropy decays because of intramolecular process which is identical for all the molecules investigated.

The mechanism of the fast anisotropy decay may be explained by considering the chemical structure of the molecule. Due to the high symmetry of the molecule, its energy states should be triple-degenerate (rotational symmetry). It is known that star-shaped molecules may have two excited states with the same energy and different electron density distribution [17]. Therefore, the molecules investigated have up to six excited states with very close energy levels. This means that the molecule, excited in one of this states, could evolve into one of the others. During this process, redistribution of the electron density occurs between the branches which leads to depolarization of the system and disappearance of the correlation between induced and final transition dipole moments. According to the experimental data, such intramolecular process of mixing of excited states takes ~0.5 ps for all three molecules.

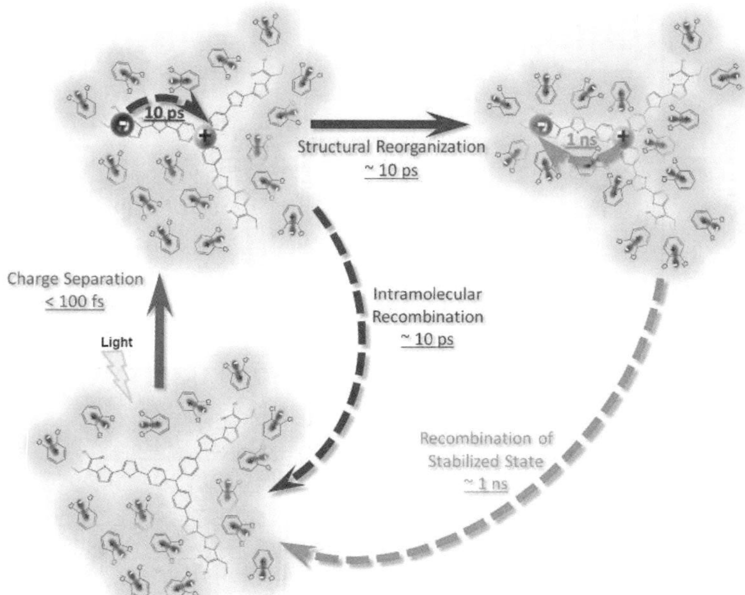

Fig. 6 An artist impression of charge generation and recombination in DCB solution of the star-shaped molecules

After the initial decay via excited state mixing, the anisotropy decays further with a much slower timescale. The relaxation time increases from 0.7 ns for the smallest T1 molecules up to 2.2 ns for T3 molecule. It is well-known that rotational relaxation for similar molecules in the same solvent increases with increasing molecular volume [28-30], so one can conclude that the slow relaxation of the anisotropy is related to the molecules' rotations.

3.5. *Charge generation and recombination*

A pictorial representation of the dynamics in the star-shaped molecules upon optical excitation is presented in Fig.6. Under the influence of the incident light, an instantaneous intramolecular charge separation occurs; the induced charges begin to recombine afterwards. At the same time, because of the appearance of the induced dipole moment of the molecules, electrostatic interactions between the material and the solvent lead to a structural reorganization of the whole system at the timescales of about 10 ps. This in turn leads to stabilization of the separated charges, and these stabilized charges recombine at the longer timescales of nanoseconds.

4. Conclusions

Here we have studied charge dynamics of a series of novel star-shaped push-pull molecules in dichlorobenzene solutions using time-resolved pump-probe technique. It was demonstrated that the charge dynamics have two significantly different timescales. The first timescale of about 10 ps is virtually independent of the star-shaped molecule's chemical structure. This timescale was assigned to intramolecular charge recombination or/and solvation effects. The second recombination channel is much slower and takes place at timescales of nanoseconds. We assume that this recombination channel appears due to interactions of the star-shaped molecules in the excited state with the solvent and consequent stabilization of the spatial separated charges.

Additional insights into the relaxation dynamics are presented by photoinduced anisotropy dynamics. About half of the photoinduced anisotropy decays because of the intramolecular process of mixing of degenerate excited states of the molecule. This process occurs at ultrafast sub-ps timescales and does not depend on the molecule's chemical structure. The anisotropy further decays because of a rotational movement of the star-shaped molecule in the solution. Characteristic timescales for this process are around nanoseconds, and the relaxation time increases with increasing the size of the molecule.

Summing up, the star-shaped small molecules provide an efficient charge separation in solutions which makes them perspective materials as donors in organic solar cells. Solvation effects observed here might play an important role in polarizing the molecule surrounding thereby providing a direct way to lower Coulomb attraction between would-be separated charges [31]. Only the first steps toward understanding charge separation and recombination processes have been made here; further investigation of charge dynamics in thin films and donor-acceptor blends is necessary to gain a deeper knowledge of photon-to-current conversion processes in organic solar cells based on small molecules.

Acknowledgments

We thank Vlad Pavelyev and Almis Serbenta (Zernike Institute for Advanced Materials, University of Groningen, The Netherlands) for valuable contribution to the experimental section, and Yoann Olivier and Jerome Cornil (Service de Chimie des Materiaux Nouveaux, Université de Mons, Belgium) for theoretical support.

References

1. You, J., L. Dou, K. Yoshimura, T. Kato, K. Ohya, T. Moriarty, K. Emery, C.-C. Chen, J. Gao, G. Li, and Y. Yang, *Nat Commun*, **4**: 1446 (2013).
2. Walker, B., C. Kim, and T.Q. Nguyen, *Chemistry of Materials*, **23**(3): 470-482 (2011).
3. Sharma, G.D., J.A. Mikroyannidis, S.S. Sharma, M.S. Roy, and K.R.J. Thomas, *Organic Electronics*, **13**(4): 652-666 (2012).
4. Seulgi, S., C. Hyunbong, K. Haye Min, K. Chulwoo, P. Sanghyun, N. Cho, S. Kihyung, L. Jae Kwan, and K. Jaejung, *Solar Energy Materials and Solar Cells*, **98**: 224-32 (2012).
5. Li, Z., G.R. He, X.J. Wan, Y.S. Liu, J.Y. Zhou, G.K. Long, Y. Zuo, M.T. Zhang, and Y.S. Chen, *Advanced Energy Materials*, **2**(1): 74-77 (2012).
6. Kim, J., N. Cho, H. Min Ko, C. Kim, J. Kwan Lee, and J. Ko, *Solar Energy Materials and Solar Cells*, **102**(0): 159-166 (2012).
7. Shang, H.X., H.J. Fan, Y. Liu, W.P. Hu, Y.F. Li, and X.W. Zhan, *Advanced Materials*, **23**(13): 1554-1557 (2011).
8. Yin, B., L.Y. Yang, Y.S. Liu, Y.S. Chen, Q.J. Qi, F.L. Zhang, and S.G. Yin, *Applied Physics Letters*, **97**(2) (2010).
9. Tamayo, A.B., B. Walker, and T.Q. Nguyen, *Journal of Physical Chemistry C*, **112**(30): 11545-11551 (2008).

10. Sun, Y., G.C. Welch, W.L. Leong, C.J. Takacs, G.C. Bazan, and A.J. Heeger, *Nature Materials*, **11**(1): 44-48 (2012).

11. van der Poll, T.S., J.A. Love, T.Q. Nguyen, and G.C. Bazan, *Advanced Materials*, **24**(27): 3646-3649 (2012).

12. Luo, J., B. Zhao, H.S. On Chan, and C. Chi, *Journal of Materials Chemistry*, **20**(10): 1932-1941 (2010).

13. Kim, K.H., Z. Chi, M.J. Cho, J.-I. Jin, M.Y. Cho, S.J. Kim, J.-S. Joo, and D.H. Choi, *Chemistry of Materials*, **19**(20): 4925-4932 (2007).

14. Lin, Z.H., J. Bjorgaard, A.G. Yavuz, and M.E. Kose, *Journal of Physical Chemistry C*, **115**(30): 15097-15108 (2011).

15. Zhang, J., J. Yu, C. He, D. Deng, Z.-G. Zhang, M. Zhang, Z. Li, and Y. Li, *Organic Electronics*, **13**(1): 166-172 (2012).

16. Cravino, A., S. Roquet, P. Leriche, O. Aleveque, P. Frere, and J. Roncali, *Chemical Communications*, (13): 1416-1418 (2006).

17. Ripaud, E., Y. Olivier, P. Leriche, J. Cornil, and J. Roncali, *Journal of Physical Chemistry B*, **115**(30): 9379-9386 (2011).

18. Dan, D., S. Suling, Z. Jing, H. Chang, Z. Zhanjun, and L. Yongfang, *Organic Electronics*, **13**: 2546–2552 (2012).

19. Min, J., T. Ameri, A. Elschner, Y. Luponosov, S. Peregudova, D. Baran, T. Heumuller, N. Li, S. Ponomarenko, and C.J. Brabec, *Organic electronics*, **14**(1): 219–229 (2013).

20. Min, J., H. Zhang, T. Stubhan, Y.N. Luponosov, M. Kraft, S.A. Ponomarenko, T. Ameri, U. Scherf, and C.J. Brabec, *Journal of Materials Chemistry A*, **1**(37): 11306-11311 (2013).

21. Min, J., Y.N. Luponosov, A. Gerl, M.S. Polinskay, S.M. Peregudova, P.V. Dmitryakov, A.V. Bakirov, M.A. Shcherbina, S.N. Chvalun, S. Grigorian, N. Kaush-Busies, S.A. Ponomarenko, T. Ameri, and C.J. Brabec, *Adv. Energy. Mater.* , DOI: 10: 1002/aenm.201301234 (2013).

22. Bakulin, A.A., D.S. Martyanov, D.Y. Paraschuk, M.S. Pshenichnikov, and P.H.M. van Loosdrecht, *Journal of Physical Chemistry B*, **112**(44): 13730-13737 (2008).

23. Gordon, R.G., *Journal of Chemical Physics*, **45**(5): 1643 (1966).

24. Hwang, I.W., C. Soci, D. Moses, Z. Zhu, D. Waller, R. Gaudiana, C.J. Brabec, and A.J. Heeger, *Advanced Materials*, **19**(17): 2307-2312 (2007).

25. Sheng, C.X., M. Tong, S. Singh, and Z.V. Vardeny, *Physical Review B*, **75**(8): 085206 (2007).

26. Rolczynski, B.S., J.M. Szarko, H.J. Son, Y. Liang, L. Yu, and L.X. Chen, *Journal of the American Chemical Society*, **134**(9): 4142-4152 (2012).

27. Tautz, R., E. Da Como, T. Limmer, J. Feldmann, H.-J. Egelhaaf, E. von Hauff, V. Lemaur, D. Beljonne, S. Yilmaz, I. Dumsch, S. Allard, and U. Scherf, *Nat Commun*, **3**: 970 (2012).

28. Kurnikova, M.G., N. Balabai, D.H. Waldeck, and R.D. Coalson, *Journal of the American Chemical Society*, **120**(24): 6121-6130 (1998).

29. Williams, A.M., Y. Jiang, and D. Ben-Amotz, *Chemical Physics*, **180**(2–3): 119-129 (1994).

30. Dote, J.L., D. Kivelson, and R.N. Schwartz, *The Journal of Physical Chemistry*, **85**(15): 2169-2180 (1981).

31. Koster, L.J.A., S.E. Shaheen, and J.C. Hummelen, *Advanced Energy Materials*, **2**(10): 1246-1253 (2012).

NONLINEAR SPECTROSCOPY OF INTERFACES AND ITS APPLICATION TO ORGANIC ELECTRONICS

SILVIA G. MOTTI, FRANCISCO C. B. MAIA[a]
and PAULO B. MIRANDA

*Instituto de Física de São Carlos, Universidade de São Paulo, CP 369
São Carlos, SP 13560-970, Brazil*

In this chapter we give a brief overview of two second-order nonlinear spectroscopy techniques that are useful to the study of interfaces: second-harmonic generation (SHG) and sum-frequency generation (SFG). SFG yields the vibrational spectrum of surface molecules, from which information about their chemical structure and interactions may be obtained, while SHG may be used to acquire the electronic spectrum of interfacial molecules. Additionally, both are intrinsically surface-specific, sensitive to molecular orientation and interfacial electric fields, and capable of probing real-time dynamics at surfaces. Examples of applications to interfaces relevant to organic electronics (LEDs, FETs, and solar cells) will be reviewed.

1. Introduction

Surfaces and interfaces are very important from the scientific point of view, since the structure, reactivity and intermolecular interactions may be very different at interfaces with respect to bulk materials. They also play a fundamental role in many technological applications,[1] such as adhesion, colloidal stability, electrochemistry and charge transfer processes in general. In the field of organic electronics, where the active semiconducting materials are small molecules or conjugated polymers, it is becoming increasingly important to understand their interfacial properties,[2,3] since the device performance may be severely affected by the surface arrangement of these molecules, or their interaction with neighboring materials. For example, nearly every device has a metallic contact for charge injection, and in organic field-effect transistors (OFETs) the current conduction occurs at a very narrow region at the semiconductor-dielectric interface.

For that reason, several experimental techniques have been used to investigate surfaces and interfaces involving organic semiconductors, with the

---

[a] Current address: Laboratório Nacional de Luz Síncrotron - LNLS, CP 6192, CEP 13083-970, Campinas - SP, Brazil

most notable ones being X-ray and ultraviolet photoelectron spectroscopy (XPS and UPS),[3] X-ray diffraction,[4] neutron reflectivity,[5,6] Fourier-transform infrared spectroscopy (FTIR)[7-9] and surface-enhanced Raman spectroscopy (SERS).[9,10] Although much important information can be obtained with these techniques, such as energy level alignment, surface composition and organization of molecules, and the occurrence of charge transfer or surface reactions, many of them have important limitations. For example, UPS and XPS of the semiconductor-metal interface must be performed with ultrathin metal films (a few Å thick), which may actually represent a different interface with respect to real devices. The same applies to SERS, which uses rough metal surfaces to obtain the signal enhancement for surface sensitivity. Grazing incidence X-ray diffraction is highly surface sensitive and gives detailed structural information, but is not applicable to amorphous interfaces. Finally, FTIR spectroscopy may have its surface sensitivity increased by modulation techniques, but residual bulk contributions are unavoidable. Therefore, probing buried solid-solid interfaces of interest to organic electronics with sufficient surface sensitivity is still a challenging task.

Here we will discuss two nonlinear optical techniques for studying interfaces that have been finding increasing applications to organic electronics. They are based on the second-order nonlinear optical processes of second-harmonic generation (SHG) and sum-frequency generation (SFG), which makes them inherently surface sensitive, since those processes can only occur where inversion symmetry is lost.[11] Therefore, the SHG and SFG signals cannot be generated in the bulk of centrosymmetric materials (isotropic solids, for example), and may only arise from the interfacial layer where the inversion symmetry is necessarily broken. In the next section we will describe the key features of these techniques, beginning with their basic theory and then highlighting the main capabilities and advantages. In section 3 we describe a few selected applications of SHG and SFG to interfaces of interest in organic electronics, including *in situ* probing of working devices, such as light-emitting diodes (OLEDs), photovoltaic cells (OPVs) and transistors (OFETs). We did not intend to make an extensive review of these applications, but instead to illustrate the potential applications that complement information which can be obtained with other techniques.

2. Probing Interfaces with SHG and SFG Spectroscopy

Sum-Frequency Generation (SFG) spectroscopy is a nonlinear optical technique that can obtain the vibrational spectrum of interfacial molecules, discriminating them from those in the bulk material. It is therefore surface-specific, with intrinsic selectivity to interfacial contributions. In SFG spectroscopy, two high-intensity laser beams at frequencies ω_1 and ω_2 overlap at an interface and generate an output beam at frequency $\omega = \omega_1 + \omega_2$ (the SFG signal) in the reflection direction, as illustrated in Figure 1. The intensity of the SFG signal is proportional to the square of the effective second-order nonlinear susceptibility of the interface, $\chi_{eff}^{(2)}(\omega = \omega_1 + \omega_2)$. Since $\chi^{(2)}$ vanishes in media with inversion symmetry (such as gases, bulk liquids, amorphous solids and most crystals) but is allowed at interfaces, where the inversion symmetry is broken, SFG spectroscopy is intrinsically sensitive to interfaces. As the process relies on breaking the inversion symmetry, only non-centrosymmetric molecules can be detected in SFG. However, if such molecules arrange at an interface with random orientations, the net SFG signal vanishes. Conversely, if there is a substantial SFG signal, it can be concluded that molecules have a net average orientation at the interface. Thus, it is possible to obtain information about the average orientational ordering of the interfacial molecules. For vibrational spectroscopy, ω_2 is tunable in the mid-infrared, (in the range of the vibrational modes of the surface molecules), while ω_1 is a fixed frequency within the visible spectrum, so that ω is also in the visible-UV range and can be detected with high sensitivity. In some cases, ω_1 may be tunable as well, yielding the electronic spectrum of interfacial molecules,[12] although this is less common due to the added complexity to the experimental setup and data analysis. SHG is a particular case of SFG, where only one input beam is used (in the near infrared or visible) to generate an output beam at the second-harmonic ($\omega = 2\omega_1$). It shares the same features with SFG spectroscopy, except that instead of performing vibrational spectroscopy, the electronic spectrum of interfacial molecules can be obtained by tuning ω_1 so that ω is scanned across the electronic resonances.

2.1. *Theoretical Background*

We will now describe the basic theory of SFG at interfaces as a vibrational spectroscopy tool. The treatment for SHG is similar and can be obtained by setting $\mathbf{E}_1 = \mathbf{E}_2$, $\omega_1 = \omega_2$ and $\omega = 2\omega_1$ in the equations below. Only the essential points to understand the basic features of these techniques will be presented. A detailed theory can be found elsewhere.[11,13,14] Tutorial reviews of SHG and SFG spectroscopy[15,16] with their applications to many fields of surface science, and recent reviews of applications to selected fields are also available.[17-19]

186

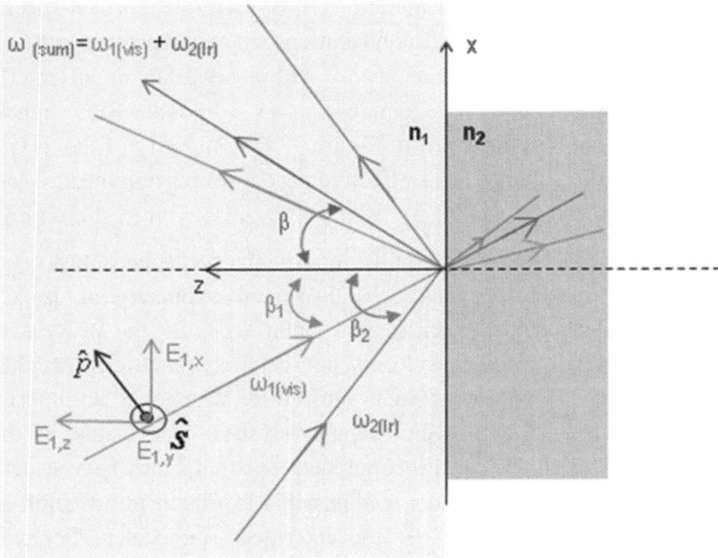

Figure 1. Experimental geometry for SFG at interfaces. Two media with refractive indices n_1 and n_2 are separated by a thin interfacial region of refractive index n′ (not shown), which is responsible for the nonlinear response. Each of the input and output beams have frequencies and incidence angles ω_i and β_i, respectively, and may be s- or p-polarized.

We generally consider an interfacial system as a three-layer system composed of two centrosymmetric media 1 and 2 (see Figure 1) and an interfacial layer (not shown in Figure 1) which is responsible for the nonlinear response. Under the irradiation of two optical fields $\mathbf{E}_1(\mathbf{r},t) = \mathbf{E}_1 e^{i(\mathbf{k}_1 \cdot \mathbf{r} - \omega_1 t)} + c.c.$ and $\mathbf{E}_2(\mathbf{r},t) = \mathbf{E}_2 e^{i(\mathbf{k}_2 \cdot \mathbf{r} - \omega_2 t)} + c.c.$ with frequencies and wavevectors ω_1, $\mathbf{k}_1$ and ω_2, $\mathbf{k}_2$ respectively, the generated effective second-order nonlinear polarization $\mathbf{P}_{eff}^{(2)}(\omega = \omega_1 + \omega_2)$ can be expressed as a multipole expansion,[20,21]

$$\mathbf{P}_{eff}^{(2)}(\omega = \omega_1 + \omega_2) = \mathbf{P}^{(2)}(\omega) - \nabla \cdot \mathbf{Q}^{(2)}(\omega) + \frac{c}{i\omega}\nabla \times \mathbf{M}^{(2)}(\omega) + \cdots \tag{1}$$

where the first, second and third terms on the right-hand side are the contributions to the second-order polarization due to electric dipole, electric quadrupole and magnetic dipole, respectively. At the interface, the optical fields have a large spatial variation. If we keep only terms to the first order in the spatial derivatives, $\mathbf{P}^{(2)}$, $\mathbf{Q}^{(2)}$ and $\mathbf{M}^{(2)}$ can be expressed as

$$\mathbf{P}^{(2)}(\omega) = \chi^{D} : \mathbf{E}_{1}(\omega_{1})\mathbf{E}_{2}(\omega_{2}) + \chi^{P_{1}} : \nabla\mathbf{E}_{1}(\omega_{1})\mathbf{E}_{2}(\omega_{2}) + \chi^{P_{2}} : \mathbf{E}_{1}(\omega_{1})\nabla\mathbf{E}_{2}(\omega_{2})$$

$$\mathbf{Q}^{(2)}(\omega) = \chi^{Q} : \mathbf{E}_{1}(\omega_{1})\mathbf{E}_{2}(\omega_{2})$$

$$\mathbf{M}^{(2)}(\omega) = \chi^{M} : \mathbf{E}_{1}(\omega_{1})\mathbf{E}_{2}(\omega_{2}) \tag{2}$$

In Eqs. (2), χ's are tensors describing the material response. If a medium has inversion symmetry, the tensors χ above should be invariant upon the inversion operation, whereas $\mathbf{P}_{eff}^{(2)}$, $\mathbf{E_i}$ and ∇ are vectors and therefore change sign upon inversion. From this we see immediately that the term proportional to χ^{D} (the local electric dipole contribution) should vanish. All other terms may be nonvanishing even in media with inversion symmetry. At the interface, however, the inversion symmetry is necessarily broken and the electric dipole contribution may also become nonvanishing. Since the electric dipole contribution is the lowest order in the expansion, it is expected to be the dominant one, which is experimentally verified in most cases.[22] If this is the case, the higher-order contributions in Eqs. (1) and (2) can be neglected and the effective second-order nonlinear polarization will be confined to the interfacial layer, being given by

$$\mathbf{P}^{(2)}(\omega = \omega_{1} + \omega_{2}) = \chi_{eff}^{(2)}(\omega = \omega_{1} + \omega_{2}) : \mathbf{E}_{1}(\omega_{1})\mathbf{E}_{2}(\omega_{2}) \tag{3}$$

where $\chi_{eff}^{(2)}(\omega = \omega_{1} + \omega_{2})$ is the effective electric dipole nonlinear susceptibility of the interface, and we have dropped the superscript D in χ for simplicity. This oscillating polarization sheet will then radiate two coherent beams at frequency ω, in the reflected and transmitted directions (see Figure 1). The output angle of the sum-frequency beam is given by phase matching in the surface plane:

$$\mathbf{k}_{1//} + \mathbf{k}_{2//} = \mathbf{k}_{//} \tag{4}$$

where $\mathbf{k}_{i//}$ is the component of the wavevector $\mathbf{k_i}$ parallel to the surface plane. For the geometry shown in Figure 1 it can be expressed as

$$\omega_{1} n_{1}(\omega_{1})\sin\beta_{1} + \omega_{2} n_{1}(\omega_{2})\sin\beta_{2} = \omega n_{1}(\omega)\sin\beta \tag{5}$$

where $n_{i}(\Omega)$ is the refractive index of medium i at frequency Ω and β_{i} is the incidence angle of the beam at frequency ω_{i}. One can then substitute the polarization given by Eq. (1) as a source in the wave equation to obtain the radiated sum-frequency field. The sum-frequency intensity in the reflected direction is given by[11,13,14]

$$I(\omega) = \frac{8\pi^{3}\omega^{2}\sec^{2}\beta}{c^{3}n_{1}(\omega)n_{1}(\omega_{1})n_{1}(\omega_{2})}\left|\chi_{eff}^{(2)}\right|^{2} I_{1}(\omega_{1})I_{2}(\omega_{2}) \tag{6}$$

where $I(\Omega)$ is the intensity of the field at frequency Ω. The effective nonlinear susceptibility $\chi_{eff}^{(2)}$ takes the form

$$\chi_{eff}^{(2)} = [\hat{\mathbf{e}}(\omega) \cdot \mathbf{L}(\omega)] \cdot \chi^{(2)} : [\mathbf{L}(\omega_1) \cdot \hat{\mathbf{e}}(\omega_1)][\mathbf{L}(\omega_2) \cdot \hat{\mathbf{e}}(\omega_2)] \tag{7}$$

with $\hat{\mathbf{e}}(\Omega)$ being the polarization unit vector and $\mathbf{L}(\Omega)$ the Fresnel factor for the field at frequency Ω. The Fresnel factor $\mathbf{L}(\Omega)$ is a diagonal tensor whose elements are the Fresnel transmission coefficients that relate the cartesian components of the field in the interfacial layer to the input field components.

In general, $\chi^{(2)}$ has 27 independent elements. However, we can use the symmetry of the interface to reduce the number of nonvanishing independent elements. In the case of an azimuthally isotropic interface, there are only four independent nonvanishing elements in $\chi^{(2)}$, all with an even number of x and y components. With the lab coordinates chosen such that z is along the interface normal and x in the incidence plane as shown in Figure 1, they are $\chi_{xxz} = \chi_{yyz}$, $\chi_{xzx} = \chi_{yzy}$, $\chi_{zxx} = \chi_{zyy}$ and χ_{zzz}. These four elements can be assessed by measuring SFG with four different input and output polarization combinations, namely, SSP (referring to s-polarized sum-frequency field, s-polarized $\mathbf{E}_1$, and p-polarized $\mathbf{E}_2$, respectively), SPS, PSS and PPP. The effective nonlinear susceptibilities under these four polarization combinations can be expressed as

$$\chi_{eff,SSP}^{(2)} = L_{yy}(\omega)L_{yy}(\omega_1)L_{zz}(\omega_2)\sin\beta_2 \; \chi_{yyz}$$

$$\chi_{eff,SPS}^{(2)} = L_{yy}(\omega)L_{zz}(\omega_1)L_{yy}(\omega_2)\sin\beta_1 \; \chi_{yzy}$$

$$\chi_{eff,PSS}^{(2)} = L_{zz}(\omega)L_{yy}(\omega_1)L_{yy}(\omega_2)\sin\beta \; \chi_{zyy}$$

$$\chi_{eff,PPP}^{(2)} = -L_{xx}(\omega)L_{xx}(\omega_1)L_{zz}(\omega_2)\cos\beta\cos\beta_1\sin\beta_2 \; \chi_{xxz}$$
$$\quad - L_{xx}(\omega)L_{zz}(\omega_1)L_{xx}(\omega_2)\cos\beta\sin\beta_1\cos\beta_2 \; \chi_{xzx}$$
$$\quad + L_{zz}(\omega)L_{xx}(\omega_1)L_{xx}(\omega_2)\sin\beta\cos\beta_1\cos\beta_2 \; \chi_{zxx}$$
$$\quad + L_{zz}(\omega)L_{zz}(\omega_1)L_{zz}(\omega_2)\sin\beta\sin\beta_1\sin\beta_2 \; \chi_{zzz} \tag{8}$$

where $L_{xx}(\Omega)$, $L_{yy}(\Omega)$ and $L_{zz}(\Omega)$ are the diagonal elements of $\mathbf{L}(\Omega)$, given by

$$L_{xx}(\Omega) = \frac{2n_1(\Omega)\cos\gamma}{n_1(\Omega)\cos\gamma + n_2(\Omega)\cos\beta}$$

$$L_{yy}(\Omega) = \frac{2n_1(\Omega)\cos\beta}{n_1(\Omega)\cos\beta + n_2(\Omega)\cos\gamma}$$

$$L_{zz}(\Omega) = \frac{2n_2(\Omega)\cos\beta}{n_1(\Omega)\cos\gamma + n_2(\Omega)\cos\beta}\left(\frac{n_1(\Omega)}{n'(\Omega)}\right)^2 \tag{9}$$

In the above equations, $n'(\Omega)$ is the refractive index of the interfacial layer, β is the incidence angle of the beam in consideration and γ is the refracted angle ($n_1(\Omega)\sin\beta = n_2(\Omega)\sin\gamma$). Since the interfacial layer is only one (or a few) monolayer thick, its refractive index can be different from that of its own bulk material and difficult to measure. It is therefore the usual practice that $n'(\Omega)$ is chosen to be equal to either $n_1(\Omega)$, $n_2(\Omega)$ or the bulk refractive index of the material at the interface. However, as noticed previously[23,24] and shown later by Zhuang et al.,[25] the determination of molecular orientation (see discussion below) is quite sensitive to the value of $n'(\Omega)$, and choosing $n'(\Omega)$ to be equal to $n_1(\Omega)$ or $n_2(\Omega)$ is not always a good approximation.

Since we will discuss applications of both SHG and SFG to organic interfaces, we now present the necessary modifications to Eq. (8) for the case of SHG. In this case, the last two sub-indices of χ_{ijk} are interchangeable,[b] since the input fields are the same. Thus there are only three nonvanishing independent χ-elements, $\chi_{xxz} = \chi_{yyz} = \chi_{xzx} = \chi_{yzy}$, $\chi_{zxx} = \chi_{zyy}$ and χ_{zzz}. They can be deduced from measurement with three different input and output polarization combinations, PS, SM and PP. Here the first and second letters denote the output and input polarization, respectively, and M refers to the polarization midway between s and p. From Eq. (7) the effective nonlinear susceptibilities take the forms

$$
\chi^{(2)}_{eff,PS} = L_{zz}(\omega)(L_{yy}(\omega_1))^2 \sin\beta \; \chi_{zyy}
$$

$$
\chi^{(2)}_{eff,SM} = L_{yy}(\omega)L_{zz}(\omega_1)L_{yy}(\omega_1)\sin\beta_1 \; \chi_{yzy}
$$

$$
\begin{aligned}
\chi^{(2)}_{eff,PP} = &+L_{zz}(\omega)(L_{xx}(\omega_1))^2 \sin\beta \; \cos^2\beta_1 \; \chi_{zxx} \\
&-2L_{xx}(\omega)L_{zz}(\omega_1)L_{xx}(\omega_1)\cos\beta \sin\beta_1 \cos\beta_1 \; \chi_{xzx} \\
&+L_{zz}(\omega)(L_{zz}(\omega_1))^2 \sin\beta \sin^2\beta_1 \; \chi_{zzz}
\end{aligned} \tag{10}
$$

In the case where the interface is composed of molecules, $\chi^{(2)}$ is related to the molecular second-order hyperpolarizability $\alpha^{(2)}$ by the coordinate transformation

$$
\chi^{(2)}_{ijk} = N_s \sum_{\xi,\eta,\zeta} \left\langle (\hat{\mathbf{i}}\cdot\hat{\boldsymbol{\xi}})(\hat{\mathbf{j}}\cdot\hat{\boldsymbol{\eta}})(\hat{\mathbf{k}}\cdot\hat{\boldsymbol{\zeta}}) \right\rangle \alpha^{(2)}_{\xi\eta\zeta} \tag{11}
$$

where N_s is the surface density of molecules, (i,j,k) and (ξ,η,ζ) are unit vectors along the lab and molecular coordinates, respectively, and the angular brackets denote an average over the molecular orientational distribution (angles θ,φ,ψ in Figure 2).

---

[b] In the case of SFG, the *first* two indices of $\chi^{(2)}$ are also interchangeable, if the sum-frequency and visible beams are away from electronic resonances, as will be mentioned later.

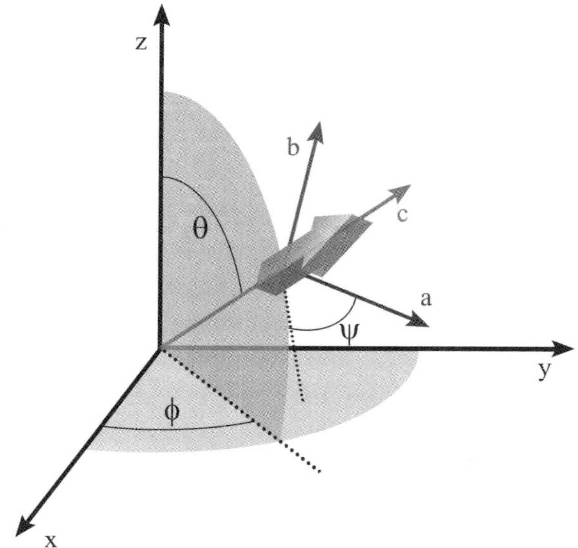

Figure 2. Laboratory (x, y, z) and molecular (a,b,c) reference frames for the coordinate transformation of Eq. (11). The orientation of the molecule is described by the angles (θ,φ,ψ).

Eqs. (8) (or (10), for SHG) and (11) show how it may be possible to obtain information about the orientation of interfacial molecules with SFG spectroscopy (or SHG). By changing the polarization of the input and output beams, all nonvanishing elements of $\chi^{(2)}$ can be determined, using the known experimental geometry and the Fresnel factors. With a model for the hyperpolarizability tensor of the molecule, $\alpha^{(2)}_{\xi\eta\zeta}$, the $\chi^{(2)}$ elements can be calculated as a function of the molecular orientation (θ,φ,ψ). Comparing the calculation to the measured $\chi^{(2)}_{ijk}$ elements, the orientation may be deduced.

As a simple example, we consider SHG from a monolayer of rod-like molecules that are cylindrically symmetric along the symmetry axis c, such that there are only two nonvanishing independent elements in $\alpha^{(2)}$, $\alpha^{(2)}_{aac} = \alpha^{(2)}_{bbc}$ and $\alpha^{(2)}_{ccc}$. For this case, we find from Eq. (11) for an azimuthally isotropic surface (ψ is not relevant due to the cylindrical symmetry of the molecule, and the average has been performed for an isotropic distribution in φ):

$$\chi_{xxz} = \chi_{yyz} = \frac{1}{2} N_s \alpha \left[\langle \cos\theta \rangle (1+r) - \langle \cos^3\theta \rangle (1-r) \right]$$

$$\chi_{xzx} = \chi_{yzy} = \chi_{zxx} = \chi_{zyy} = \frac{1}{2} N_s \alpha \left(\langle \cos\theta \rangle - \langle \cos^3\theta \rangle \right)(1-r)$$

$$\chi_{zzz} = N_s \alpha \left[r\langle \cos\theta \rangle + \langle \cos^3\theta \rangle (1-r) \right] \tag{12}$$

where $\alpha = \alpha_{ccc}$, $r = \alpha_{aac} / \alpha_{ccc}$, and θ is the polar angle of the symmetry axis c with respect to the lab z-axis. The χ elements can be deduced from SHG measurements with three different input and output polarization combinations, as shown by Eq. (10). Since an absolute determination of $N_s\alpha$ is usually not of interest, we can determine from the measurements two ratios of independent nonvanishing χ elements. Then, from Eqs. (12) we can find the orientation θ and the hyperpolarizability ratio r of the molecule by assuming a very narrow (δ-function) distribution for θ. It should be noted that Eqs. (12) also highlight another interesting feature of SHG and SFG spectroscopies: the phase of $\chi^{(2)}$ contains information about the absolute orientation of surface molecules. This can easily be seen in Eq. (12), since the sign of $\cos\theta$ (and therefore of $\chi^{(2)}$) changes if the molecules are pointing upward or downward at the interface.

For the analysis of the vibrational spectrum by SFG, we must examine the microscopic expressions for $\alpha^{(2)}$, which can be calculated by standard second-order perturbation theory [Ref. 11, pg 17]. From the microscopic expressions it can be seen that if one of the frequencies ω, ω_1 or ω_2 coincides with a vibrational or electronic transition, energy denominators become small and resonant enhancement of $\alpha^{(2)}$ is attained (Eqs. (13)). In the case of IR-visible SFG, we are concerned with vibrational resonances of the surface molecules. With the IR frequency (ω_2) near a vibrational transition from ground to first exited vibrational level, $\alpha^{(2)}$ (and therefore $\chi^{(2)}$, by substituting $\alpha^{(2)}$ in Eq. (11)) can be written as[26]

$$\alpha^{(2)} = \alpha^{(2)}_{NR} + \sum_q \frac{\alpha_q}{\omega_2 - \omega_q + i\Gamma_q}$$

$$\chi^{(2)} = \chi^{(2)}_{NR} + \sum_q \frac{\chi_q}{\omega_2 - \omega_q + i\Gamma_q} \tag{13}$$

where the subscript NR refers to a nonresonant contribution due to all other resonances away from the IR range in consideration, and α_q (χ_q), ω_q, and Γ_q denote the strength, resonant frequency and width of the qth vibrational mode, respectively. The strength of each resonance α_q can be expressed in the usual notation of Raman and IR spectroscopy as

$$\alpha_{q,\xi\eta\zeta} = \frac{1}{2\omega_q} \frac{\partial \alpha_{\xi\eta}^{(1)}}{\partial Q} \frac{\partial \mu_\zeta}{\partial Q} \Delta\rho_q \tag{14}$$

where $\dfrac{\partial \alpha_{\xi\eta}^{(1)}}{\partial Q}$ and $\dfrac{\partial \mu_\zeta}{\partial Q}$ are the Raman polarizability derivative and IR dipole moment derivative for the normal mode Q, respectively, and $\Delta\rho_q$ is the population difference between ground and excited vibrational states for the qth mode. Eqs. (13) and (14) have important consequences. First of all, the resonant enhancement of $\alpha^{(2)}$ as an infrared laser is scanned over vibrational resonances gives the surface vibrational spectrum. Since the SFG signal is proportional to $|\chi^{(2)}|^2$, Eq. (13) also demonstrates a feature of nonlinear spectroscopic methods that are very different from their linear counterparts: there is *interference* of the resonant contribution with the nonresonant background, leading to changes in the spectral lineshape which depend on both the presence of nearby resonances and also on the magnitude and phase of $\chi_{NR}^{(2)}$ – this nonresonant background may be complex, in general, although it should become real if ω and ω_1 are away from resonances in any medium. This interference leads to a more complicated lineshape than in the case of usual IR or Raman spectroscopy. An example is schematically illustrated in Figure 3 for real and positive χ_q and χ_{NR}. Therefore, in general one can only extract parameters such as resonant frequencies, linewidths and mode strengths by fitting the spectrum to a theoretical expression for $\chi^{(2)}$ such as Eq. (13). In particular, the peak SFG intensity only occurs at the resonance ($\omega_2 = \omega_q$) and is proportional to the square of the mode amplitude χ_q if the nonresonat background is negligible. Furthermore, if a mode is broadened but keeps the same amplitude χ_q, the area under the peak is *not* constant, as it would be in conventional IR or Raman spectroscopy. Another consequence of Eq. (14) is that a vibrational mode must be both IR- and Raman-active in order to be observable in the SFG spectrum. This is equivalent to the requirement that a molecule must not have an inversion center to have nonvanishing $\alpha^{(2)}$.[27]

Finally, if both ω and ω_1 are away from electronic resonances, $\dfrac{\partial \alpha_{\xi\eta}^{(1)}}{\partial Q}$ is symmetric upon interchange of indices ξ and η,[27] resulting that the first two indices of $\chi_{ijk}^{(2)}$ are interchangeable.

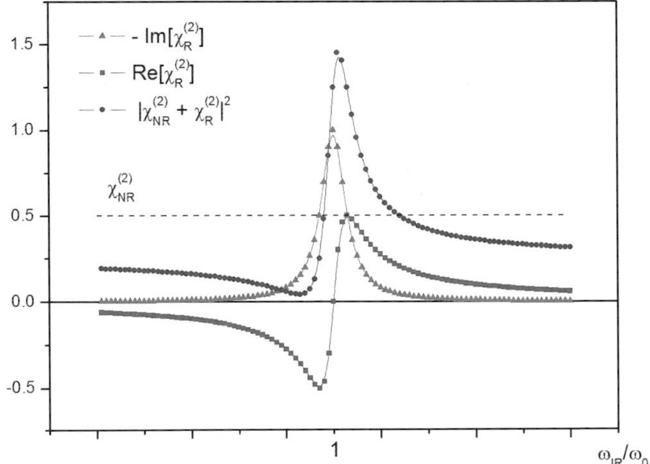

Figure 3. Plot of a simulated lineshape for a vibrational resonance in SFG spectroscopy with real and positive χ_q and χ_{NR} values. χ_{NR} is shown as a horizontal dashed line, together with the real ($\blacksquare$) and imaginary ($\blacktriangle$) parts of the resonant contribution, χ_R, and the SFG intensity ($\bullet$). ω_0 is the resonance frequency of the vibrational mode.

2.2. *Capabilities and Advantages*

The theoretical framework described in the previous section indicated a few features of SHG and SFG spectroscopy that make them useful tools for investigating interfaces. In this section we summarize them, point out other capabilities of these techniques and highlight their advantages with respect to other methods.

- *Surface specificity*. If the electric dipole contribution dominates (the great majority of experimental systems), the SFG or SHG signals come only from the interface, where the inversion symmetry is broken. Therefore, these techniques intrinsically discriminate the interfacial contribution from that of the bulk material. Consequently, they are among the few techniques available to investigate the free surface of a neat liquid.[28] One question that always arises is how deep into the interface would SHG/SFG spectroscopy be probing. Since the techniques rely on breaking the inversion symmetry, the answer depends on the particular system investigated. For instance, it could be just a monolayer, as in the case of the surface of a surfactant solution, where the adsorbed monolayer is oriented with chains away from the liquid, but the molecules in solution have random orientations. However, for charged interfaces in contact with an electrolyte, the E-field within the

194

electrical double layer breaks the inversion symmetry near the interface, and the whole depth of the double layer (typically up to tens of nm) may contribute to the signal. This has been clearly demonstrated in the case of the OH stretches of interfacial water.[29] Such a feature is interesting for probing the interfacial electric fields in organic devices, and will be further discussed below. Finally, for a solution of chiral molecules there is no inversion symmetry, since inversion would change the handedness of the molecules. In this case, the SFG signal may be generated from the bulk solution (up to the coherence length for the particular experimental configuration – usually several hundred nm, for the reflection geometry).[30]

- *Vibrational or electronic spectroscopy.* As outlined in the previous section, when one of the frequencies of input or output beams is resonant, the surface nonlinearity $\chi^{(2)}$ is enhanced, so that tuning the resonant beam yields a surface spectrum. For SFG spectroscopy, one of the beams is tuned in the mid-infrared to obtain a vibrational spectrum of surface molecules, with the other input beam in the visible or near infrared region, but usually off electronic resonances, as shown schematically in Figure 4(a). As in any vibrational spectroscopy, information on the molecular arrangement at the interface is obtained indirectly by the interpretation of SFG spectra. For instance, the frequency/linewidth of vibrational resonances could be related to the interaction of molecular moieties with neighboring molecules, such as in the case of H-bonding. The SFG output may be further enhanced if the other input beam is also resonant with electronic transitions (Figure 4(b)). In this case, the doubly-resonant (DR-SFG) spectrum shows different enhancements for each vibrational mode, revealing information on the coupling between electronic and vibrational transitions at the surface. Lastly, the electronic spectrum of the interfacial molecules may be obtained with SHG spectroscopy, where usually the input beam is tuned so that the

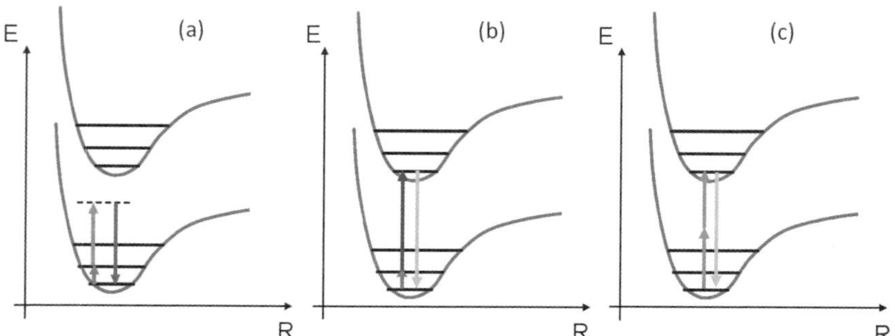

Figure 4. Schematic energy diagram showing the transitions involved in (a) SFG, (b) DR-SFG and (c) SHG spectroscopies.

second-harmonic output scans through the electronic resonances, as sketched in Figure 4(c).

- *Sub-monolayer sensitivity.* Since usually only a monolayer of molecules can contribute to SFG or SHG, in most examples in the literature these techniques are detecting one, or a fraction of a full monolayer. However, the usual dependence of the signal intensity on the *square* of the surface density makes it in most cases difficult to detect a small fraction of a monolayer. However, the sensitivity may be enhanced by using the interference with χ_{NR} (from the sample or introduced externally in a heterodyne detection scheme),[17] which makes the spectral feature depend linearly on surface density for a small resonant contribution.[31] However, it is usually quite challenging to detect coverages of the order of a few percent.

- *Determination of molecular orientation.* As discussed in section 2.1, Eq. (11) shows that with a model for the hyperpolarizability tensor of the molecule, $\alpha_{\xi\eta\zeta}^{(2)}$, the $\chi_{ijk}^{(2)}$ elements can be calculated as a function of the molecular orientation (θ,φ,ψ). Comparing the calculation to the measured $\chi_{ijk}^{(2)}$ elements, the orientation may be deduced. In the case of SFG, each vibrational mode has its own resonant amplitude tensor α_q, and based on the symmetry of the vibration it is possible to list its nonvanishing independent elements. As shown in Eq. (12), their ratios are important for a quantitative determination of molecular orientation, and they can be found from the Raman depolarization ratio or from theoretical calculations. Each mode may be associated with a particular moiety on the molecule. If $(\chi_q)_{ijk}$ can be obtained by fitting the SFG spectrum to Eq. (13), and $(\alpha_q)_{\xi\eta\zeta}$ is known, then according to Eq. (11) the average orientation of that moiety may be deduced. By selecting different vibrations, it may be possible to map the orientation of different parts of the molecule,[25] which may give more confidence on the deduced molecular orientation, or further indicate the molecular conformation at the interface (see discussion below). An additional input needed for such a quantitative analysis are the refractive indices of the monolayer and of the surrounding media (n', n_1 and n_2, respectively), at the input and output frequencies, as they determine the Fresnel factors L_{ii} that relate the measured effective susceptibilities to the $\chi_{ijk}^{(2)}$ elements. Since they may also be subject to some uncertainty, such a quantitative analysis has to be considered with care. However, the same critique applies to other optical techniques, such as infrared spectroscopy, which is sometimes used to obtain quantitative information on molecular orientation at interfaces.[32,33] In that case, the optical field at the interface is also related to the input field by the same Fresnel factors, a fact that is usually neglected in the orientation analysis using infrared spectroscopy. Finally, one unique capability of SFG and SHG is the possibility of determining the absolute orientation of an asymmetric molecule, since $\chi^{(2)}$ changes sign if the molecule has the same tilt angle at the interface, but points on average upward or downward. Interference techniques[17,34] can be used to determine the phase of $\chi^{(2)}$, and using a sample of known orientation, the absolute orientation of other samples can be deduced.

- *Probing molecular conformation.* One particularly interesting aspect of SFG spectroscopy in regards to monolayers of long-chain molecules is the ability to probe (at least qualitatively) the conformation of alkyl chains ($R-(CH_2)_n-CH_3$), a major component of surfactants, lipids and other biomolecules. As shown by Guyot-Sionnest *et al.*[35] for a monolayer of pentadecanoic acid ($CH_3(CH_2)_{13}COOH$) at the air/water interface, when it is compressed to a highly condensed phase, where the alkyl chains are known to be in the *all-trans* conformation and nearly vertical, the SFG spectrum in the CH stretch range is dominated by the terminal CH_3 group, even though there is usually an overwhelmingly larger number of CH_2 groups along the alkyl chain (see Figure 5(a)). The CH_2 groups are arranged in a zigzag configuration with adjacent groups pointing in opposite directions, leading to inversion symmetry of their arrangement and cancelation of their contribution. On the other hand, the CH_3 groups are well ordered, pointing away from the subphase and giving a strong contribution to the SFG spectrum. As the monolayer is expanded (Figure 5(b) and 5(c)), thermally activated *gauche* conformations appear, breaking the symmetry of the CH_2 arrangement and making the CH_3 more widely distributed. This leads to an increase of the CH_2 stretches and a reduction of the CH_3 symmetric stretches ($<\cos \theta>$ in Eq. (12) decreases). Therefore, the ratio of their amplitudes can be used to quantify the relative conformational order of alkyl chains.[36,37] We should note that in the limit of a large number of gauche defects in the alkyl chains the orientations of both CH_2 and CH_3 groups may become random, so that all peaks should become vanishingly small. From this discussion, it is clear that even such qualitative analysis of the SFG spectra of alkyl chains in the CH stretch region can give very important information about chain conformation at interfaces.

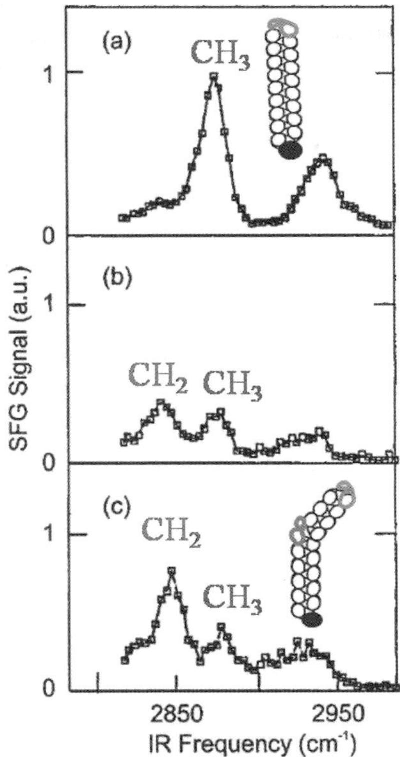

Figure 5. SFG spectra for a pentadecanoic acid Langmuir film at various surface densities: (a) 22 Å^2/molecule; (b) 34 Å^2/molecule; (c) 47 Å^2/molecule. The polarization combination is SSP. The inset indicates an (a) *all-trans* conformation of the alkyl chain and (c) a chain with one *gauche* defect. Adapted from reference [35].

- *Sensitivity to interfacial electric fields.* An important aspect of SHG and SFG to applications in organic electronics is their ability to probe interfacial electric fields. They are always present near charged surfaces immersed in an electrolyte (electrochemical interfaces, lipid membranes, etc), or may arise due to charge transfer at a metal/semiconductor interface. Such E-fields are usually very strong (10^6 to 10^8 V/m) and confined to a thin region near the interface (a few tens on nm). This electric field breaks the inversion symmetry within the interfacial region, leading to an enhanced effective second-order susceptibility, $\chi_{eff}^{(2)}$. There are two mechanisms that may contribute to such an enhancement:[18] i) a net orientation (alignment) of dipolar molecules by the interfacial electric field, and ii) a coupling of the DC interfacial field E_0 with the third-order susceptibility of the bulk material, $\chi^{(3)}$, which is allowed even in centrosymmetric media. The first may be important in electrolytes, while in a solid the latter dominates. The second-order polarization can then be written as

$$\mathbf{P}^{(2)}(\omega = \omega_1 + \omega_2) = \chi_I^{(2)} : \mathbf{E}_1(\omega_1)\mathbf{E}_2(\omega_2) + \chi_F^{(2)}\left(\mathbf{E}_0\right) : \mathbf{E}_1(\omega_1)\mathbf{E}_2(\omega_2)$$
$$+ \chi^{(3)} : \mathbf{E}_1(\omega_1)\mathbf{E}_2(\omega_2)\mathbf{E}_0 \tag{15}$$

where the $\chi^{(2)}$ term was separated in a field-independent part, $\chi_I^{(2)}$, and another contribution $\chi_F^{(2)}$ due to molecular alignment, which is dependent on the interfacial field E_0 (assumed to be proportional to E_0 in the small field regime). The generated SFG (or SHG) signal will be proportional to the square of the total effective second-order polarization $\mathbf{P}_{eff}^{(2)}$, which should be integrated throughout the interfacial region. If its thickness is much less than the coherence length , this integration leads to a dependence of $\mathbf{P}_{eff}^{(2)}$ on the surface potential Φ_0 with respect to the bulk as shown by Eq. (16), with $\chi_{eff}^{(3)}$ including the contributions of both $\chi_F^{(2)}$ and $\chi^{(3)}$.

$$\mathbf{P}_{eff}^{(2)} = \chi_I^{(2)} : \mathbf{E}_1(\omega_1)\mathbf{E}_2(\omega_2) + \chi_{eff}^{(3)} : \mathbf{E}_1(\omega_1)\mathbf{E}_2(\omega_2)\Phi_0 \tag{16}$$

Therefore, both SHG and SFG can be used to probe interfacial electric fields, a feature that is quite interesting for applications in physical chemistry[38] and organic devices.[39] In the latter case, both in-plane and normal components of E_0 may be probed by selecting the appropriate polarizations for the input and output beams.

- *Wide applicability.* As optical techniques, SHG and SFG spectroscopies have wide applicability and can be used to investigate all interfaces accessible by light, including solid/liquid, solid/solid and liquid/liquid interfaces. The only requirement is that the beams must reach the interface. However, since usually at least one of them is resonant and experiences high absorption, one of the media must be transparent to that beam. Although this imposes some restrictions (for example, applications of SFG spectroscopy to liquid/liquid interfaces are rather limited,[28] due to strong infrared absorption in liquids and the fact that the liquid layers cannot be made very thin), they are not as severe as in the case of UHV and scanning probe techniques, which usually cannot be applied to liquid surfaces or solid/solid interfaces.

- *Ultrafast dynamics at interfaces.* As we will see in the next section, usually ultrashort laser pulses are used in the implementation of these techniques. Therefore, it is also possible to investigate surface dynamics in the picosecond or femtosecond timescales, both for electronic[40,41] and vibrational processes.[42] A few examples will be shown in section 3.

In summary, SHG and SFG vibrational spectroscopy have some unique advantages and capabilities as surface analytical tools. They are intrinsically surface specific and therefore able to obtain surface spectra even when the bulk is resonantly absorbing, as in the study of a neat liquid surface. They also have submonolayer sensitivity and as optical techniques can probe any interface accessible by light, including buried interfaces. Quantitative information on molecular orientation at interfaces can also be deduced, and their sensitivity to interfacial electric fields is particularly relevant to organic electronics. In the case of vibrational spectroscopy of alkyl chains, we have shown how SFG provides very direct qualitative information about chain conformation. This kind of information is not easily obtained by other techniques in such a direct way. We will now briefly outline the experimental requirements for SHG and SFG spectroscopies.

2.3. *Experimental Implementation*

SHG experiments are usually not focused on obtaining (electronic) spectral information, but mainly intend to detect how the broken symmetry at the interface (due to molecular alignment or the presence of DC electric fields, for example) changes with experimental parameters. In such case, only a fixed incidence wavelength is needed, so that the experiments can be performed with one pulsed excitation laser at an appropriate wavelength. Usually ultrashort pulses are used (from 100 fs to tens of ps) to attain high peak intensities at the sample (typically of the order o 0.1 to 10 GW/cm^2), but with a low repetition rates (tens of Hz to a few kHz) to avoid average heating and damage to the samples (typical average powers of a few mW only).

For SFG experiments, a laser system capable of producing synchronized pulses in the visible and mid-infrared (IR) is needed. This is usually accomplished by combining harmonic generation of the laser output and pumping an optical parametric generator/amplifier (OPG/OPA) system for infrared generation. The most popular laser systems used to date can be generally divided in two categories: (i) ps pulse duration and low repetition rate (~10 Hz), where the SFG spectrum is obtained by scanning the frequency of the mid-IR pulse generated by the OPG/OPA and measuring the intensity of the SFG signal, and (ii) ~100 fs pulse duration and higher repetition rates (~1kHz), where the bandwidth of the mid-IR pulse produced by the OPA is large enough to allow measuring the whole SFG spectrum at once by mixing the mid-IR pulse with a ps visible pulse and detecting the SFG signal on a multichannel spectrometer. These latter systems have the advantage that laser intensity fluctuations do not affect the quality of the SFG spectra, generally yielding higher signal-to-noise data. In our experiments we used a commercial ps SFG spectrometer (Ekspla, Lithuania). The readers are referred to the cited references to obtain details of their particular SHG and SFG setups.

3. Applications to Organic Electronics

In this section we will briefly describe selected applications of SHG and SFG spectroscopy to investigate several interfaces that are relevant to organic electronics, including studies of functional devices under operation. This highlights the capability of *in situ* and real time probing of interfaces by these nonlinear optical techniques. First we discuss investigations of interfaces involving materials useful for organic electronics, but not in a device architecture. In the following sections we then describe studies of organic devices, such as diodes, solar cells and field-effect transistors (OFETs).

3.1. *Structure and Charge Transfer at Interfaces*

Understanding the structure of conjugated polymers or small molecules at dielectric and metal interfaces is quite important to rationally improve the performance and lifetime of organic devices. Several recent studies using SHG and SFG spectroscopy try to correlate the interfacial structure to device performance, particularly in the case of OFETs. These will be discussed in more detail in section 3.4. Here we describe a few studies that focused on getting information on the structure of conjugated molecules at various interfaces, but without necessarily building a device and relating the molecular structure to its performance.

Geiger and Marowsky[43] have observed SHG from thin films of oligothiophenes. Although the molecules are centrosymmetric, they demonstrated that the signal was generated only by the first monolayer of molecules, probably due to their interaction with the silica substrate, which breaks the inversion symmetry of the molecular charge density. They also showed that $\chi^{(2)}$ increases with the third power of the conjugation length and that the orientation of the oligomers at the interface depend on whether they are "end-capped" or not. Chou *et al.*[44] have used doubly-resonant IR-vis SFG to obtain the spectrum of the electronic transitions coupled to the C-C stretches of the conjugated backbone of MEH-PPV, both at the air/polymer and polymer/solid(CaF_2) interfaces. Modeling the electronic spectra by a contribution from a Gaussian distribution of oligomers, they have obtained an average conjugation length of 5.8 and 5.1 at the polymer/solid and air/polymer interfaces, respectively. Such a difference was attributed to a more rigid surface confinement at the polymer/ solid interface.

Vibrational SFG spectroscopy was recently used by several groups to probe the orientation of interfacial layers in conjugated polymer and small molecule films. Massari and co-workers[45] have qualitatively investigated the orientation of the perylene derivative PTCDI-C8 vapor-deposited on bare silica and on silica covered by an alkylsilane monolayer. They found that the SFG signal had contributions from both the outer air/film and the inner film/substrate interfaces. Their analysis showed that the as-deposited films on both substrates are similar. However, upon thermal annealing the outer surfaces became significantly reordered, the inner interface was not affected very much for the monolayer-covered silica substrate, while for the bare silica the PTCDI-C8 molecules became more inclined towards the substrate. Benderskii and colleagues[46] have also investigated the effect of thermal annealing on the orientation of the conjugated polymer P3HT spin-coated on silica or aluminum films (partially oxidized - AlO_x). As shown in Figure 6, they concluded that the thiophene rings adopt a more edge-on orientation upon annealing on the silica substrate, while on aluminum the rings are initially more upright, tilting towards the substrate after annealing. We have also investigated the structure of the conjugated polymers poly(9,9'-dioctylfluorene) – PF8 and P3HT in contact with metals (Au and Al).[47] Two different fabrication methods for the metallic contact were investigated: metal thermally evaporated over spin-coated polymer film, and polymer spin-coated over metal film previously evaporated on glass. We found significant differences in the molecular orientation at both interfaces, and discussed how the molecular structure at the interface may affect the spontaneous charge transfer to the metal. Another interesting example is the study of films from PEDOT:PSS,[48] which is a water-soluble conducting polymer complex that is largely used as a transparent conductor and hole-injection layer in polymeric devices. They have determined quantitatively the orientation of the phenyl groups of PSS in pristine films and in complexes with PEDOT, finding both its tilt and twist angles, and concluded that the orientations were quite similar in both samples.

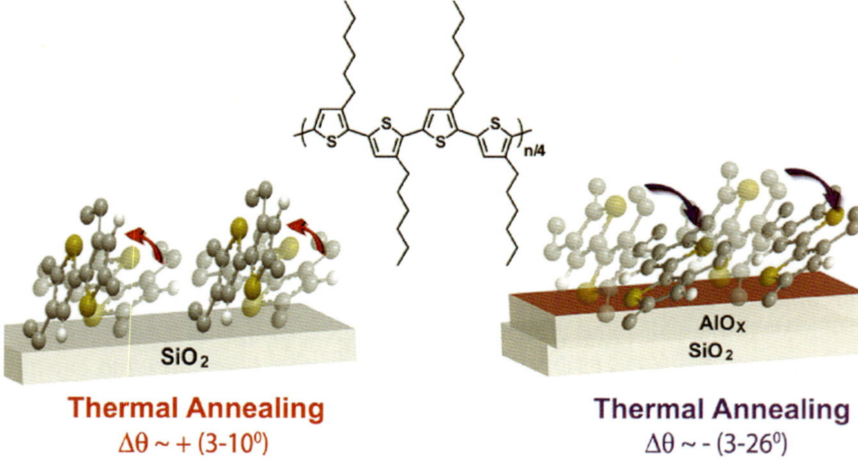

Figure 6. Annealing-induced changes in the ring orientation of P3HT molecules on silica substrates (left) and on partially oxidized aluminum films over silica (right). The light gray and dark-colored molecules represent the orientation before and after annealing, respectively. Adapted from reference [46].

Another very important system in organic electronics are self-assembled monolayers (SAMs). They can be used as molecular wires, if conjugated, or insulating layers, in the case of non-conjugated molecules such as long chain alkylthiols or alkylsilanes. These non-conjugated SAMs may also be used as conditioning layers on substrates, affecting the structure of subsequent active layers of devices.[4,45,49] The structure and conformational order of these SAMs can be sensitively probed by SFG spectroscopy.[50] However, we should stress that SFG can probe them even at buried solid/solid interfaces.[49] For example, Asanuma *et al.*[51] have investigated the conformation of alkyl SAMs at the Si/gold interface, forming a metal-insulator semiconductor (MIS) junction. They concluded that gold overlayers introduce significant *gauche* defects within the initially highly oriented alkyl chains, and the effect depends on the gold deposition method (vacuum evaporation vs sputtering) and the alkyl chain length.

Regarding the doping or charge transfer to conjugated molecules, as early as in 1993 Aktsipetrov *et al.*[52] have detected an increase in SHG signal from polyaniline films near the metal-insulator transition, which they attributed to either a doping-induced resonant SHG or to a break in molecular inversion symmetry by the formation of bipolarons. More recently, SFG spectroscopy was used to investigate the surface of polyaniline films in its conducting form emeraldine salt (ES).[53] It was found a strong SFG signal, indicating that the chains are non-planar and nearly vertical to the surface, based on the polarization dependence of the SFG spectra. Resonances characteristic of benzenoid rings and polarons could be detected.

SHG has also been used to probe ultrafast charge transfer at donor-acceptor interfaces. X-Y. Zhu and co-workers[41] have used a combination of femtosecond time-resolved SHG and two-photon photoemission (2PPE) to probe the formation and relaxation of charge-transfer (CT) excitons at a copper phtalocyanine-fullerene (CuPc/C_{60}) interface, as a model system for organic photovoltaics. These CT excitons are believed to be intermediate states between the photoexcited singlet exciton (S_1) in the donor molecule (CuPc, in this case) and the free charges in the materials (hole in CuPc and electron in C_{60}), with a possible dissociation mechanism being due to excess energy after photoexcitation (electronic and/or vibrational), the so-called hot CT excitons. However, a definite experimental proof of this state, the nature of its relaxation and the role of excess energy in charge generation has been elusive, in part because of difficulties of femtosecond transient absorption spectroscopy in assigning spectral features to CT excitons and attaining enough surface sensitivity to discriminate this state from bulk signals. In their work, Zhu and colleagues used time-resolved 2PPE, which is also highly surface sensitive, to probe the energy spectrum and the dynamics of formation and relaxation for the CT exciton in very thin bilayer systems (CuPc/C_{60} and C_{60}/CuPc). In turn, a definite proof of the CT nature of this state came from time-resolved SHG measurements, which probed the interfacial electric field induced by charge separation after photoexcitation. The results are shown in Figure 7(a), together with a scheme of such experiment. Changes induced by pump pulse in the SHG intensity of the probe beam are measured as a function of the pump-probe delay, therefore measuring the formation dynamics of the CT state. It was found that for excitation within the S_1 absorption band of CuPc, the CT excitons are formed in a timescale $\tau \cong 80$ fs, while instantaneous (< 20 fs) direct excitation of the CT exciton is possible just below the onset of the S_1 absorption. Similar results were obtained for a C_{70} acceptor, with $\tau \cong 150$ fs. It is interesting to note that there is no appreciable decay of the induced SHG signal in a several ps, indicating that charge recombination is negligible in this timescale. The inset in Figure 7 (b) shows a scheme of the energy levels and the transitions involved in the charge photogeneration for this system, indicating the timescales for CT formation (~ 80 fs) and for CT energy relaxation (~ 1 ps) , as measured by 2PPE (Figure 7(b)).

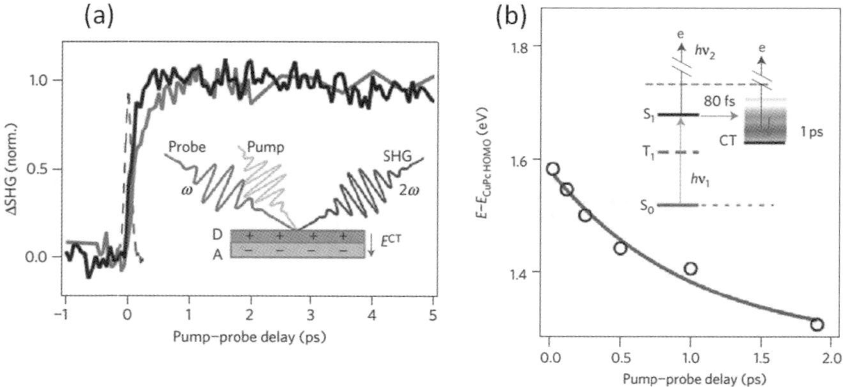

Figure 7. (a) Time-resolved SHG measurements for the CuPc/fullerene interface, with excitation at the S_1 band of CuPc. Dark curve is with C_{60} and lighter one is with the C_{70} acceptor. The inset illustrates the concept of this experiment. (b) Position of the CT state measured by time-resolved 2PPE as a function of the pump-probe delay. The inset shows the scheme of the 2PPE experiment (excitation by hv_1 and ionization by hv_2), and the energy levels and transitions involved in the charge photogeneration process. Adapted from reference [41].

The same group has also applied time-resolved 2PPE and SHG to investigate the route from singlet state fission to multiple electron transfer in pentacene/C_{60} bilayers.[54] Using 2PPE they followed the energetics and dynamics of the multiple exciton (ME) state formation after photoexcitation of S_1. This ME state can decay into a CT state or a pair of triplet states (T_1), which in turn also decay into CT exciton state (see Figure 8). As in the previous example, SHG was crucial in assigning the observed spectral features in 2PPE to the formation of a CT state, which produced an interfacial electric field that enhanced the SHG signal.

We should emphasize that these examples evidence the advantage of SHG and SFG to probe buried (solid/solid) interfaces, which are critical to organic electronic devices. In contrast, X-ray scattering has been used to investigate the organic semiconductor/dielectric interface in OFETs,[4,49,55] but in fact it probes the whole (crystalline) part of the organic thin film. This has been pointed out recently by a study of a free-standing film of a molecular semiconductor, combining X-ray scattering and X-ray absorption (NEXAFS).[56] The latter has been measured both in transmission (bulk sensitive) and by electron yield (surface sensitive), with different molecular orientations obtained in each case, and only the bulk orientation was consistent with X-ray scattering data. Therefore, the nonlinear optical techniques have the advantage of selectively

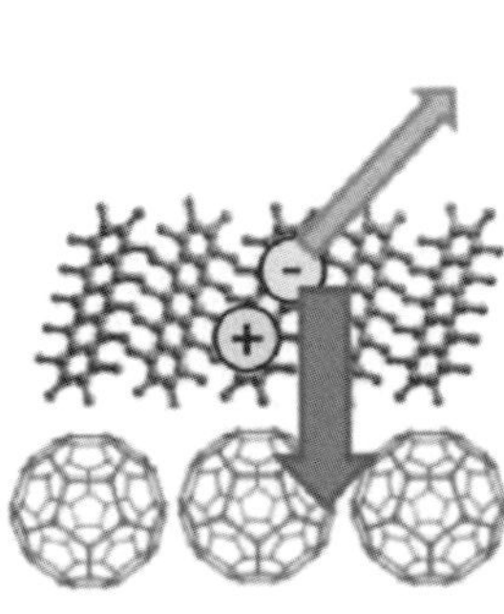
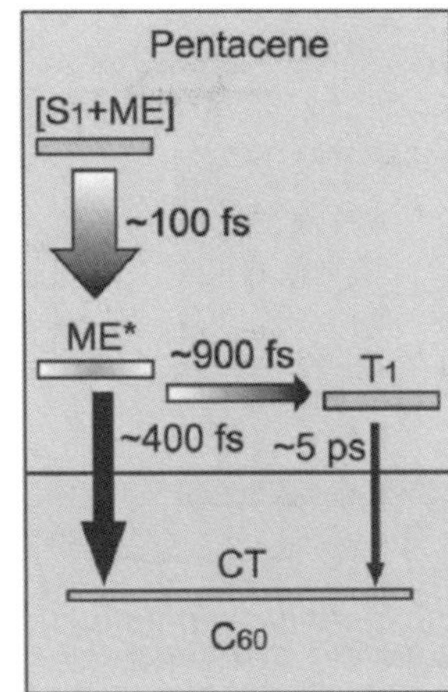

Figure 8. Energy levels and timescales for the transitions involved in the singlet fission process, leading to multielectron transfer in pentacene/C_{60} bilayers. ME^* is the multiple exciton state, T_1 is the triplet state and CT is the charge transfer state. Adapted from reference [54].

probing the interfacial molecules even at buried interfaces, without any bulk contribution, and recent studies are bringing new insights into how the control of interfacial structure may affect device performance (see section 3.4).

3.2. *Interfaces in Organic Diodes*

The Iwamoto group pioneered the application of Electric-Field Induced SHG (EFISHG, as described at the end of section 2.2) to probe electric fields within organic devices. By choosing the appropriate excitation wavelength so that the second-harmonic is resonant with molecular electronic transitions, different molecular layers in a device can be probed selectively, and it is therefore possible to investigate how the E-field changes with bias within each layer of an OLED[57,58] or a diode with blocking insulating layers.[59,60] When a double-layer diode is biased and a steady state current flows through the device, each layer behaves as an RC circuit and charges accumulate at the electrodes and at the interface between two dielectrics (Q_s), which is known as the Maxwell-Wagner (MW) effect, as illustrated in Figure 9. The characteristic time for charging Q_s is the Maxwell-Wagner time constant τ_{MW}, which depends on the dielectric properties of the two materials.[61] As the device current varies, Q_s and the electric fields within each organic layer change, and this can be detected by EFISHG.

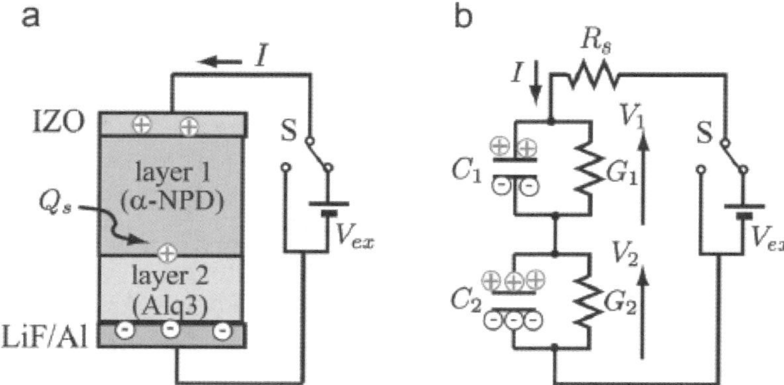

Figure 9. (a) Schematic view of a double-layer diode, indicating the accumulated charges at the electrodes and at the interface between the organic layers, Q_s. (b) Equivalent circuit model for the MW effect. Adapted from reference [61].

They have also developed a time-resolved (TR-SHG) setup to observe the dynamics of the electric fields in devices after switching the external bias, in the ns to ms range. As sketched in Figure 10(a) for an OLED,[57,61,62] the excitation laser pulses are synchronized with a square wave voltage that is applied to the device. The dynamics of the electric fields in the device are obtained by measuring the EFISHG intensity as a function of the time delay t_d between the square wave and the laser pulse train. A typical result is shown in Figure 10(b), where the charging and discharging of an OLED are followed in real time from a few ns to a few ms after bias changes. From the measured time-dependent SHG signal it is possible to obtain the electric field in the α-NPD hole-transporting layer and the charge density Q_s accumulated at the α-NPD/Alq$_3$ interface. The charging and discharging times of Q_s correlate well with the dynamics of electroluminescence (EL) in the device, explaining the reason for the asymmetric temporal behavior during EL turn on and turn off. In another report,[62] Iwamoto's group has proposed and verified a model for explaining two regimes of EL in a similar OLED, at low and high frequencies of operation. The EFISHG results show that accumulated holes at the α-NPD/Alq$_3$ interface (Q_s) suppress hole injection at low frequencies, reducing the EL due to recombination of injected electrons and holes from opposite electrodes. At high frequencies, Q_s assists electron injection from the cathode, yielding an increase in EL intensity due to their recombination with the accumulated holes at the interface.

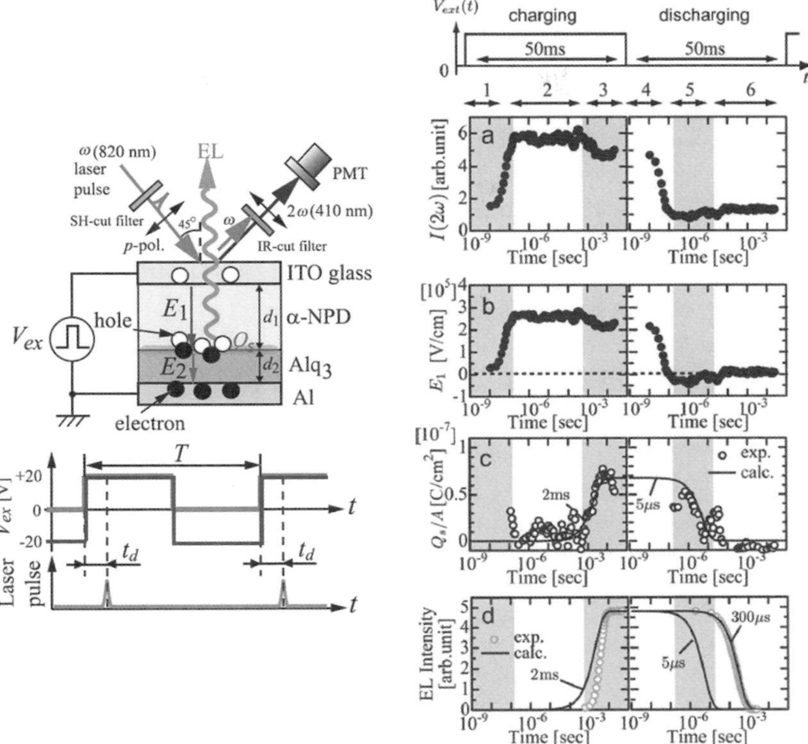

Figure 10. (left) Experimental setup for TR-SHG. The SHG intensity is measured as a function of the time delay t_d between the bias voltage rise and the incoming laser pulse. It is also possible to simultaneously measure the EL for the OLED. (right) SHG transient (a) on the OLED shown on left, from which is possible to deduce the electric field in the α-NPD layer (b) and the accumulated charge density at the α-NPD/Alq$_3$ interface (c). Curve (d) is EL transient. Adapted from references [61] and [62].

The TR-SHG technique has also been applied to diode structures containing an organic semiconductor and an insulating blocking layer, such as ITO/polyimide/pentacene/Au, where polyimide (PI) is the blocking layer. Both the dynamics of charging and discharging[59,63] and the change of field distribution upon bias[60] have been studied. In an interesting observation, Lim *et al.*[59] have observed that freshly prepared samples yielded SHG transients showing charging at the electrodes and at the semiconductor/insulator interface, while aged samples only displayed electrode charging. They attributed the absence of interfacial charging (Q_s) to reduced carrier injection in the semiconductor due to increased charge trapping. This shows the potential of this technique to probe interfacial degradation in organic devices. Very recently, Taguchi *et al.*[64] have investigated the hysteresis in the capacitance-voltage characteristics of IZO/PT/C_{60}/pentacene/Au diodes, where polyterpenol (PT) acts as hole-transporting and electron-blocking layer. The TR-SHG transients showed that charges accumulate at the PT/C_{60} (electrons) and at the C_{60}/pentacene (holes) interfaces, and the resulting space charge field is responsible for the hysteresis loop observed in the capacitance measurements.

Finally, SFG vibrational spectroscopy has also been applied to investigate an OLED device under operation.[65] In this case, the SFG signal is doubly resonant (DR-SFG) and the intensity is also enhanced by the $\chi^{(3)}$ contribution, giving rise to an electric-field-induced-SFG spectroscopy, the vibrational analogue of EFISHG. It was therefore used to probe how the electric filed distribution changed with bias in a complex device, comprised of 5 organic layers and two electrodes. By means of the characteristic vibrational spectrum of each organic semiconductor, it was possible to discriminate among their contributions and follow the field concentration change from one layer to another upon reversing the bias.

3.3. *Interfaces in Solar Cells*

In the field of organic solar cells, TR-SHG has recently been used by the Iwamoto group to investigate the internal electric field, interface charging and carrier relaxation times in pentacene/C_{60}[66-68] and copper-phtalocyanine/C_{60}[69] double-layer organic solar cells. By choosing the appropriate laser wavelength, the SHG was resonant only with C_{60} molecules, so that EFISHG in these cases was selectively probing the electric field within the C_{60} layer. For the copper-phtalocyanine/C_{60} device, it was found that the organic heterojunction acquired a negative interfacial charge Q_s upon illumination, and it decayed in the dark via a two-step process, one with a time-constant of the order of tens of μs, and another in the tens of ms range. In contrast, for the pentacene/C_{60} solar cell, Q_s was positive under photoillumination and produced an interfacial voltage of 0.16 V that deformed the potential profile within the device.[68] This contribution of the interfacial charges to the field inside the C_{60} layer is opposite to that generated by the charges collected on the electrodes due to the photovoltaic effect, and its dependence on the external load resistance was investigated.[66] A combination of impedance spectroscopy and TR-SHG[67] has shown that the interface charge relaxation time decreased 70 times under illumination due to increased carrier density and, as expected, applying to the solar cell a bias equal to the open-circuit voltage suppressed the interfacial charging.

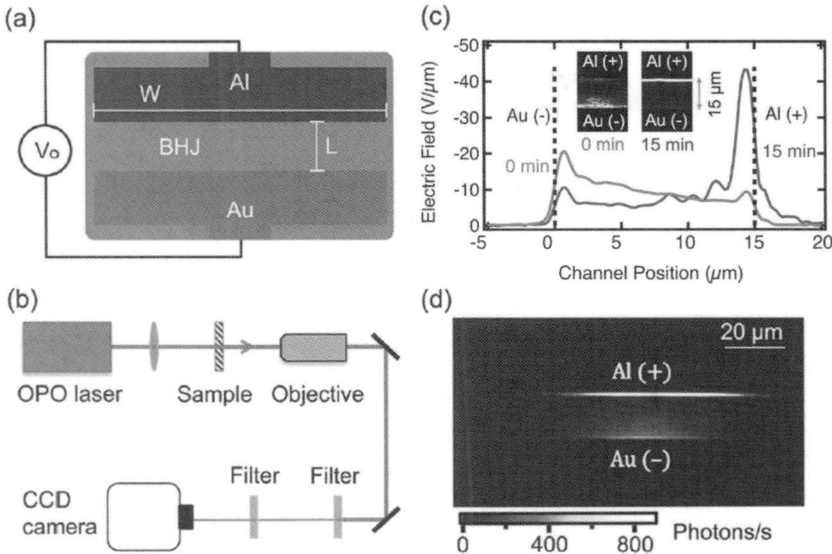

Figure 11. (a) Schematic top view of the lateral BHJ device. (b) Optical setup for EFISHG microscopy in transmission. (c) Cross section of the electric field distribution along the channel length L obtained from the SHG microscopy image in (d) for a P3HT:PCBM device. Two curves are shown, at the beginning of the measurement (0 min) and after 15 min of laser irradiation. Adapted from reference [70].

Very recently, Zhu and coworkers[70] have employed SHG microscopy to map the electric field distribution in a model bulk heterojunction (BHJ) photovoltaic device. They fabricated a lateral BHJ which consisted of a thin film of a blend of a conjugated polymer donor (P3HT or PSBTBT) and the C_{60} derivative PCBM as an electron acceptor. Phase separation within the blend forms the BHJ, where charges are photogenerated and driven to the electrodes by the built-in field produced by the difference in the work functions of the metallic contacts (or by an external bias). The electrodes are evaporated metal strips with a micrometric gap exposing the active BHJ (see Figure 11(a)). The experimental setup for SHG microscopy in the transmission configuration is sketched in Figure 11(b). The measurement in Figure 11(d) shows an image of the EFISHG signal, which is directly mapping the spatial distribution of the in-plane electric field in the BHJ film of P3HT:PCBM with a bias of − 200 V. The field distribution is plotted in Figure 11(c) for a pristine device and after 15 min of measurements, showing the buildup of a marked space charge near the Al electrode. This has been attributed to the creation of electron traps due to photo-oxidation of the polymer, which results in the formation of a space charge that screens the field near the electron-collecting electrode.

3.4. *Interfaces in Field-Effect Transistors*

Organic field-effect transistors (OFETs) are the most studied devices by the nonlinear optical techniques, partly due to their importance in the field of organic electronics, but also because the device architecture is simpler, with fewer organic layers to be probed, allowing therefore a more direct interpretation of the results. We will first describe EFISHG studies of OFETs by the Iwamoto group, and then illustrate recent applications of SFG spectroscopy and microscopy to probe organic FETs.

In 2005, Manaka *et al.*[71] have observed EFISHG from the channel of a pentacene FET. Using the fundamental and SHG beams polarized mostly along the channel direction (drain-source direction), the EFISHG signal was most sensitive to the in-plane bias field. They demonstrated that the SHG signal increased upon applying a V_{DS} bias in the off-state ($V_{GS} = 0$), but decreased when the device was turned on by applying a negative V_{GS}. This is because in the on-state, a conducting channel is formed by carrier injection, leading to space charge formation and a redistribution of the electric field within the FET channel. Interestingly, they also showed that treating the Si/SiO_2 substrate with UV/ozone was necessary to obtain devices with good electric characteristics and SHG signal modulation. The direct measurement by EFISHG of the electric field distribution within the device channel upon bias was reported in subsequent papers, first with a low spatial resolution (~ 20 μm) by scanning the laser spot across the channel,[39] and later with high resolution using an SHG microscopy setup similar to that shown in Figure 11 (but detecting the SHG signal in reflection).[72] In this case, they also mapped the conductivity within the channel using a near-field scanning microwave microprobe.

A time-resolved SHG microscopy technique has also been developed to visualize the carrier motion in FET channels.[73,74] It is a variant of the TR-SHG technique described in Section 3.2 (see Figure 10), except that instead of measuring the total SHG intensity, the signal is imaged onto a sensitive CCD camera by a microscope objective (as in Figure 11(b), but with the detection in the reflection direction). A scheme of the TR-SHG microscopy is shown in Figure 12(a) and (b). Varying the delay τ of the laser pulse with respect to the rise of the FET bias pulse and collecting the EFISHG image, it is possible to follow the time evolution of the electric field distribution in the device, from

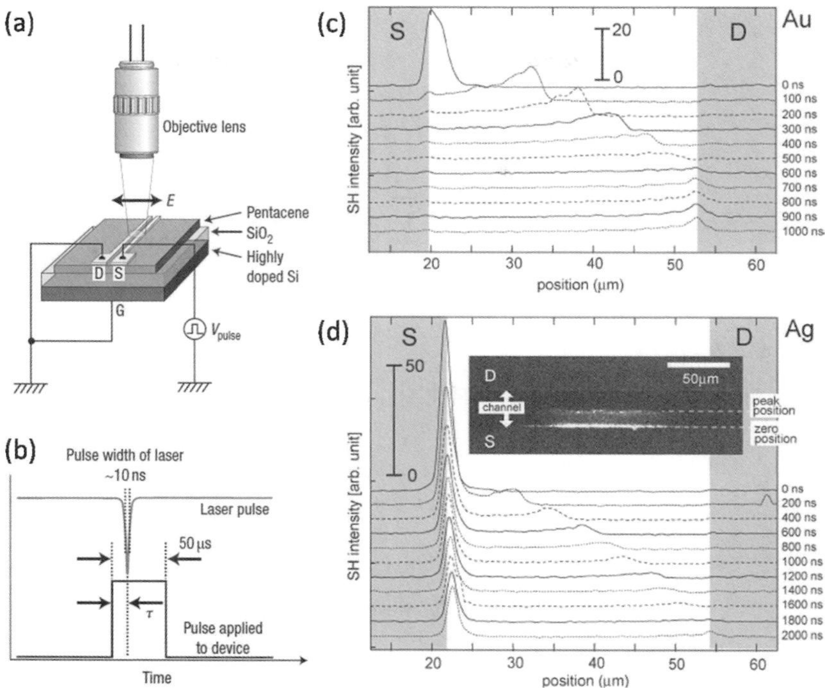

Figure 12. (a) Optical setup for TR-EFISHG microscopy in reflection and OFET device structure. (b) Timing between the applied bias pulse and the laser excitation, showing the variable delay τ. (c) and (d), cross section of the SH intensity along the channel length L from the images obtained at the delays indicated on the right, for Au and Ag electrodes, respectively, for a pentacene OFET. The inset in (d) shows an SHG image from which one curve in the graph is obtained (cross section along the channel). Adapted from references [73,74].

which is possible to deduce the motion of charge density within the FET channel. Results for a pentacene FET on Si/SiO_2 is shown in Figure 12 (c) and (d) for devices with Au and Ag electrodes, respectively. Two features are worth noting. First, there is a peak in the SHG intensity at a certain distance from the source electrode, which propagates to the drain electrode, broadens and loses intensity as time increases. This locates the edge of the carrier sheet injected at the source electrode (where the E-field has a maximum value), and indicates that carriers are moving towards the drain by a drift-diffusion mechanism. The mobilities extracted by modeling this carrier motion are independent of the electrode material. However, the second important observation is that at very early times there is a peak in the electric field (or the sqrt(SHG intensity)) near the source electrode, which rapidly disappears for Au electrodes, but reduces initially and then saturates after ~ 2 μs for the Ag electrode. This reflects the fact that hole injection is barrierless with Au, but occurs with ~0.8 eV barrier at the pentacene/Ag interface. A large E-field must be sustained at this interface to keep carrier injection during device operation.

The effect of carrier trapping is crucial to the performance of organic devices, and this has also been addressed by SHG experiments. The combination of capacitance-voltage characteristics and SHG measurements in pentacene FETs allowed investigating the origin of hysteresis behavior in the capacitance measurements after bias stress.[75] SHG spatial profiles along the channel length helped to identify that electrons trapped in pentacene after bias stress assist hole injection from Au electrodes, leading to hysteresis in the capacitance-voltage measurements. In another study,[76] TR-SHG microscopy (similar to Figure 12) as a function of gate bias could identify the contribution of trapped charges to the electric field in the channel. From the propagation of the charge sheet in the channel, the carrier mobility could be measured, and it increased as traps were filled.

SFG vibrational spectroscopy has also been used to investigate OFETs under operation. The first reports by Ye et $al.$[77,78] for OFETs based on pentacene or phenyl-thiophene oligomers with hexyl end-chains showed a correlation between the gate bias (in the hole accumulation regime and with $V_{DS} =0$) and the nonresonant background of the SFG spectra ($\chi_{NR}^{(2)}$ in Equation (13)), due to the field-induced electronic contribution to $\chi_{NR}^{(2)}$, as in EFISHG. They also observed a change in the relative intensity of the CH_2 and CH_3 symmetric stretches of the hexyl end-chains, implying that interface charging induced a mechanical stress on the active layer, which in turn led to an increase in conformational disorder of the alkyl chains of the phenyl-thiophene oligomer. Similar results were obtained by Anglin et $al.$[79] for a P3HT FET, except that they did not observe any conformational change of the hexyl side chains, and $\chi_{NR}^{(2)}$ changed with both positive and negative bias, although the electrical characteristics showed only unipolar behavior (negative bias FET operation). They argued that positive bias was leading to trapped electrons at the interface, which changed the nonresonant background due to the interfacial electric field, but did not contribute to current in the device. More recently, the same group has obtained SFG spectra in the vibrational fingerprint region (1150 to 2300 cm^{-1}) for gate-biased FETs with a poly(triarylamine) – PTAA – active layer.[80] Figure 13(a-c) displays the SFG spectra with positive, negative and no gate bias. Three main features can be distinguished in the spectra. For positive or no gate bias, the electric transfer curve (Figure 13(d)) suggests no carrier accumulation at the interface, and the

SFG spectrum consists of mainly one resonance at ~1600 cm^{-1}, attributed to the C=C stretches of the aromatic rings of PTAA. For negative bias, holes accumulate at the interface, inducing a sharp and strong peak at 1260 cm^{-1} and a broad feature at high wavenumbers, of which only a tail is shown for frequencies above 1800 cm^{-1}. They are both a signature of polarons in the polymer which accumulate at the interface, the first associated with doping-induced vibrational modes (IRAV modes) and the latter due to an electronic transition associated to polarons. Figure 13(d) also shows that the transfer curve of the device correlates well with the intensity of the IRAV peak at 1260 cm^{-1}. Interestingly, they found that different surface treatments of the Si/SiO$_2$ substrate led to ambipolar charge accumulation, as detected by SFG spectra, but not ambipolar transport in the device. Another interesting work was that of Nakai et al.,[81] who used SFG microscopy to probe a pentacene FET with an additional insulating layer of polyvinyl phenol (PVP) over the Si/SiO$_2$ substrate. They have also observed the increase of the nonresonant background upon gate bias, but a vibration assigned to PVP could also be detected. During device operation (with negative V_{GS} and V_{DS}), the SFG image shows a redistribution of the charge density at the interface, indicating a gradual channel formation.

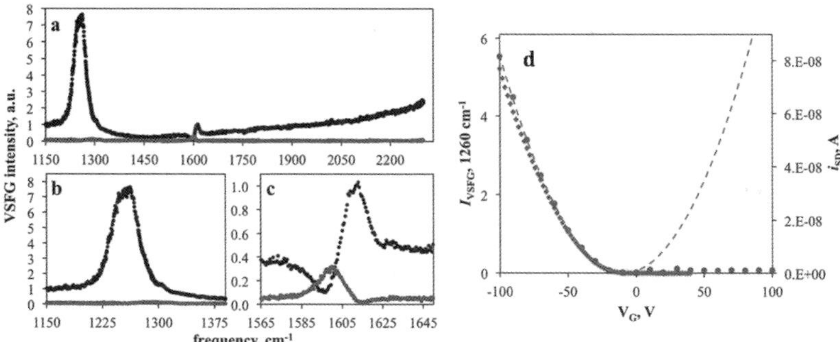

Figure 13. SFG spectra for a PTAA FET with +100V (dark gray), 0V (light gray) and − 100V (black) gate bias (V_{DS} = 0). The spectra with +100 V and 0 V bias are nearly identical. Panel (a) shows the spectra in a wide range, while panels (b) and (c) show in more detail the IRAV and ring mode ranges. (d) Overlaid transfer curve (♦) and SFG intensity (•) of the peak at 1260 cm^{-1} as a function of the gate bias. The dashed line is a parabolic fit to the SFG intensity at negative bias. Adapted from reference [80].

The intrinsic surface specificity of SFG spectroscopy was used to correlate device performance and interfacial properties of the organic semiconductor at the dielectric interface. Geiger and coworkers[49] have recently used X-ray scattering to probe the crystalline order of the organic semicondutor p-BTDT in an FET constructed on octadecylsilane (OTS) modified SiO_2 dielectrics. SFG spectroscopy was used to probe the conformational order of the OTS monolayer on Si/SiO_2, by measuring the amplitude ratio of the CH_2 and CH_3 stretch vibrations. It was found that different assembling protocols led to OTS monolayers with significant changes in conformational order, which dramatically affected the carrier mobility of the devices. However, X-ray and AFM measurements indicate that these changes in mobility are not due to changes in local orientation of the organic molecules at the dielectric interface, but to a change in the growth mode and crystallinity of the organic film on the modified substrates. In another study, Anglin et al.[82] have investigated P3HT FETs fabricated on Si/SiO_2 substrates with different surface functionalizations of the SiO_2 dielectric. SFG was used to study the orientation of the backbone rings of P3HT on the functionalized surfaces, showing that higher mobilities are achieved on low surface energy dielectrics, which lead to a more edge-on orientation of the polymer rings. It should be emphasized that this was done on non-annealed polymer films, probing both crystalline and amorphous regions (the major component), which could not be done with X-ray diffraction.

4. Conclusion

In summary, we presented here an overview of two second-order nonlinear optical techniques to study interfaces: second-harmonic generation (SHG) and sum-frequency generation (SFG) vibrational spectroscopy. Initially we described their basic theory and then highlighting their main capabilities and advantages, such as: intrinsic surface specificity, submonolayer sensitivity, wide applicability (including liquid surfaces and buried interfaces), possibility of obtaining quantitative information on molecular orientation at interfaces, and their sensitivity to interfacial electric fields and alkyl chain conformation (for SFG spectroscopy). After that, a few selected applications of SHG and SFG to interfaces of interest in organic electronics were discussed, including *in situ* probing of working devices, such as light-emitting diodes (OLEDs), photovoltaic cells (OPVs) and transistors (OFETs).

Most applications up to now concentrate on investigating the electric field and charge distributions within the devices by using electric-field-induced SHG (or SFG), with variants such as time-resolved SHG or SHG/SFG microscopy. In these cases, the techniques are not intrinsically sensitive to interfaces, but probe all regions within the bulk that have an electric-field breaking the symmetry. However, a few examples have shown that SHG and SFG have great potential to investigate the microscopic molecular structure at interfaces, bringing up structural information that can be correlated to the device performance. It is likely that in the near future these techniques will provide important insights on the role of interfaces to the performance and stability of organic electronic devices.

Acknowledgments

This work was supported by the Brazilian agencies CNPq, Fapesp, and Finep. SGM acknowledges a graduate fellowship from Capes.

References

1. A. W. Adamson, A. P. Gast, *Physical Chemistry of Surfaces*, 6th ed., Wiley, New York (1997).
2. N. Koch, *ChemPhysChem* **8**, 1438 (2007).
3. S. Braun, W. R. Salaneck, M. Fahlman, *Adv. Mater.* **21**, 1450 (2009).
4. R. J. Kline, M. D. McGehee, M. F. Toney, *Nature Mater.* **5**, 222 (2006).
5. W. J. Mitchell, P. L. Burn, R. K. Thomas, G. Fragneto, *Appl. Phys. Lett.* **82**, 2724 (2003).
6. K. H. Lee *et al.*, *Adv. Mater.* **23**, 766 (2011).
7. V. Parente *et al.*, *Adv. Mater.* **3**, 319 (1998).
8. Y. Furukawa, J. Yamamoto, D.-C. Cho, T. Mori, *Macromol. Symp.* **205**, 9 (2004).
9. Y. Furukawa et al., *Vibrational Spectroscopy* **60**, 5 (2012).
10. J. He, H. Zhou, F. Wan, Y. Lu, G. Xue, *Vibrational Spectroscopy* **31**, 265 (2003).
11. Y. R. Shen, *The Principles of Nonlinear Optics*, Wiley-Interscience, New York (1984), chapter 25.
12. M. B. Raschke, M. Hayashi, S. H. Lin, Y. R. Shen, *Chem. Phys. Lett.* **359**, 367 (2002).
13. Y. R. Shen, *Surf. Sci.* **299/300**, 551 (1994).
14. R.W. Boyd, *Nonlinear Optics*, 2nd ed. San Diego: Academic Press. pg. 117, 2003.
15. C.T. Williams, D. A. Beattie, *Surf. Sci.* **500** 545 (2002).
16. A. G. Lambert, P. B. Davies, D. J. Neivandt, *Appl. Spectr. Rev.* **40**, 103 (2005).
17. Y. R. Shen, *Annu. Rev. Phys. Chem.* **64**, 129 (2013).
18. F. M. Geiger, *Annu. Rev. Phys. Chem.* **60**, 61 (2009).
19. A. M. Jubb, W. Hua, H. C. Allen, *Annu. Rev. Phys. Chem.* **63**, 107 (2012).
20. P. Guyot-Sionnest, W. Chen and Y. R. Shen, *Phys. Rev. B* **33**, 8254 (1986).
21. P. Guyot-Sionnest and Y. R. Shen, *Phys. Rev. B* **38**, 7985 (1988).
22. G. A. Sefler, Q. Du, P. B. Miranda, Y. R. Shen, *Chem. Phys. Lett.* **235**, 347 (1995).

23. R. Braun, B. D. Casson, and C. D. Bain, *Chem. Phys. Lett.* **245**, 326 (1995).

24. T. G. Zhang, C. H. Zhang and G. K. Wong, *J. Opt. Soc. Am. B* **7**, 902 (1990).

25. X. Zhuang, P. B. Miranda, D. Kim, Y. R. Shen, *Phys. Rev. B* **59**, 12632 (1999).

26. Y. R. Shen, *Proc. Natl. Acad. Sci USA* **93**, 12104 (1996).

27. D. A. Long, *Raman Spectroscopy*, McGraw-Hill, New York (1977).

28. P. B. Miranda, Y. R. Shen, *J. Phys. Chem. B* **103**, 3292 (1999).

29. Q. Du, E. Freysz, Y.R. Shen, *Phys. Rev. Lett.* **72**, 238 (1994).

30. M. A. Belkin, T. A. Kulakov, K.-H. Ernst, L. Yan, Y. R. Shen, *Phys. Rev. Lett.* **85**, 4474, (2000).

31. T. Nishida *et al.*, *Phys. Rev. Lett.* **96**, 077402 (2006).

32. K. Seto, J. Pham, Y. Furukawa, *Chem. Phys. Lett.* **529**, 31 (2012).

33. J. F. Rabolt, F. C. Burns, N. E. Schlotter, and J. D. Swalen, *J. Chem. Phys.* **78**, 946 (1983).

34. R. Superfine, J. Y. Huang, Y. R. Shen, *Opt. Lett.* **15**, 1276 (1990).

35. P. Guyot-Sionnest.; J. H. Hunt, Y. R. Shen, *Phys. Rev. Lett.* **59**, 1597 (1987).

36. F. J. Pavinatto *et al.*, *Langmuir* **25**, 10051 (2009).

37. P. B. Miranda, V. Pflumio, H. Saijo, Y. R. Shen, *Thin Solid Films* **327–329**, 161 (1998).

38. P. L. Hayes, J. N. Malin, D. S. Jordan, F. M. Geiger, *Chem. Phys. Lett.* **499**, 183 (2010).

39. T. Manaka, E. Lim, R. Tamura, D. Yamada, M. Iwamoto, *Appl. Phys. Lett.* **89**, 72113 (2006).

40. J. A. McGuire, M. B. Raschke, Y. R. Shen, *Phys. Rev. Lett.* **96**, 087401 (2006).

41. A. E. Jailaubekov *et al.*, *Nat. Mater.* **12**, 66 (2013).

42. J. A. McGuire, Y. R. Shen, *Science* **313**, 1945 (2006).

43. F. Geiger, G. Marowsky, *Chem. Phys. Lett.* **254**, 179 (1996).

44. Q. Li, R. Hua, K. C. Chou, *J. Phys. Chem. B* **112**, 2315 (2008).

45. D. B. O'Brien, T. C. Anglin, A. M. Massari, *Langmuir* **27**, 13940 (2011).

46. P. Dhar, P. P. Khlyabich, B. Burkhart, S. T. Roberts, S. Malyk, B. C. Thompson, A. V Benderskii, *J. Phys. Chem. C* **117**, 15213 (2013).

47. F. C. B. Maia, P. B. Miranda, *manuscript in preparation*.

48. X. Lu, S. A. Spanninga, C. B. Kristalyn, Z. Chen, *Langmuir* **26**, 14231 (2010).

49. S. R. Walter et al., *J. Am. Chem. Soc.* **134**, 11726 (2012).

50. J. Löbau, A. Rumphorst, K. Galla, S. Seeger, K. Wolfrum, *Thin Solid Films* **286**, 272 (1996).

51. H. Asanuma, H. Noguchi, Y. Huang, K. Uosaki, H.-Z. Yu, *J. Phys. Chem. C.* **113**, 21139 (2009).

52. O. Aktsipetrov, A. Ermushev and A. Petukhov, L. Daikhin, M. Levi, V. Vereta, *Synth. Met.* **54**, 327 (1993).
53. G. Niaura, Z. Kuprionis, V. Kod, R. Mažeikienė, A. Malinauskas, *Materials Science (Medžiagotyra)* **9**, 436 (2003).
54. W-L. Chan, M. Ligges, A. Jailaubekov, L. Kaake, L. Miaja-Avila, X-Y. Zhu, *Science* **334**, 1541 (2011).
55. H. Dong, et al., *J. Phys. Chem. B* **113**, 4176 (2009).
56. T. Schuettfort, L. Thomsen, C. R. McNeil, *J. Am. Chem. Soc.* **135**, 1092 (2013).
57. D. Taguchi, M. Weis, T. Manaka, M. Iwamoto, *Appl. Phys. Lett.* **95**, 263310 (2009).
58. E. Lim, T. Manaka, M. Iwamoto, *Chem. Phys. Lett.* **516**, 254 (2011).
59. E. Lim, D. Taguchi, M. Iwamoto, *Chem. Phys. Lett.* **515**, 306 (2011).
60. E. Lim, D. Taguchi, M. Iwamoto, *Chem. Phys. Lett.* **561-562**, 97 (2013).
61. D. Taguchi, S. Inoue, L. Zhang, J. Li, M. Weis, T. Manaka, M. Iwamoto, *J. Phys. Chem. Lett.* **1**, 803 (2010).
62. A. Sadakata, K. Osada, D. Taguchi, T. Yamamoto, M. Fukuzawa, T. Manaka, M. Iwamoto, *J. Appl. Phys.* **112**, 083723 (2012).
63. L. Zhang, D. Taguchi, J. Li, T. Manaka, M. Iwamoto, *J. Appl. Phys.* **108**, 093707 (2010).
64. D. Taguchi, T. Manaka, M. Iwamoto, K. Bazaka, M. V. Jacob, *Chem. Phys. Lett.* **572**, 150 (2013).
65. T. Miyamae, N. Takada, T. Tsutsui, *Appl. Phys. Lett.* **101**, 73304 (2012).
66. X. Chen, D. Taguchi, K. Lee, T. Manaka, M. Iwamoto, *Chem. Phys. Lett.* **511**, 491 (2011).
67. X. Chen, D. Taguchi, T. Shino, T. Manaka, M. Iwamoto, *J. Appl. Phys.* **110**, 74509 (2011).
68. D. Taguchi, T. Shino, L. Zhang, J. Li, M. Weis, T. Manaka, M. Iwamoto, *Appl. Phys. Express* **4**, 21602 (2011).
69. D. Taguchi, T. Shino, X. Chen, L. Zhang, J. Li, M. Weis, T. Manaka, M. Iwamoto, *Appl. Phys. Lett.* **98**, 133507 (2011).
70. J. D. Morris *et al.*, *Appl. Phys. Lett.* **102**, 033301 (2013).
71. T. Manaka, E. Lim, R. Tamura, M. Iwamoto, *Appl. Phys. Lett.* **87**, 222107 (2005).
72. A. Babajanyan, H. Melikyan, J. Kim, K. Lee, M. Iwamoto, B. Friedman, *Organic Electronics* **12**, 263 (2011).
73. T. Manaka, E. Lim, R. Tamura, M. Iwamoto, *Nat. Photonics* **1**, 581 (2007).
74. T. Manaka, F. Liu, M. Weis, M. Iwamoto, *J. Phys. Chem. C* **113**, 10279 (2009).
75. E. Lim, T. Manaka, R. Tamura, M. Iwamoto, *J. Appl. Phys.* **101**, 094505 (2007).
76. Y. Tanaka, T. Manaka, M. Iwamoto, *Chem. Phys. Lett.* **507**, 195 (2011).

77. H. Ye, A. Abu-Akeel, J. Huang, H. E. Katz, D. H. Gracias, *J. Am. Chem. Soc.* **128**, 6528 (2006).
78. H. Ye, J. Huang, J. R. Park, H. E. Katz, D. H. Gracias, *J. Phys. Chem. B* **111**, 13250 (2007).
79. T. C. Anglin, D. B. O'Brien, A. M. Massari, *J. Phys. Chem. C* **114**, 17629 (2010).
80. T. C. Anglin, Z. Sohrabpour, A. M. Massari, *J. Phys. Chem. C* **115**, 20258 (2011).
81. I. F. Nakai, M. Tachioka, A. Ugawa, T. Ueda, K. Watanabe, Y. Matsumoto, *Appl. Phys. Lett.* **95**, 243304 (2009).
82. T. C. Anglin, J. C. Speros, A. M. Massari, *J. Phys. Chem. C* **115**, 16027 (2011).

PHOTOINDUCED CHARGE TRANSFER DYNAMICS AT HYBRID GAAS/P3HT INTERFACES

JUN YIN, MANOJ KUMAR, MAJID PANAHANDEH-FARD AND ZILONG WANG

Division of Physics and Applied Physics, School of Physical and Mathematical Sciences, Nanyang Technological University, 21 Nanyang Link, Singapore 637371

FRANCESCO SCOTOGNELLA

Dipartimento di Fisica, Politecnico di Milano, Piazza Leonardo da Vinci 32, 20133 Milano, Italy

Center for Nano Science and Technology@PoliMi, Istituto Italiano di Tecnologia, Via Giovanni Pascoli 70/3, 20133 Milano, Italy

CESARE SOCI[*]

Division of Physics and Applied Physics, School of Physical and Mathematical Sciences, Nanyang Technological University, 21 Nanyang Link, Singapore 637371

Centre for Disruptive Photonic Technologies, Nanyang Technological University, 21 Nanyang Link, Singapore 637371

1. Introduction

Organic-inorganic hybrid photovoltaic systems, for instance conjugated polymers blended with inorganic single crystals, thin films, or functionalized nanostructures (nanoparticles, nanorods or quantum dots) [1-2], offer great flexibility for designing novel solar cells with high power conversion efficiency (PCE) [3-5]. The overall photocurrent of hybrid photovoltaics is primarily affected by the competition between interfacial charge separation and charge recombination [6-8]. Typically, the energetic offsets between the frontier orbitals of conjugated polymers and conduction bands (CB) and valence bands (VB) of inorganic nanocrystals make inorganic parts act as electron acceptor,

---

[*] Corresponding Author: csoci@ntu.edu.sg

promoting dissociation of the photogenerated excitons and preventing charge recombination [9]. Recently, significant research efforts were directed toward hybrid photovoltaics using different inorganic nanostructure composites of group IV (Si [10] and Ge [11]), group IV-VI (PbS [12] and PbSe [13]), group II-VI (CdSe [14-15] and CdS [16]) and metal oxides (TiO_2 [1] and ZnO [17-18]). Specifically, mainstream III-V semiconductors (*e.g.* GaAs and InP) with high carrier mobility and direct bandgap absorption well-overlapped with the solar irradiance are rapidly emerging as exceptional photovoltaic material for thin film technologies [19], including dye or polymer sensitized hybrid solar cells [20-23]. Early demonstration of hybrid photovoltaics based on GaAs/quaterthiophene and GaAs/octithiophene bilayer yielded 1.7% [20] and 4.2% PCE [24]; more recently, use of GaAs nanowires either blended in P3HT bulk heterojunctions [21] or in a bilayer thin-film configuration [22, 25] allowed achieving PCEs of >2.3%.

Despite very promising device performances, current understanding of the electronic properties leading to charge transfer at organic-inorganic interfaces like GaAs/poly(3-hexylthiophene-2,5-diyl) (P3HT) is limited [26-29]. Moreover, the nature of charged photoexcitations at the interface of highly delocalized inorganic crystals and more localized, disordered conjugated systems is of great fundamental interest [30-34]. Based on above perspectives, better understanding of the nature of primary photoexcitations and fundamental physical processes at III-V/polymer heterointerfaces are desirable from both experimental and theoretical standpoints.

Several recent studies of hole and electron transfer mechanisms in III-V/conjugated polymer systems provided useful insights into interfacial processes of these composites and guidance for the design of efficient photovoltaic devices. Blackburn et al. [15] investigated electron and hole transfer from colloidal InP quantum dots to organic hole-transfer material by photoluminescence and transient absorption spectroscopy. In such system hole transfer induces a rapid ~4 ps component to the transient absorption (TA) decay, which is manifested in the dynamics of both visible photobleaching and mid-IR absorbance. Lanzani et al. [35] found strong photoinduced absorption signatures in oligothiophene cations upon photoexcitation of a GaAs/oligothiophene bilayer, indicating that charge transfer involves exciton diffusion from oligothiophene bulk to the GaAs/oligothiophene interface. Subsequently, a detailed picture of photoinduced electron transfer at the GaAs/CuPcF$_{16}$ interface was also provided [29]. Very recently, Zhu and coworkers revealed the presence of competing charge separation processes at the GaAs/CuPc interface by resolving distinct channels of charge carrier separation driven by the delocalized space charge field in GaAs and charge transfer from GaAs to localized CuPc due to an energetic driving force at the interface [8]. In theory, Giustino and coworkers investigated an idealized GaAs(10-10)/P3HT interface by DFT modeling, showing that the resulting interfacial dipole can lower the highest occupied molecular orbital (HOMO) of the conjugated polymer until the whole system attains equilibrium [36]. Our group demonstrated that the polar GaAs(111)B surface tends to facilitate hole transfer from the VB states to the HOMO of P3HT compared to the nonpolar surface GaAs(110) [37].

Here we compare theoretical predictions regarding the charge carrier photogeneration and photoinduced interfacial charge transfer at the heterointerface of a prototype bilayer formed by hole-conducting polymer (P3HT) and n-type GaAs substrates with different orientation and polar surfaces by combining steady-state and ultrafast spectroscopy measurements with density functional theory (DFT) calculations. Experimental evidence of efficient electron transfer at GaAs/P3HT interfaces is found in photoluminescence (PL) and transient absorption spectra and dynamics with excitation energy above the P3HT optical gap by the enhancement of polaron formation rate and increase of polaron lifetime. On the other hand, selective excitation with energy below the optical gap of P3HT allows isolating the contribution of hole injection from GaAs to P3HT. All these findings are supported by DFT calculations which confirm that, irrespective of the crystal orientation, P3HT acts as an efficient electron donor. On the other hand, the intrinsic surface dipole moment of GaAs surfaces is enhanced by induced charge displacement, facilitating hole transfer to P3HT in the case of GaAs(111)B.

2. Discussions

2.1. *Electronic Charge Rearrangement – Density Function Theory Predictions*

The efficiency of electron transfer across GaAs/P3HT interfaces, which ultimately determines charge generation efficiency for photovoltaic applications, is subject to the interfacial energy alignment between the conjugated polymer and the inorganic substrate, the electronic couplings between them, and the change of the polymer geometric and electronic structure due to the thermal fluctuations [38]. Recent theoretical work by Prezhdo *et al.* described the interfacial charge separation and relaxation processes in hybrid organic-inorganic systems by time-domain atomistic simulations [39]. In the case of alizarin/TiO$_2$ interface, the adiabatic mechanism dominates over non-adiabatic ones due to the strong electronic coupling across the interface [40]. For graphene/TiO$_2$ interface, electron injection is found to be ultrafast due to the strong electronic coupling between graphene and TiO$_2$, and both electron injection and energy transfer accelerate for photoexcited states that are delocalized between the two subsystems [41].

The energy alignment and electronic coupling between a polymer and a semiconductor substrate can be obtained via the analysis of the total density of states (DOS) and of the projected density of states (PDOS) after charge redistribution [42]. For the quantification of the actual charge transfer and separation processes between polymer and semiconductor substrates, the charge redistribution upon P3HT adsorption was evaluated, and the changes of interfacial dipole moment and induced work function estimated [43]. In organic/inorganic heterojunctions, adsorption of organic molecules on inorganic substrates is always accompanied by a certain degree of electronic charge rearrangement across the hybrid interface; as a consequence, the generated dipole moment between the molecule and the inorganic semiconductor caused by charge transfer, offsets the interfacial potential and reduces the energy level mismatch [44-45]. Thus, the nature of the polarity of inorganic surface has substantial effects on the formation of interfacial dipoles and may strongly influence charge separation.

Here the charge transferred from the P3HT molecule to the GaAs substrate was calculated by the difference $\Delta\rho(r) = \rho_{GaAs/P3HT} - [\rho_{GaAs} + \rho_{P3HT}]$ (r is the position vector within the computational cell; ρ_{GaAs} is the charge density of the GaAs(110) and (111)B slabs; ρ_{P3HT} is the charge density of P3HT layer without substrate; and $\rho_{GaAs/P3HT}$ is the electronic charge density of the GaAs/P3HT interface). Figures 1(a,a') show that the overall DOS of the GaAs and P3HT hybrid system (dotted line) is largely ruled by the PDOS of GaAs (shaded grey area) while the PDOS of P3HT (solid line) only slightly perturbs the edge of GaAs VB. As compared to the relatively small overlap of the HOMO of P3HT (H_{P3HT}) and the GaAs VB state, the states of P3HT molecule extend over a broad energy range and there is large electronic overlap between the lowest unoccupied molecular orbital (LUMO) level of P3HT (L_{P3HT}) and the GaAs CB state, which generates strong electronic coupling. The characterization of orbital overlap across the interface is performed by the analysis of the frontier orbitals of the P3HT molecule interacting with the GaAs substrate. Figures 1(b,b') and 1(c,c') show the spatial electronic distribution at the HOMO and LUMO energies of P3HT. In the case of GaAs(110)/P3HT, electrons may be efficiently transferred to the bulk of the GaAs crystal due to the highly overlapped electron clouds distribution at the interface, indicating the possible formation of hybrid delocalized states. On the other hand, hole transfer is unfavorable and in both case would be confined to surface states of GaAs(110) because of the poor overlap between the corresponding CB and HOMO. In the case of the polar GaAs(111)B surface, the DOS is severely affected by the presence of surface dangling bonds: As-4p states of the top surface layer and Ga-4p states of the bottom surface layer lie right at the Fermi level inside the bandgap, leading to a large number of intragap states and significant narrowing of the energy bandgap upon lowering CB energy [46]. Hence the lowering of GaAs(111)B CB has dramatic effects on the relative overlap between GaAs and P3HT states, and changes the resulting electron density distribution. Specifically, the CB of GaAs(111)B acquires large overlap with the HOMO of P3HT, which favors the hole transfer from GaAs(111)B to P3HT. While electron transfer from P3HT to GaAs(111)B may be significantly reduced compared to the previous case and somehow confined to the GaAs(111)B surface, holes are allowed to delocalize from the thiophene ring over to the GaAs(111)B surface states and deeply into the bulk thanks to the larger coupling.

For both GaAs(110) and GaAs(111)B/P3HT interfaces, adsorption of P3HT onto the substrate induces significant charge transfer, together with the formation of distinct charge accumulation layers and substantial charge reorganization at the interface. This indicates the significance of electrostatic interaction between the polymer and the substrates [47-48]. The one-dimensional plane-averaged charge density difference ($\Delta\rho$) along z-direction shown in Figures 1(d,d') provides quantitative estimation of electron ($\Delta\rho<0$) and hole ($\Delta\rho>0$) accumulation layer, suggesting a much larger charge redistribution in the case of the polar GaAs(111)B as compared to the nonpolar GaAs(110) surface (up to $\sim5.3\times10^{-3}$ vs $\sim0.8\times10^{-3}$ e/Å^3). The main difference between such two surfaces is the type of charges accumulated at the interface: holes accumulate both above and below the GaAs surface in nonpolar GaAs(110), while a small electron accumulation layer appears on the top As monolayer in polar GaAs(111)B. This electron accumulation layer reduces electrostatic screening due to the large interfacial holes density and facilitates hole transfer from GaAs to P3HT comparing with the case of nonpolar GaAs(110).

Both bare GaAs(110) and GaAs(111)B surfaces have distinct polarization properties; although GaAs(110) is usually considered as a nonpolar surface, a small intrinsic dipole moment is found due to the tendency of surface Ga atoms to sink toward the bulk, leaving behind an As-terminated surface. The unbalanced charge at the surface forms a negative dipole moment pointing toward the bulk which, in our simulations, results in a surface charge density of 2.2×10^{13} e/cm^2. Unlike the nonpolar GaAs(110) surface, a large intrinsic dipole

224

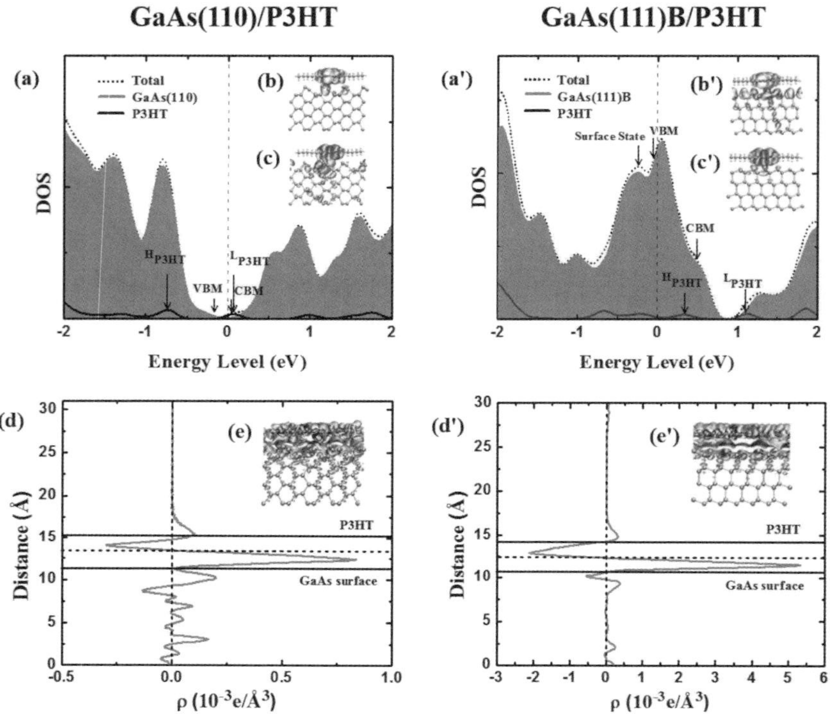

Figure 1. Density of states, electronic orbital distribution and charge redistribution in GaAs(110)/P3HT (left) and GaAs(111)B/P3HT (right) hybrid systems: (a, a') density of states (the dashed line in indicates the position of the Fermi energy); charge distribution of the electron (b, b') and hole (c, c') orbitals; one-dimensional plane-averaged charge density difference upon P3HT adsorption (d, d'); three-dimensional representation of the charge density difference with an isovalue of ±0.005 e/Å^3 (e, e'). The solid lines in (d, d') indicate the average positions of the GaAs surface and the P3HT plane, while the horizontal dashed line shows the interfacial distance at which charge depletion converts into charge accumulation.

moment exists in GaAs(111)B thanks to the alternating As and Ga terminal layers. In this case, we obtained a larger surface charge density of 7.4×10^{13} e/cm^2 [49]. Upon charge redistribution with adsorbed P3HT, the intrinsic surface dipole moment of GaAs surfaces is enhanced by the induced charge displacement. To understand the origin of the interfacial dipole moment, a Löwdin charge analysis of the charge density was carried out for these hybrid systems [50]. By comparing the sum of the Löwdin charge on the GaAs and P3HT molecule before and after formation of the interface a total charge (ΔQ) of 0.207e and 0.209e for GaAs(110)/P3HT and GaAs(111)B/P3HT is found to be transferred between P3HT and GaAs. These charge values are lower than typical organic molecules on metals (Au(111)/Naphthalocyanine~0.7e, Cu(110)/Petencene~0.8e) [51-52], but of the same order of inorganic metal-oxide/polymer interfaces (ZnO/P3HT~0.3e) [53], and metal-oxide/graphene interface (ZnO/Graphene~0.4e) [54], which are known to induce significant charge transfer.

2.2. *Energy Alignment – XPS and UPS Measurements*

DFT simulations were validated performing XPS and UPS measurements of the energy alignment of GaAs/P3HT interfaces. According to the integer charge transfer (ICT) model [55], the energy level alignment of organic/inorganic systems with weak interfacial interactions can be determined from the change of work function upon adsorption of the organic molecule. Removing charges from conjugated polymer can induce substantial geometric and electronic relaxation effects, which leads to localized positive polaronic states ($p+$) [56]. If the work function of substrate (Φ_{SUB}) is larger than the energy of the polaronic states (E_{p+}), electrons will spontaneously transfer from the organic layer into the inorganic substrate, creating a dipole that reduces the vacuum level, where the interfacial dipole energy (Δ) caused by charge redistribution can be obtained from the energy difference between the Φ_{SUB} and E_{p+}.

The position of the HOMO level of P3HT is determined to be 1.10 and 0.64 eV below the Fermi energy from the Fermi-edge regions of UPS spectra of P3HT-coated GaAs(110) and GaAs(111)B substrates. Meanwhile, from the cutoff of the UPS spectra, the work function of P3HT on GaAs(100) and GaAs(111)B substrates ($\Phi_{P3HT/SUB}$) are found to be 3.78 and 4.00 eV, respectively. To determine the energetics of the bare substrates, the thin P3HT films were removed *in-situ* by Ar ion sputtering, until exposure of clean GaAs surfaces. GaAs(110) and GaAs(111)B substrates show the valence band maxima (VBM) of 0.69 and 0.68 eV and work functions of 4.76 and 4.95 eV. The LUMO level of P3HT is determined by adding the optical gap energy (1.9 eV) to the HOMO; similarly, the CBM of GaAs is the sum of VBM and GaAs optical gap energy (1.42 eV).

The overall picture of energy level alignment determined by the above measurements is sketched in Figures 2(a,a'). A nested configuration (type I) is obtained at the GaAs(110)/P3HT interface whereas a staggered band alignment (type II) is observed in the GaAs(111)B/P3HT case, in good agreement with the calculated energy alignment in Figure 1(a,a'). Both configurations favor electron transfer from P3HT to GaAs substrates thanks to the barrier between conduction band minimum (CBM) of GaAs and the LUMO of P3HT film. The higher work function of bare GaAs substrates compared to the hybrid GaAs/P3HT systems could be attributed to the built-up of interfacial dipole barrier of -0.98 and -0.95 eV, respectively, which result from the displacement of negative charge from P3HT film to GaAs substrates. On the other hand, only in the case of GaAs(111)B the valence level offset is favorable for the hole injection from the P3HT layer (E_{HOMO}-E_{VBM}=0.04 eV), suggesting that P3HT could act as a 'hole acceptor'; opposite behavior is expected for the GaAs(110) surface. The induced vacuum level shift by the interfacial dipole, $\Delta\Phi$, can be calculated from the comparison of the electrostatic potential between the GaAs surface and the P3HT molecular plane using the Helmholtz equation $\Delta\Phi = \mu n/\varepsilon_0$ [43] (μ is the interface dipole moment, i.e. the amount of excess of charge obtained from Löwdin charge analysis multiplied by the interfacial distance; $n = 1/A$, while A is the surface area of the interface). Since only one monolayer of P3HT was considered in the simulations, the values of $\Delta\Phi = 0.769$ and 0.868 eV obtained for the GaAs(111)B and GaAs(110), respectively, are slightly lower than the interfacial dipole barriers observed experimentally.

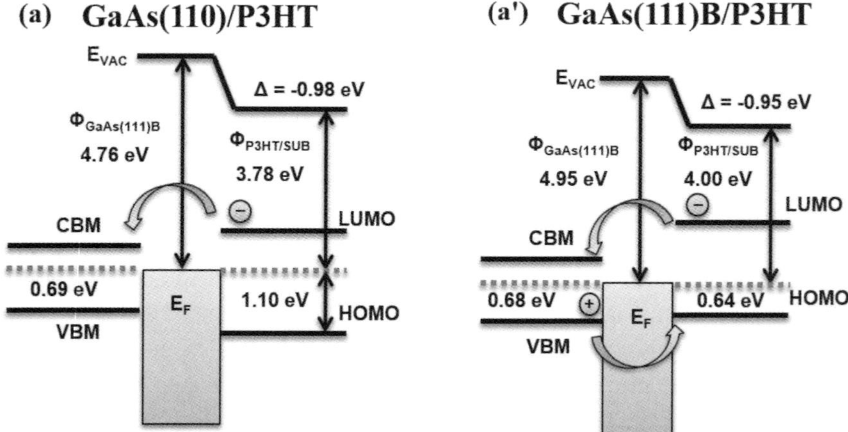

Figure 2. Schematic of the energy diagrams determined from UPS measurements of GaAs(110)/P3HT (a) and GaAs(111)B/P3HT (a') heterointerfaces.

2.3. *Charge Transfer Efficiency – Steady-State Spectroscopy Measurements*

The effects of charge transfer across the hybrid interface are illustrated by the optical absorption and photoluminescence spectra of the combined systems. In the top panel of Figure 3(a) the reflectance of GaAs(111)B substrate is superimposed to the absorption of P3HT film: the reflectance spectrum of GaAs/P3HT is nearly flat in the entire visible region and the typical vibronic features of pristine P3HT film [57] can be clearly resolved from the reflectance spectrum. Quenching of the PL intensity at steady-state provides a clear evidence of interfacial charge transfer. As shown in the bottom panel of Figure 3(a), the PL spectrum of the GaAs/P3HT shows main peaks originating from P3HT with vibronic features at 660 and 720 nm [58] and from GaAs substrate with emission peak at 875 nm. However, the PL intensity of P3HT in GaAs/P3HT is reduced compared to pristine P3HT with the same film thicknesses. Quenching of the P3HT PL suggests the reduction of exciton population upon charge transfer [58-59]. Both emission and absorption spectra of GaAs/P3HT bilayer suggest that GaAs acts both as effective electron acceptor and light absorber in the combined system.

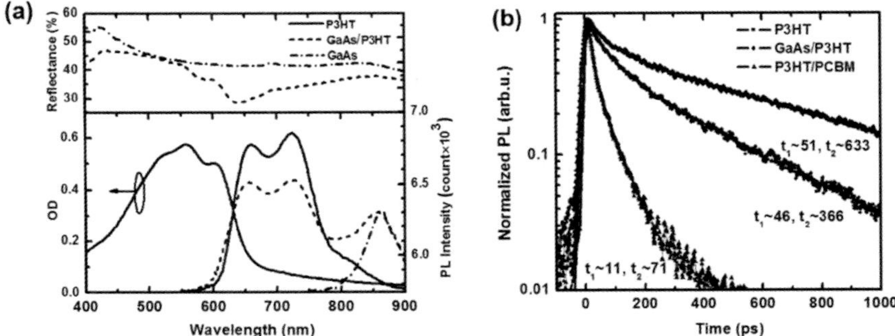

Figure 3. (a) Top panel: reflectance of GaAs (dash-dotted line) and GaAs/P3HT (dashed line); Bottom panel: absorption and steady-state photoluminescence spectra of P3HT (solid lines), and photoluminescence spectra of GaAs/P3HT bilayer (dashed line), and GaAs substrate (dash-dotted line). (b) Time resolved PL of pristine P3HT film (squares), P3HT/PCBM blend film (triangles), and GaAs/P3HT bilayer (circles). The time resolved PL spectra were detected at 660 nm with photoexcitation at 500 nm. The PL decay dynamics were fitted to a bioexponential function; t_1 and t_2 are the extracted lifetimes. Thickness of the P3HT films used in both absorption and PL measurements was 40 nm.

The observed reduction of PL lifetime of P3HT is due to effective exciton dissociation at the GaAs/P3HT interface and photoinduced charge transfer between P3HT and GaAs. The charge transfer efficiency can be estimated using the exciton lifetimes extracted from Figure 3(b). The relative charge transfer efficiency of the combined system can be estimated from the ratio of the fast components of the radiative decay to that of P3HT/PCBM [58]. The estimated charge transfer rate of GaAs/P3HT is found to be ~3% of that of P3HT/PCBM blend film, which is known to have efficiency of ~100% [60]. This shows that the GaAs/P3HT system also facilitates dissociation of photogenerated excitons and subsequently interfacial transfer by the formation of hybrid interfacial states.

2.4. *Photogenerated Carrier Dynamics – Ultrafast Spectroscopy Measurements*

Femtosecond pump-probe transient absorption measurements were performed in reflection geometry in the range of 500-900 nm to investigate the prompt dynamics of photogenerated species with different excitation energies. Here the negative features in differential transient reflectance (TR) spectra (note $-\Delta R/R$ on the vertical axis) correspond to photobleaching (PB) or stimulated emission (SE) from excited states and positive features correspond to photoinduced absorption (PIA) mainly due to polarons [61].

Transient absorption dynamics at three significant probe wavelengths (namely 500, 650 and 900 nm) are reported in Figures 4 and 5. The dynamics in Figure 4 were obtained using the second harmonic of a Ti:Sapphire amplified system (λ_{ex}=400 nm), while those in Figure 5 using the fundamental of the Ti:Sapphire (λ_{ex}=800 nm) as excitation. Samples were excited with energy per pulse of 0.3 µJ. Both Figures 4 and 5 show a comparison between the dynamics of GaAs(110)/P3HT (left) and GaAs(111)B/P3HT (right) hybrid systems to highlight the differences arising from the two crystallographic orientations (surface polarity) of GaAs.

Dynamic traces obtained with excitation wavelength at 400 nm in the different probe spectral regions (Figure 4) are attributed based on transient reflectance (TR) spectra (not shown here). GaAs(110) and GaAs(111)B/P3HT systems show similar dynamic features: PB and PIA signals are long-lived (>500 ps), consistent with the formation of charged polaron states in P3HT. The rise time of PIA correlates well with the fast decay time of PB signals due to the depletion of the ground state upon charge transfer between P3HT and GaAs and dissociation of singlet excitons. The prolonged PB signal also indicates small repopulation of excited states following charge transfer. The main difference between the dynamics of the two hybrid systems appears in the characteristic PB signal at 500 nm. In the case of GaAs(111)B/P3HT the PB signal at 500 nm decays quickly within the first 200 fs and then remains almost unchanged up to the ns time scale. The initial decay correlates well with the rise of the long-lived PIA signal at 650 nm due to the formation of polarons in P3HT. In the case of GaAs(110)/P3HT, the signal at 500 nm nearly instantaneously decays to ~17% of its initial amplitude and becomes positive after approximately 1 ps due to the predominant contribution from GaAs. These observations correlate well with our theoretical prediction that electron transfer from P3HT to GaAs is expected in both hybrid systems, but largely favored in the case of GaAs(111)B polar surface.

Ultrafast spectroscopy data and energy considerations suggest that prompt formation of charges at the GaAs/P3HT heterointerfaces may be induced by either electron transfer from P3HT to GaAs or by hole injection from GaAs to P3HT. To address this issue, the excitation energy was reduced to below the optical gap of P3HT. In this case, selectively excitation of GaAs allows isolating the contribution of hole injection from GaAs to P3HT. Figure 5 shows the transient dynamics obtained with excitation wavelength of 800 nm. One- and two-photon absorption in P3HT was ruled out at these optical fluences from the absence of any photoluminescence or SE from the polymer. As a result very weak signals are observed at 650 and 900 nm, while the PB previously seen at 500 nm is absent. The TR spectra of GaAs/P3HT bilayers excited at 800 nm show similar features as for photoexcitation at 400 nm (not shown here). Notably, even without direct excitation of the polymer, PIA features of P3HT are clearly visible. In the case of GaAs(111)B, the long-lived (up to ns) charged polarons corresponding to the PIA signals probed at 650 strongly indicate interfacial charge generation upon hole injection from GaAs(111)B to P3HT, with a final state that coincides with the electron transfer state from P3HT to GaAs(111)B. Therefore, upon selective excitation of GaAs(111)B, holes are efficiently injected into P3HT within a time span of t<1 ps. Conversely, GaAs(110)/P3HT shows smaller PIA at 650 nm, with somehow faster decays. This indicates that hole transfer is unfavorable compared to the case of the polar GaAs(111)B orientation, consistent with the theoretical predictions and UPS measurements discussed earlier.

230

GaAs(110)/P3HT   GaAs(111)B/P3HT

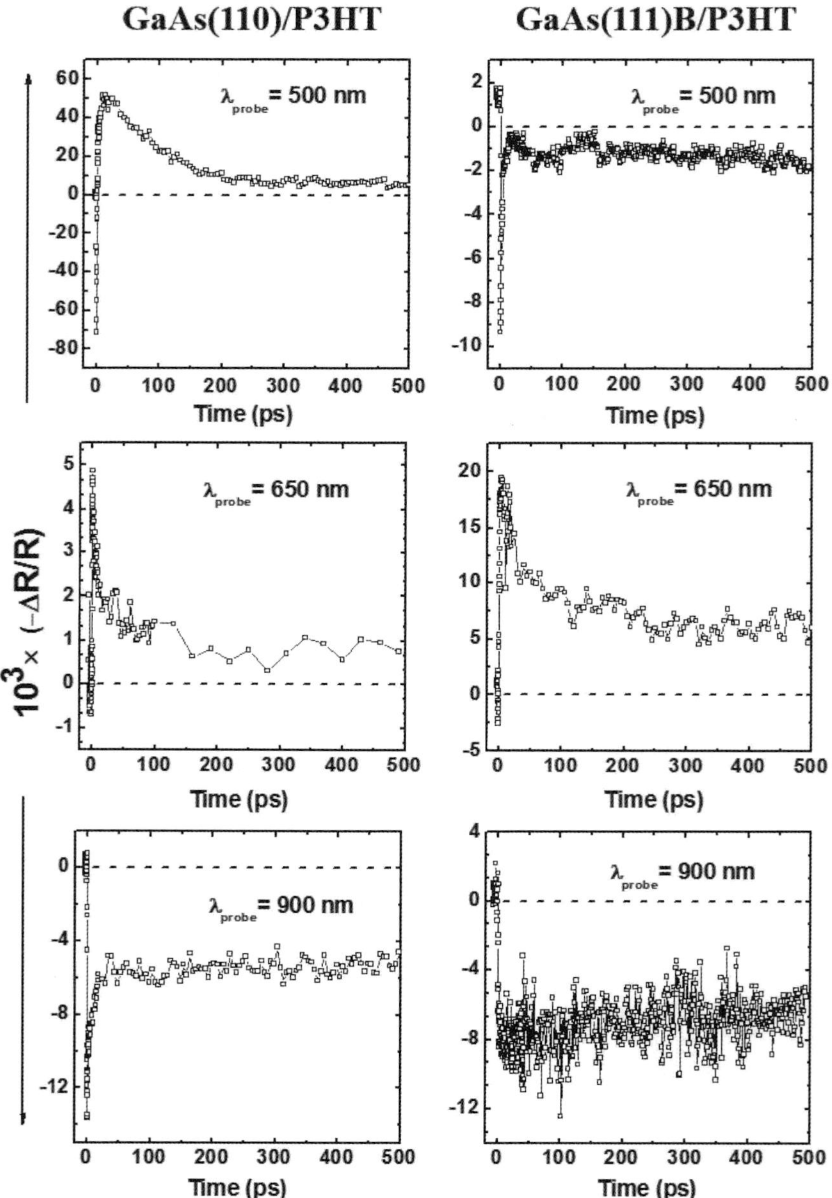

Figure 4. Transient reflection decay profiles of GaAs(110)/P3HT (left) and GaAs(111)B/P3HT (right), at variuos probe wavelengths: photobleaching (PB) at 500 nm, photoinduced absorption (PIA) at 650 nm, and PB at 900 nm. Excitation wavelength: 400 nm.

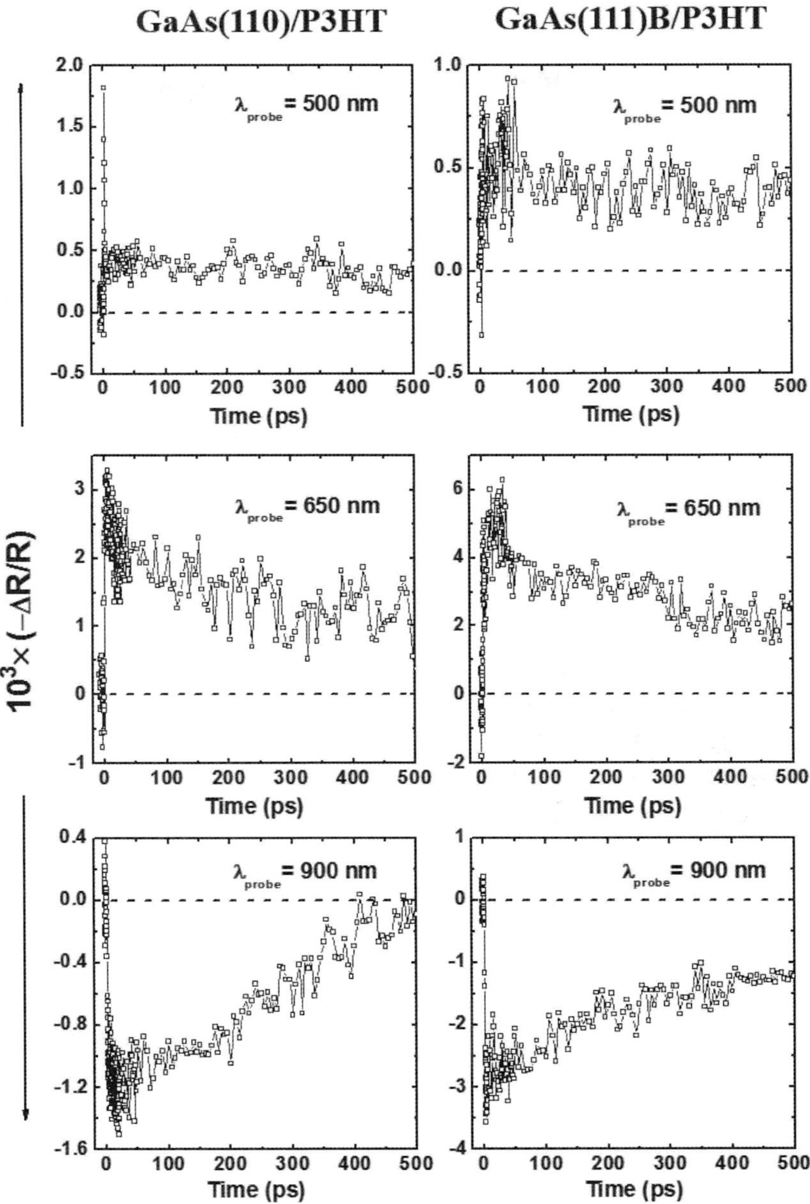

Figure 5. Transient reflection decay profiles of GaAs(110)/P3HT (left) and GaAs(111)B/P3HT (right), at variuos probe wavelengths: photobleaching (PB) at 500 nm, photoinduced absorption (PIA) at 650 nm, and PB at 900 nm. Excitation wavelength: 800 nm.

3. Conclusions

We investigated charge carrier photogeneration and transfer in hybrid GaAs/P3HT bilayers with different surface orientation (polarity) by using DFT calculations, photoemission spectroscopy, and ultrafast spectroscopy measurements. The combination of experimental data and theoretical predictions provides clear evidence for interfacial charge transfer of both electrons and holes at the heterointerface. DFT calculations show that the different polarity of the GaAs surface affects the electronic orbital and charge redistribution properties. Both GaAs(110)/P3HT and GaAs(111)B/P3HT favor electron transfer from P3HT to GaAs substrates. Substantial surface dipole moments ($\Delta Q \approx 0.21e$) may be induced on the GaAs surfaces, which promotes charge transfer across the interface. The experimental transient absorption measurements of GaAs/P3HT bilayers show clear signatures of both electron and hole transfer through the presence of large populations of long-lived polarons in PIA signal. Selective excitation of GaAs in GaAs(111)B/P3HT bilayers shows that hole injection from GaAs(111)B to P3HT induces long-lived (~ns) PB and PIA, which is reduced in the case of GaAs(110).

Our theoretical predictions are in agreement with the experimental findings, and provide a robust framework to understand and design hybrid photovoltaic systems. To increase charge separation and reduce charge recombination, large interfacial areas in contact with the organic materials will be required when designing hybrid solar cells. Low-dimensional compounds, such as GaAs nanowires or quantum dots may then be effectively incorporated in hybrid organic-inorganic devices, providing a large extent of flexibility for surface, bandgap, and DOS engineering.

Acknowledgments

We are grateful to Dr. Shi Chen and Prof. Alfred Huan for their help with photoelectron spectroscopy, and to Jefri Teguh for time-resolved photoluminescence measurements. Thanks also to Prof. Guglielmo Lanzani and Dr. Dmitri Migas for the fruitful discussions regarding this work. C.S. acknowledges financial support from the Nanyang Assistant Professorship, F.S. from the Tan Chin Tuan Fellowship, and M.P.F. from the Yousef Jameel Scholarship.

References

1. W. U. Huynh; J. J. Dittmer; A. P. Alivisatos, *Science* **295**, 2425 (2002).
2. Y. Yin; A. P. Alivisatos, *Nature* **437**, 664 (2005).
3. C. Y. Liu; Z. C. Holman; U. R. Kortshagen, *Nano Lett.* **9**, 449 (2009).
4. J. A. Chang; J. H. Rhee; S. H. Im; Y. H. Lee; H. J. Kim; S. I. Seok; M. K. Nazeeruddin; M. Gratzel, *Nano Lett.* **10**, 2609 (2010).
5. S. Dayal; N. Kopidakis; D. C. Olson; D. S. Ginley; G. Rumbles, *Nano Lett.* **10**, 239 (2010).
6. Y. Vaynzof; D. Kabra; L. H. Zhao; P. K. H. Ho; A. T. S. Wee; R. H. Friend, *Appl. Phys. Lett.* **97**, 033309 (2010).
7. *Chemsuschem* **3**, 1103 (2010).
8. H. Park; M. Gutierrez; X. Wu; W. Kim; X. Y. Zhu, *J. Phys. Chem. C* **117**, 10974 (2013).
9. P. Vanlaeke; A. Swinnen; I. Haeldermans; G. Vanhoyland; T. Aernouts; D. Cheyns; C. Deibel; J. D'Haen; P. Heremans; J. Poortmans et al., *Sol. Energy Mater. Sol. Cells* **90**, 2150 (2006).
10. L. N. He; Rusli; C. Y. Jiang; H. Wang; D. Lai, *IEEE Electron Device Lett.* **32**, 1406 (2011).
11. A. Du Pasquier; D. D. T. Mastrogiovanni; L. A. Klein; T. Wang; E. Garfunkel, *Appl. Phys. Lett.* **91**, 183501 (2007).
12. S. Gunes; K. P. Fritz; H. Neugebauer; N. S. Sariciftci; S. Kumar; G. D. Scholes, *Sol. Energy Mater. Sol. Cells* **91**, 420 (2007).
13. D. Cui; J. Xu; T. Zhu; G. Paradee; S. Ashok; M. Gerhold, *Appl. Phys. Lett.* **88**, 183111 (2006).
14. Y. Zhou; F. S. Riehle; Y. Yuan; H.-F. Schleiermacher; M. Niggemann; G. A. Urban; M. Kruger, *Appl. Phys. Lett.* **96**, 013304 (2010).
15. J. L. Blackburn; D. C. Selmarten; R. J. Ellingson; M. Jones; O. Micic; A. J. Nozik, *J. Phys. Chem. B* **109**, 2625 (2005).
16. S. Ren; L.-Y. Chang; S.-K. Lim; J. Zhao; M. Smith; N. Zhao; V. Bulovic; M. Bawendi; S. Gradecak, *Nano Lett.* **11**, 3998 (2011).
17. W. J. E. Beek; M. M. Wienk; R. A. J. Janssen, *Adv. Funct. Mater.* **16**, 1112 (2006).
18. S. D. Oosterhout; M. M. Wienk; S. S. van Bavel; R. Thiedmann; L. J. A. Koster; J. Gilot; J. Loos; V. Schmidt; R. A. J. Janssen, *Nat. Mater.* **8**, 818 (2009).
19. E. Yablonovitch; O. D. Miller; S. R. Kurtz; Ieee. *The Opto-Electronic Physics That Broke the Efficiency Limit in Solar Cells.* Ieee: New York, 2012; p 1556.
20. J. Ackermann; C. Videlot; A. El Kassmi, *Thin Solid Films* **403**, 157 (2002).
21. S. Q. Ren; N. Zhao; S. C. Crawford; M. Tambe; V. Bulovic; S. Gradecak, *Nano Lett.* **11**, 408 (2011).
22. J. J. Chao; S. C. Shiu; C. F. Lin, *Sol. Energy Mater. Sol. Cells* **105**, 40 (2012).

23. G. Mariani; R. B. Laghumavarapu; B. T. de Villers; J. Shapiro; P. Senanayake; A. Lin; B. J. Schwartz; D. L. Huffaker, *Appl. Phys. Lett.* **97**, 013107 (2010).

24. J. Ackermann; C. Videlot; A. El Kassmi; R. Guglielmetti; F. Fages, *Adv. Funct. Mater.* **15**, 810 (2005).

25. L. Yan; W. You, *Acs Nano* **7**, 6619 (2013).

26. C. K. Yong; K. Noori; M. Gao; H. J. Joyce; H. H. Tan; C. Jagadish; F. Giustino; M. B. Johnston; L. M. Herz, *Nano Lett.* **12**, 6293 (2012).

27. H. Park; M. Gutierrez; X. Wu; J. W. Kim; X. Zhu, *J. Phys. Chem.C,* **117**, 10974 (2013).

28. J. Cabanillas-Gonzalez; A. Gambetta; M. Zavelani-Rossi; G. Lanzani, *Appl. Phys. Lett.* **91**, 122113 (2007).

29. J. Cabanillas-Gonzalez; H.-J. Egelhaaf; A. Brambilla; P. Sessi; L. Duò; M. Finazzi; F. Ciccacci; G. Lanzani, *Nanotechnology* **19**, 424010 (2008).

30. F. Bassani; G. C. La Rocca; D. M. Basko; V. M. Agranovich, *Phys. Solid State* **41**, 701 (1999).

31. S. Blumstengel; S. Sadofev; C. Xu; J. Puls; F. Henneberger, *Phys. Rev. Lett.* **97**, 237401 (2006).

32. Q. Zhang; T. Atay; J. R. Tischler; M. S. Bradley; V. Bulovic; A. V. Nurmikko, *Nat. Nanotechnol.* **2**, 555 (2007).

33. S. Nizamoglu; X. W. Sun; H. V. Demir, *Appl. Phys. Lett.* **97**, 263106 (2010).

34. C. K. Yong; H. J. Joyce; J. Lloyd-Hughes; Q. Gao; H. H. Tan; C. Jagadish; M. B. Johnston; L. M. Herz, *Small* **8**, 1725 (2012).

35. J. Cabanillas-Gonzalez; A. Gambetta; M. Zavelani-Rossi; G. Lanzani, *Appl. Phys. Lett.* **91**, (2007).

36. C. K. Yong; K. Noori; Q. Gao; H. J. Joyce; H. H. Tan; C. Jagadish; F. Giustino; M. B. Johnston; L. M. Herz, *Nano Lett.* **12**, 6293 (2012).

37. J. M. Yin, D. B.; Panahandeh-Fard, M.; Chen, S.; Wang ZL.; Lova P.; and Soci, C. , *J. Phys. Chem. Lett.* **4**, 3303 (2013).

38. S. Dag; L. W. Wang, *Nano. Lett.* **8**, 4185 (2008).

39. A. V. Akimov; A. J. Neukirch; O. V. Prezhdo, *Chem. Rev.* **113**, 4496 (2013).

40. W. R. Duncan; C. F. Craig; O. V. Prezhdo, *J. Am. Chem. Soc.* **129**, 8528 (2007).

41. R. Long; N. J. English; O. V. Prezhdo, *J. Am. Chem. Soc.* **134**, 14238 (2012).

42. S. Nenon; R. Mereau; S. Salman; F. Castet, *J. Phys. Chem. Lett.* **3**, 58 (2012).

43. N. Sai; R. Gearba; A. Dolocan; J. R. Tritsch; W. L. Chan; J. R. Chelikowsky; K. Leung; X. Y. Zhu, *J. Phys. Chem. Lett.* **3**, 2173 (2012).

44. H. Ishii; K. Sugiyama; E. Ito; K. Seki, *Adv. Mater.* **11**, 972 (1999).

45. X. Crispin; V. Geskin; A. Crispin; J. Cornil; R. Lazzaroni; W. R. Salaneck; J. L. Bredas, *J. Am. Chem. Soc.* **124**, 8131 (2002).

46. L. Lin; J. Robertson, *Appl. Phys. Lett.* **98**, 082903 (2011).

47. H. Nemec; J. Rochford; O. Taratula; E. Galoppini; P. Kuzel; T. Polivka; A. Yartsev; V. Sundstrom, *Phys. Rev. Lett.* **104**, 197401 (2010).

48. O. L. A. Monti, *J. Phys. Chem. Lett.* **3**, 2342 (2012).

49. F. Bernardini; V. Fiorentini; D. Vanderbilt, *Phys. Rev. B* **56**, 10024 (1997).

50. P. Giannozzi; S. Baroni; N. Bonini; M. Calandra; R. Car; C. Cavazzoni; D. Ceresoli; G. L. Chiarotti; M. Cococcioni; I. Dabo et al., *J. Phys.: Condens. Matter* **21**, 395502 (2009).

51. A. Terentjevs; M. P. Steele; M. L. Blumenfeld; N. Ilyas; L. L. Kelly; E. Fabiano; O. L. A. Monti; F. Della Sala, *J. Phys. Chem. C* **115**, 21128 (2011).

52. K. Muller; A. P. Seitsonen; T. Brugger; J. Westover; T. Greber; T. Jung; A. Kara, *J. Phys. Chem. C* **116**, 23465 (2012).

53. K. Noori; F. Giustino, *Adv. Funct. Mater.* **22**, 5089 (2012).

54. Z. Geng. W., X. F. , Liu. H. X. , Yao. X. J., *J. Phys. Chem. C* **117**, 10536 (2013).

55. S. Braun; W. R. Salaneck; M. Fahlman, *Adv. Mater.* **21**, 1450 (2009).

56. Z. Xu; L. M. Chen; M. H. Chen; G. Li; Y. Yang, *Appl. Phys. Lett.* **95**, 013301 (2009).

57. P. Salek; O. Vahtras; J. D. Guo; Y. Luo; T. Helgaker; H. Agren, *Chem. Phys. Lett.* **374**, 446 (2003).

58. D. Jarzab; K. Szendrei; M. Yarema; S. Pichler; W. Heiss; M. A. Loi, *Adv. Funct. Mater.* **21**, 1988 (2011).

59. J. Piris; T. E. Dykstra; A. A. Bakulin; P. H. M. van Loosdrecht; W. Knulst; M. T. Trinh; J. M. Schins; L. D. A. Siebbeles, *J. Phys. Chem. C* **113**, 14500 (2009).

60. J. M. Guo; H. Ohkita; H. Benten; S. Ito, *J. Am. Chem. Soc.* **132**, 6154 (2010).

61. F. Deschler; A. De Sio; E. von Hauff; P. Kutka; T. Sauermann; H. J. Egelhaaf; J. Hauch; E. Da Como, *Adv. Funct. Mater.* **22**, 1461 (2012).

THE FIRST STEP IN VISION: VISUALIZING WAVEPACKET MOTION THROUGH A CONICAL INTERSECTION

DARIO POLLI, DANIELE BRIDA, CRISTIAN MANZONI and GIULIO CERULLO

IFN-CNR, Dipartimento di Fisica, Politecnico di Milano, P.za L. da Vinci 32, 20133 Milano, Italy

PIERO ALTOE' and MARCO GARAVELLI

Dipartimento di Chimica "G. Ciamician", Università di Bologna, V. F. Selmi 2, 40126 Bologna, Italy

OLIVER WEINGART

Lehrstuhl für theoretische Chemie, Universität Duisburg-Essen, Universitätsstr.5, 45117 Essen, Germany

KATELYN SPILLANE and PHILIPP KUKURA

Department of Chemistry, University of Oxford, Physical and Theoretical Chemistry Laboratory, Oxford OX1 3QZ

RICHARD A. MATHIES

Chemistry Department, University of California at Berkeley, Berkeley, CA 94720, USA

The primary photochemical event of visual excitation is the ultrafast conversion of 11-*cis* retinal to its all-*trans* form in rhodopsin. It has been postulated theoretically that rhodopsin's unique reactivity is mediated by a conical intersection connecting the ground and excited electronic states, but experimental verification has thus far been lacking due to the extremely rapid nature of the process. Here, we followed rhodopsin isomerisation using ultrafast optical spectroscopy with simultaneous sub-20-fs time resolution and broad spectral coverage from the visible to the near infrared. We tracked coherent wavepacket motion from the photoexcited Franck-Condon region to the photoproduct by monitoring the loss of reactant emission and the subsequent appearance of photoproduct absorption. The excellent agreement between our experimental observations and state-of-the-art molecular dynamics calculations exhibiting a true electronic states crossing provides compelling evidence for the existence and importance of conical intersections in visual photochemistry.

1. Introduction

Retinal proteins, also known as opsins, are a widespread family of photo-sensory proteins which are found in all domains of living organisms. They are trans-membrane proteins sharing a common structure consisting of seven α-helices connected to each other by protein loops. Inside the protein pocket, the chromophore retinal (vitamin-A aldehyde) is attached through a Schiff base linkage to a Lysine residue in the seventh helix. Opsins can be broadly classified in Type I, or bacterial opsins, found in prokaryotes, and type II opsins, found in animals possessing image resolving eyes (mollusks, arthropods and vertebrates). Type I opsins are used by bacteria to harvest light energy in order to carry out metabolic processes. The prototypical one is bacteriorhodopsin, working as a pump for proton transfer across the cell membrane; the ensuing proton gradient is then converted into chemical energy. Type II opsins are found in the photoreceptor cells (rod and cones) of the retina, where rod opsins (rhodopsins) are highly-sensitive receptors used for night and peripheral vision while cone opsins (photopsins) are pigments used for color vision.

In visual opsins the retinal pigment is in its 11-*cis* form. Light absorption by the retinal triggers an ultrafast isomerization to the all-*trans* form [1], which gives rise to a cascade of slower reactions, ultimately leading to the detachment of retinal from the protein pocket. The high quantum yield of rhodopsin isomerization (0.65) [2], the ultrafast 200 fs production of the primary ground state photoproduct [3-6], and the high-energy storage in the first stable bathorhodopsin intermediate [7] all evidence a paradoxically fast and efficient but one-way photoactivated reaction.

Isomerization of rhodopsin is one of the most studied photochemical reactions. Experiments have focused on the goal of defining the structure of the 11-*cis* retinal protonated Schiff base chromophore at various points along the reaction pathway. Early resonance Raman intensity analysis probed the initial dynamics out of the Franck-Condon (FC) region, revealing significant initial motion along both hydrogen out-of-plane wag and torsional internal coordinates that extrapolate to roughly a 100 fs isomerization internal conversion time [8]. This prediction was consistent with the 200 fs photoproduct production time [3] as well as the observation of coherent nuclear dynamics in the ground state photoproduct that survive the internal conversion process [9]. Time-resolved structural studies using transient Raman probed into the sub-ps domain [10], but always revealed a bathorhodopsin structure that was identical to that found in the low-temperature trapped species. The time-bandwidth limitation was transcended with the introduction of femtosecond stimulated Raman spectroscopy (FSRS), which was used to probe the chromophore's vibrational structure with fs time resolution and revealed the conformational relaxation of the nascent high-energy photorhodopsin species, formed on the ground state surface after 200 fs [11]. However, the structural and electronic dynamics that occur in the critical time domain relevant to photochemical reactivity, between the 10's of fs probed by Raman intensities and 200 fs where vibrational probing by FSRS is effective, have remained experimentally unexplored.

Theoretical investigations predict that rhodopsin's unique reactivity is mediated by a conical intersection (CI) connecting the ground and excited electronic states [12, 13]. CIs have become ubiquitous features in theoretical descriptions of organic photochemistry and photobiology [14, 15] and are thought to be responsible for triggering radiationless decay and efficient and ultrafast conversion of photon energy into chemical energy. CIs are collections of molecular geometries in which two (or more) electronic states are isoenergetic, forming a multi-dimensional "seam", any point of which may serve as a doorway through which a molecule may pass to reach a lower energy electronic state. Depending on the topography around the crossing point, CIs can be classified as "peaked" or "sloped" [16]. In peaked CIs, the molecule is funneled towards the point of intersection regardless of the direction from which it approaches; in contrast, sloped topographies guide the molecule to the CI less efficiently, leading to less efficient transitions and a wider variety of reaction products.

It has been gradually accepted that CIs are a critical part of rhodopsin photochemistry [12, 13, 17-19] but experimental observation of CIs has remained elusive because the energy gap between the ground and excited electronic states of a molecule changes significantly over an ultrashort timescale. Direct observation of CIs thus calls for the combination of high temporal resolution and broad spectral coverage to enable the real-time tracking of wavepacket motion along the reaction pathway. In this work, we used ultra-broadband few-optical-cycle femtosecond pulses, tuneable from the visible to the near-infrared (NIR), to monitor the photoisomerization of the retinal chromophore in rhodopsin. These unique experimental capabilities allowed us to directly observe, for the first time, the coherent motion of the wavepacket from the FC region along the excited state surface, through the conical intersection all the way to the formation of the primary photoproduct [20]. Combined with new, highly accurate molecular dynamics, these results generate a comprehensive and dynamic picture of this important process and provide a general model for fast and efficient photochemistry mediated by CIs.

2. Experiments

2.1. *Sample preparation*

Rhodopsin molecules were extracted from the rod outer segments of bovine retinae and purified by sucrose flotation followed by sucrose density gradient centrifugation, as described elsewhere [21]. The OD280/OD500 absorbance ratio was less than 1.8. Fresh NH_2OH was added to a final concentration of 2 mM.

2.2. *Pump-probe spectroscopy*

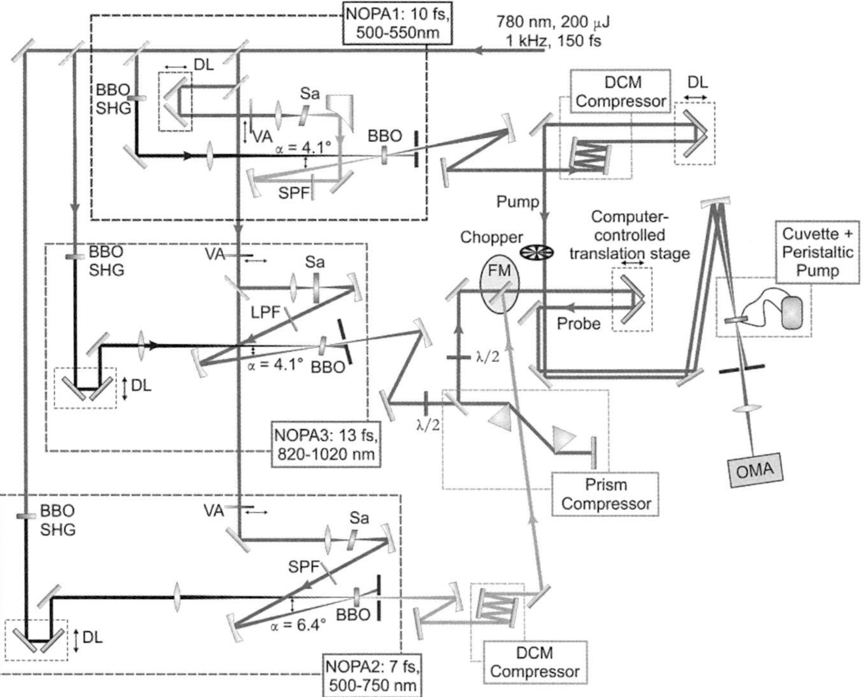

Figure 1: Experimental set up. NOPA: non-collinear optical parametric amplifier; DL: delay line; Sa: sapphire plate; VA: variable attenuator; DCM: double-chirped mirrors; SHG: second-harmonic generation; SPF: short-pass filter; LPF: long-pass filter; OMA: optical multichannel analyzer; FM: flip mirror.

For our experiments we employ a home-built two-colour pump-probe spectrometer which provides sub-20-fs temporal resolution over a broad spectral range in the visible and infrared [22]. Fig. 1 shows the schematic experimental setup. It starts with a regeneratively-amplified mode-locked Ti:Sapphire laser system (Clark-MXR model CPA-1) delivering pulses with 150-fs duration, 500-μJ energy, at 1-kHz repetition rate and 780-nm central wavelength. The system drives three independent non-collinear optical parametric amplifiers (NOPAs) [23] in the visible and in the NIR, which generate the pump and the probe pulses for the time-resolved experiments. The first NOPA (NOPA1) generates the pump pulses in the 500- to 550-nm wavelength region in order to selectively excite the rhodopsin molecule in resonance with its ground-state absorption. The second and third NOPAs are used alternatively to generate the probe pulses in the visible (NOPA2, from 500 nm to 750 nm) or in the NIR frequency range (NOPA3, from 820 nm to 1020 nm). NOPA1 and NOPA2 are compressed to their transform-limited (TL) duration by multiple bounces on custom-designed chirped mirrors, while NOPA3 is compressed to the TL by a Brewster-cut fused-silica prism pair. The pulse durations, retrieved using the Frequency-Resolved Optical Gating (FROG) technique, are 10 fs for NOPA1, 7 fs for NOPA2 and 13 fs for NOPA3. Fig. 2 shows typical spectra from the three NOPAs.

The sample is contained in a cuvette with 250-μm-thick fused-silica windows and ~300-μm optical path. The sample thickness ultimately determines the experimental temporal resolution due to group-delay mismatch between the pump pulse (in the blue) and the probe pulse (in the red/NIR): when the pump and probe pulses are perfectly synchronized (time zero) at the front face of the sample, they will be delayed at the sample exit, thus smearing out the measured time traces. The calculated group delay mismatch is 12.7 fs at 700nm and 17.4 fs at 900nm (considering the dispersion relations of water).

After the sample, the probe beam is selected by an iris and focused onto the entrance slit of a spectrometer equipped with a 1024-pixel linear photodiode array and electronics specially designed for fast read-out times and low noise [24]. The spectral resolution of the spectrograph (about 2 nm) is more than sufficient for our experiments. A fast analog-to-digital conversion card with 16-bits resolution enables single-shot recording of the probe spectrum at the full 1-kHz repetition rate. By recording pump-on and pump-off probe spectra, one can calculate the $\Delta T/T$ spectrum at the specific probe delay τ as:

$$\Delta T/T(\lambda,\tau) = [T_{on}(\lambda,\tau)-T_{off}(\lambda)]/T_{off}(\lambda)$$

By repeating this procedure for a few hundred milliseconds and averaging the resulting signals, it is thus possible to achieve a high enough S/N ratio to detect $\Delta T/T$ spectra down to the 10^{-4} level over the whole probe wavelength range. By moving the translation stage we record $\Delta T/T$ spectra at different probe delays, thus obtaining a complete 2D map: $\Delta T/T=\Delta T/T(\lambda,\tau)$.

The absorption spectrum of Rhodopsin, peaking at ~500 nm, is plotted in Fig. 2 as gray curve. To prevent the accumulation of photo-product in the irradiated volume, the sample was continuously re-circulated by means of a peristaltic pump in a closed-loop circuit with ≈4 ml total volume to ensure complete replacement of the sample in the focal volume for each consecutive laser shot. The excitation fluence was chosen so as to be in a perturbative regime, with maximum $|\Delta T/T|$ of the order of 1-2%.

To ensure that the signal is only due to the rhodopsin molecule, we also performed pump-probe experiments, in exactly the same experimental conditions, on two different samples: (i) the buffer alone and (ii) the fully bleached rhodopsin solution obtained by stopping the flow. In both cases, the transient signal disappeared except for the coherent artifact around time zero, which was subtracted from the data.

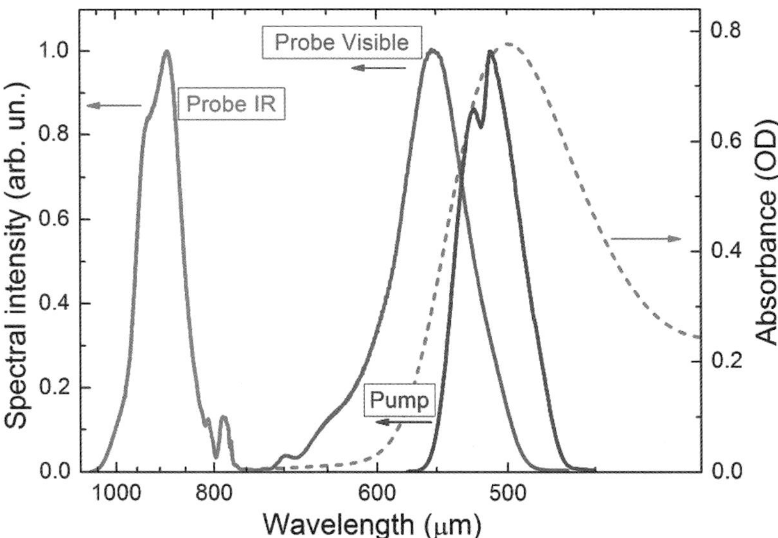

Figure 2: Pump and probe pulse spectra (solid lines) together with rhodopsin absorption spectrum in a 1-mm cuvette (dashed curve).

3. Numerical simulations

To simulate the isomerization dynamics in rhodopsin we use advanced quantum mechanics(QM)/molecular mechanics (MM) simulations. CASPT2//CASSCF/ 6-31G*/AMBER QM/MM [25] computations are performed using the

COBRAMM interface [26], using a hydrogen link-atom scheme [27] and linking the Molpro, MOLCAS-6.0 and AMBER8 [28] packages. Details of the approach have been presented elsewhere [26]. Geometry optimizations, frequency calculations and molecular dynamics involve *ab initio* CASSCF/6-31G* computations [29] for the QM region. The AMBER8 software and the *ff99* force field [28] are used for the MM region and its charges. During all calculations the retinal chromophore, the region between Schiff base nitrogen and the C_β-carbon of Lys-296, and the two neighbouring water molecules are mobile. The remaining atoms are fixed at their positions in the protein crystal (1U19 pdb code). Two different chromophore QM/MM-models have been used here, a full chromophore setup for optimizations and static isomerization path mapping, and a reduced chromophore model for molecular dynamics simulations. In the full chromophore the QM-MM frontier is placed at the C_δ-C_ε bond of the Lys-296 side chain and a full CASSCF active space of 12 electrons and 12 π orbitals (CAS 12/12) is applied. The reduced model is simplified by excluding the non-π-conjugated beta ionone ring part and employing a reduced CASSCF 10/10 active space.

The protein setup employed in this work is the same used by some of the authors in a very recent static investigation of rhodopsin photoisomerization (see [19] for the details). While the protonation state of Glutamate 181 is still a matter of debate, it was left unprotonated in the current simulations. Notably, due to its location above the center of the retinal chromophore, this residual charge does not affect the photoisomerization path and the energy of the photoactive charge transfer state S_1, as shown by Tomasello *et al.* [19] and Cembran *et al.* [30]. Therefore, this choice should not influence the computed dynamics.

To validate the reductions in orbital space and QM region, the static excited state reaction path along C11=C12 torsion was computed by constrained geometry optimization and compared to the corresponding reaction path of the full chromophore from ref. [19]. A two root state averaged wave function was employed in both calculations. We find that both the excited state energies and the S_1/S_0 energy differences in CAS(12/12) are almost perfectly reproduced by CAS(10/10). From these results we conclude that the simplifications applied are reasonable and allow a calculation of several trajectories at significantly reduced computational cost, still preserving the accuracy found in the unreduced QM-model/active space. This conclusion is further supported by dynamical computations.

CAS(10,10)/6-31G*/AMBER QM/MM numerical frequencies have been calculated on the rhodopsin ground state minimum (optimized at the same level). These are used to generate initial conditions for the dynamics by thermally sampling the vibrational modes at 300 K [31] (including zero-point energy corrections: high frequency C-H, N-H and O-H modes are excluded). Note that the temperature distribution is related only to the mobile part, while the protein stays fixed during thermal sampling and MD runs. The fixed crystal structure of the protein is considered as a mean representation of the experimental positions of the atoms, although at low (100 K) temperature. This approximation is justified by the extremely short timescale of the rhodopsin primary photochemical event investigated here (ca. 100 fs decay excited state decay time and 200 fs photoproduct appearance time) that prevents protein thermalization (i.e., equilibration). This is further supported by the results of Hayashi *et al.* [18], showing that excited state dynamics in rhodopsin are not affected by protein mobility.

Trajectories were started after FC excitation to the first excited singlet state using the COBRAMM interface, which couples Molpro-CASSCF and Amber. The wavefunction was averaged over the first three singlet states throughout the entire calculation. Newton's equations of motion were integrated on the fly using the velocity Verlet algorithm [32] and a time step of 0.5 fs. In order to properly detect possible decay points, the time step was reduced to 0.25 fs when the S_1-S_0 energy difference was below 8 kcal/mol. When the scalar product of the S_1 and S_0 state coefficients indicated that a surface crossing had been passed, the trajectories were brought to the ground state [33]. The energy gap at the hops was always below 2.2 kcal/mol.

In order to partially account for the deficiencies in CASSCF calculated spectral data, linear scaling factors were estimated to match the more accurate values obtained at the CASPT2 level of theory [34]. As applied in the dynamics, a CAS(10,10)/6-31G* three root state averaged description of the wavefunction was used for the reduced system to optimize the complete static photoisomerization path from first excited to ground state. The three root state averaged approach allowed us to estimate the contribution of both S_0-S_1 and S_1-S_2 transitions; the latter may have a non-negligible influence on time-resolved emission/absorption spectra.

The CASSCF(10/10) optimized reaction path of the reduced model (full lines) has been re-evaluated at the CASPT2 level employing a constant imaginary level shift [35] of 0.2 and according to a full chromophore approach: the QM region and active space have been enlarged on top of the reduced chromophore structures to embrace the whole chromophore and π orbital space, respectively. Remarkably, the vertical S_0->S_1 CASPT2(12,12) computed energy is 58.9 kcal/mol, in fair agreement with the recorded absorption (57.6 kcal/mol). As expected, S_1 is less steep at the CASPT2 level, and the S_2 energy is significantly reduced when considering dynamic electron correlation. Scaling factors for the stimulated emission (SE) signal were obtained by averaging the energy difference quotients from Min-S_1 to 60°. It is worth noting that, in order to increase the accuracy of the scaling factors employed in spectra simulations, both the FC structure and the region in close vicinity to the CI (i.e., 70°-110° twisted points that all have a rather small S_1/S_0 energy gap) have not been taken into account in the fitting, as they do not enter the probe wavelength window (610-720 nm and 820-1020 nm) of the experiments. Scaling factors for the photoinduced absorption (PA) signal were estimated by an analogous procedure, using the optimized ground state path from 120° toward the photoproduct.

We find that CASSCF systematically overestimates excited state energies all along the photoisomerization path including the starting (FC) and ending (CI) points, leading to a moderate change in the gradient. More specifically, both profiles are barrierless and only the very initial part (FC→Min-S_1) of the CASSCF relaxation path (involving C-C bonds order inversion, yet not the isomerization) is slightly steeper. Afterwards, when C11=C12 rotation becomes active, the two surfaces run almost parallel to each other. Due to this overestimation of the surface steepness, a somewhat faster process is expected in our simulations and, indeed, our initial relaxation out of the FC region is slightly faster than observed. However, the C11=C12 torsional mode is driven by a correct gradient and force field, and the timing error is comparable to the 20-fs time resolution of the experiments. For this reason we preferred not to apply any scaling factor to the simulated time scale, since it already agrees reasonably well with the experiments. The scaling factor was instead applied to the transition energies to gain experimentally accurate values.

Having calculated the trajectories, for each instant of time the transient absorption spectrum was calculated as a superposition of $S_1{\rightarrow}S_0$ SE and $S_1{\rightarrow}S_2$ PA before and $S_0{\rightarrow}S_1$ PA after the hop to the ground state. Each of these three signals was represented by a Gaussian function centred at the corresponding transition energy averaged over all the trajectories with half-width equal to the standard deviation of the collected data and height equal to the mean oscillator strength. To account for the finite laser pulsewidths used in the experiments, we convolved the calculated transient signals with a 20-fs full-width at half maximum Gaussian function.

4. Results and discussion

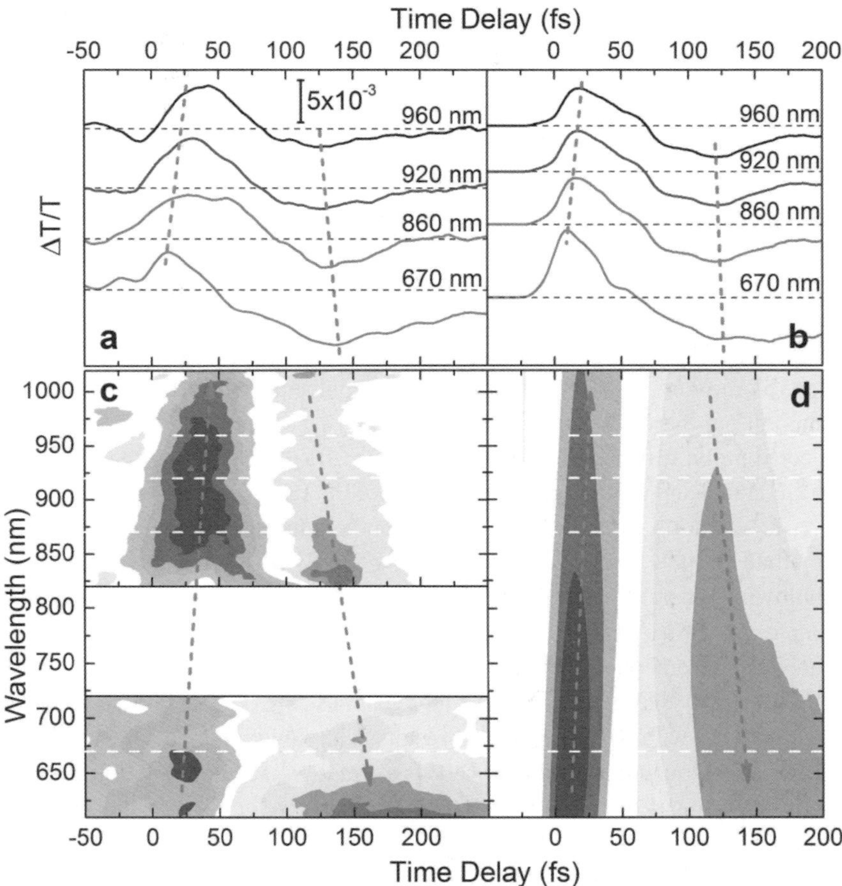

Figure 3: Wavepacket dynamics through the rhodopsin conical intersection. Experimental (a) and simulated (b) time traces at selected probe wavelengths (as indicated). Experimental (c) and simulated (d) differential transmission (ΔT/T) map as a function of probe delay and wavelength in the visible and NIR spectral regions. Gray lines are a guide to the eye highlighting the shifts of the SE and PA signals in time.

The photochemical reaction in purified bovine rhodopsin was initiated by excitation of the retinal chromophore with 10-fs pulses centred at 520 nm (with 55 nm bandwidth, pump pulses in Fig. 2), resonant with the ground-state absorption. The photo-induced dynamics were then probed by delayed pulses either in the visible wavelength region (from 500 to 720 nm, with ~7 fs duration, visible probe pulses in Fig. 2) or in the NIR (from 820 to 1020 nm, with ~13 fs duration, IR probe pulses in Fig. 2), both generated by broadband OPAs [22].

Fig. 3(c) shows a two-dimensional $\Delta T/T$ map as a function of probe wavelength and delay. Immediately following excitation, we observed a positive signal (blue in the figure) with maximum intensity at ~650 nm, which can be assigned to SE from the excited state, due to the negligible ground-state absorption in this wavelength range. The SE signal rapidly shifts to the red while losing intensity and disappearing to wavelengths longer than 1000 nm within ~75 fs. At this time, the $\Delta T/T$ signal changes sign and turns into a weak PA signal (red in the figure), which first appears at 1000 nm and then gradually blue shifts and increases in intensity. For delays longer than 200 fs, the PA signal stabilizes to a long-lived band peaking at 560 nm, indicating the presence of the all-*trans* photoproduct. We emphasize that the use of a sub-20 fs NIR probe is the key to observing the reacting molecule with a remarkably small temporal gap between the excited and ground electronic states. Time traces at selected probe wavelengths are shown in Fig. 3(a), highlighting the red shift of the SE signal and the subsequent blue shift of the PA.

To extract a dynamic model of the photoinduced process from these data we simulated the transient signals ($S_1 \rightarrow S_0$ SE, $S_1 \rightarrow S_2$ PA and $S_0 \rightarrow S_1$ PA) by employing accurate MRMP2-like transition energies over a room temperature sample of QM(CASSCF)/MM(Amber) trajectories tracking the rhodopsin-embedded chromophore excited state/ground state evolution, including excited-state decay. Virtual spectroscopies and dynamics (Figs. 3b and 3d) agree almost quantitatively with the experimental results. The same holds for the 61% photo-isomerization quantum yield and ~110 fs average $S_1 \rightarrow S_0$ hopping time extracted from the simulations, thus further supporting the reliability of the theoretical approach.

The ultrafast complex spectral evolution observed both experimentally and theoretically can be understood qualitatively with the help of Figure 4. Here, the motion of the wavepacket is depicted along the ground and excited state potential energy surfaces of the rhodopsin chromophore as a function of the isomerization coordinate. The wavepacket initially created in the FC region of the excited state of the 11-*cis* reactant rapidly evolves along the reaction pathway towards the CI, and the SE progressively shifts to the red as the band gap between the excited and ground states narrows. Near the CI region, which is reached in ~80 fs according to both experiments and simulations, the SE signal vanishes as the two surfaces approach each other and the transition dipole moment decreases. Following the "jump" to the hot ground state of the photoproduct, a symmetric PA signal is formed. This PA band rapidly shifts to the blue as the surfaces move away from each other energetically and the wavepacket relaxes to the bottom of the photoproduct well, reflecting the redistribution of the excess energy deposited in the molecule and the final torsional movement to the all-*trans* configuration.

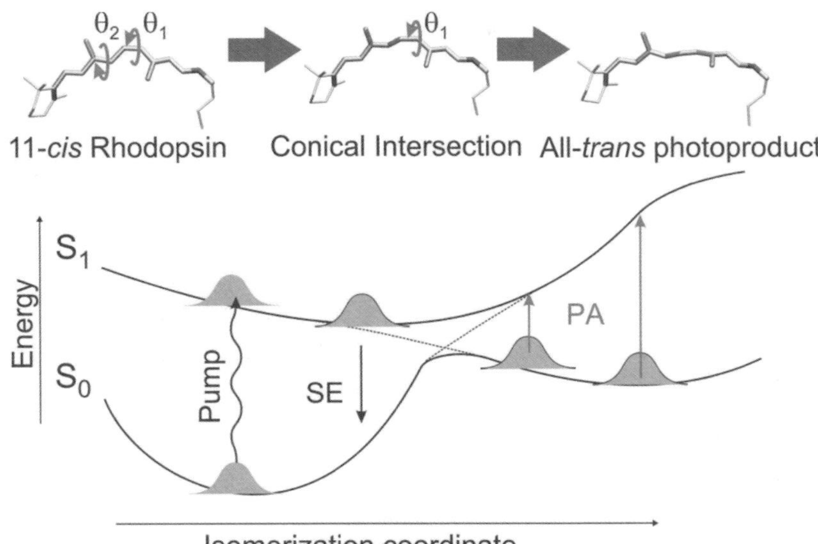

Figure 4: Isomerization potential energy surfaces of rhodopsin. Sketch of the ground- and excited-state potential energy surfaces of the chromophore in rhodopsin as a function of the isomerization coordinate. It shows that stimulated emission (SE) from the excited state of the parent molecule and photo-induced absorption (PA) from the hot photoproduct can monitor the wavepacket dynamics through the CI. On top averaged structures of the chromophore at the initial 11-cis CI and final all-trans configurations; gray/yellow colours indicate the chromophore QM/MM mobile regions in the simulations.

To complete the description of photo-induced dynamics in rhodopsin, we report the portion of the $\Delta T/T$ map probing the response of the system in the visible region, from 495 nm to 610 nm (Fig. 5(a)). In agreement with previous studies [3, 9], we observed the delayed formation of the PA band of the photorhodopsin photoproduct, which peaks at 560 nm and is complete within 200-250 fs. The signal does not display exponential build-up dynamics, but rises rather abruptly starting at ~150 fs (see Fig. 5(b)), which is the time needed for the wavepacket to cross the CI and enter the probed wavelength window on the photoproduct side. The blue region of the spectrum is dominated by the photo-bleaching (PB) signal from the ground state of the parent rhodopsin molecule, peaking at ~510 nm. These two spectral signatures partially overlap so that the

PB band shrinks in time as the PA signal forms and blue shifts. The PB amplitude also decreases due to the return of the 35% unsuccessful excitations back to the ground state of the reactant. At early probe delays, a PA band peaking at ~500 nm is evident, due to a transition from the FC excited state to a higher-lying S_n state [3], having a greater dipole moment than the ground state PB. This band disappears within ~50 fs, due to the wavepacket motion out of the FC region.

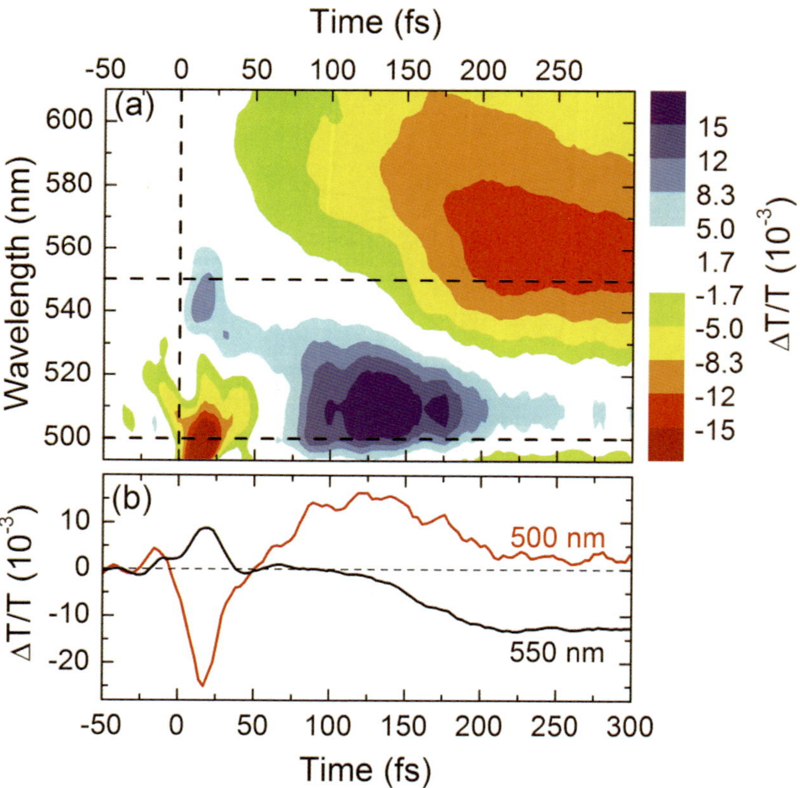

Figure 5: Rhodopsin isomerization probed in the visible spectral range. (a) $\Delta T/T$ map in the visible spectral region. (b) Time traces at 500 nm and 550 nm probe wavelengths.

Finally, The experimental $\Delta T/T(\lambda,\tau)$ maps for rhodopsin were also recorded for longer time delays with respect to the 200-fs window over which the photoproduct is formed. We observe the PA signal from the ground state of the all-trans photoproduct, peaking around 550nm. The signal is modulated by a low frequency oscillation. Figure 6(a) shows the $\Delta T/T(\lambda,\tau)$ map of the oscillatory component of the signal (after subtraction of the slowly-varying background). The oscillation, with $\approx$550 fs period, shows a phase jump and an amplitude minimum around 540nm (see Fig. 6(b)), in proximity of the PA peak of the photoproduct. The probe-wavelength dependence of the amplitude and phase of the oscillation indicates that the wavepacket oscillates in the ground state of the photoproduct, as previously observed by Wang *et al.* [9]. The fact that vibrational coherence, initiated by the pump pulse in the FC region of the reactant, is preserved through the CI and in the photoproduct, is consistent with the extreme speed of excited state wavepacket evolution.

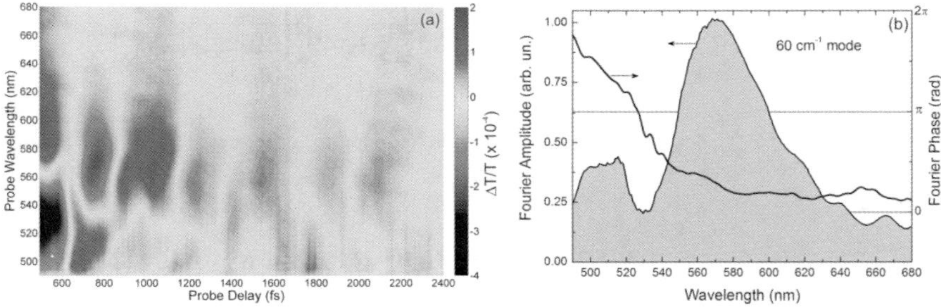

Figure 6 (a): Map of the oscillatory component of the $\Delta T/T$ signal, as a function of probe delay and wavelength; (b) Plot of the Fourier amplitude (filled area) and phase (solid blue line) of the 60 cm^{-1} mode as a function of probing wavelength.

The extremely high experimental time resolution combined with the remarkable agreement with theory makes it possible to confidently predict the real-time structural changes occurring within the first 200 fs of retinal isomerization in rhodopsin. Here, we focus on the key structures, namely the 11-*cis* reactant, the structure of retinal at the conical intersection and the all-*trans* photoproduct (Fig 4). Comparison of the retinal structure in the ground state with the structure at the CI (Fig. 4) reveals dramatic changes only in the vicinity of the isomerizing bond. We found that the average $C_{11}=C_{12}$ dihedral angle at the CI is $\theta_1 \sim -87.8°$, ranging from -75° to -105° for individual trajectories, emphasizing the relevance of the $C_{11}=C_{12}$ twist in triggering internal conversion (Fig. 7(a)). At the same time, the $C_9=C_{10}$ torsional angle θ_2 also exhibits a large change (by ~45°) on the excited-state surface as the wavepacket moves towards the CI (Fig. 7(b)).

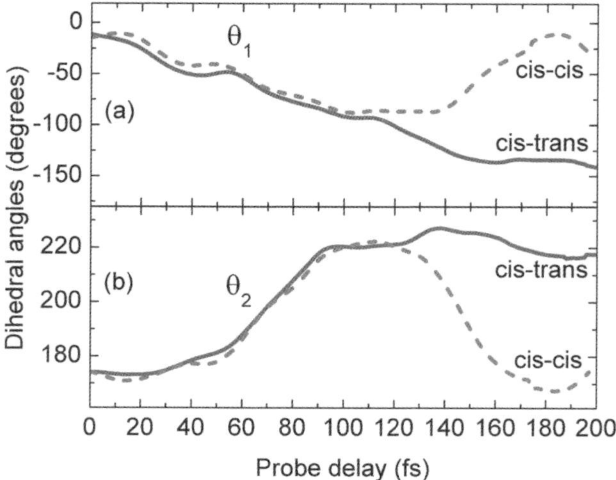

Figure 7: Torsional angles θ_1 and θ_2 during isomerizing trajectories (solid line) and in non-isomerizing paths (dashed lines).

This torsion also persists on the photoproduct side, accounting for chromophore ends that are motionless during the isomerization process in agreement with structures derived from FSRS [11] and reminiscent of Warshel's bicycle pedal proposal [36, 37].

5. Conclusions

Taken together, these data allow us to finally expose the complete molecular dynamics of this classical photochemical process and thereby understand how rhodopsin achieves its unique reaction speed. The structural evolution from the reactant towards the CI is restricted almost exclusively to the atoms in the centre of the molecule. This behaviour is promoted by the tight binding pocket provided by the protein to the chromophore, which restricts the possible motion at its ends through non-covalent interactions [38]. Therefore, retinal can use all of the incident photon energy to drive minimal atomic displacements in the relevant region and reach the CI region within 80 fs resulting in a very high reaction speed. Once the local isomerization has taken place, the overall highly strained structure can then relax in the photoproduct well to result in the more *trans*-like structure, which completes the primary isomerization reaction in rhodopsin.

These observations are consistent with a CI having a strongly peaked topography, leading to a very efficient ballistic transfer of the molecular wavepacket from the 11-*cis* reactant to the all-*trans* photoproduct and to the very high quantum efficiency of the visual response. The combined experimental-computational approach presented in this study can now be extended to investigate CIs with different topography, such as those in rhodopsin mutants or analogs (such as 9-*cis* isorhodopsin [39]).

References

1.   T. Yoshizawa and G. Wald, *Nature* **197**, 1279 (1963).
2.   J. E. Kim, M.J. Tauber and R.A. Mathies, *Biochemistry* **40**, 13774 (2001).
3.   R. W. Schoenlein, , L. A. Peteanu, , R. A. Mathies, and C. V. Shank, *Science* **254**, 412 (1991).
4.   G. Haran, E.A. Morlino, J. Matthes, R.H. Callender, and R.M. Hochstrasser, *J. Phys. Chem.* **A 103**, 2202 (1999).
5.   H. Chosrowjan *et al.*, *J. Am. Chem. Soc.* **120**, 9706 (1998).
6.   H. Kandori et al. *Chem. Phys. Lett.* **334**, 271 (2001).
7.   G. A. Schick, T. M. Cooper, R. A. Holloway, L. P. Murray, and R. R. Birge, *Biochemistry* **26**, 2556 (1987).
8.   G.R. Loppnow and R.A. Mathies, *Biophys. J.* **54**, 35 (1988).
9.   Q. Wang, R.W. Schoenlein, L.A. Peteanu, R.A Mathies, and C.V. Shank, *Science* **266**, 422 (1994).
10. J.E. Kim, D.W. McCamant, L. Zhu, and R.A. Mathies, *J. Phys. Chem.* **B 105**, 1240 (2001).
11. P. Kukura, D.W. McCamant, S. Yoon, D.B. Wandschneider, and R.A. Mathies, *Science* **310**, 1006 (2005).
12. M. Garavelli, P. Celani, F. Bernardi, M.A. Robb, and M. Olivucci, *J. Am. Chem. Soc.* **119**, 6891 (1997).
13. R. González-Luque *et al.*, *Proc. Natl. Acad. Sci. USA* **97**, 9379 (2000).
14. M. Klessinger and J. Michl, *Excited States and Photochemistry of Organic Molecules*, VCH Publishers, New York (1994).
15. B.G. Levine and T.M. Martinez, *Annu. Rev. Phys. Chem.* **58**, 613 (2007).
16. T.J. Martinez, *Nature* **467**, 412 (2010).
17. L. M. Frutos, T. Andruniow, F. Santoro, N. Ferre, and M. Olivucci, *Proc. Natl. Acad. Sci. USA* **104**, 7764 (2007).
18. S. Hayashi, E. Tajkhorshid, and K. Schulten, *Biophys. J.* **96**, 403 (2009).
19. G. Tomasello et al. *J. Am. Chem. Soc.* **131**, 5172 (2009).
20. D. Polli *et al.*, *Nature* **467**, 440 (2010).
21. W.J. DeGrip, F.J.M. Daemen, and S.L. Bonting, *Methods in Enzymology* **67**, 301 (1980).
22. D. Brida, C. Manzoni, G. Cirmi, D. Polli, and G. Cerullo, *IEEE J. J. Select. Topics Quantum Elect.* **18**, 329 (2012).
23. C. Manzoni, D. Polli, and G. Cerullo, *Rev. Sci. Instrum.* **77**, 023103 (2006).

24. D. Polli, L. Lüer, and G. Cerullo, *Rev. Sci. Instrum.* **78**, 103108 (2007).
25. H. Lin and D.G. Truhlar, *Theor. Chem. Acc.* **117**, 185 (2006).
26. P. Altoè, M. Stenta, A. Bottoni, and M. Garavelli, *Theor. Chem. Acc.* **118**, 219 (2007).
27. U.C. Singh and P.A. Kollman, *J. Comput. Chem.* **7**, 718 (1986).
28. D.A. Case *et al. J. Comput. Chem.* **26**, 1668 (2005).
29. B. O. Roos, in *Ab Initio Methods in Quantum Chemistry - II* Vol. 69 (Ed.: K. P. Lawley), 1987, p. 399.
30. A. Cembran, F. Bernardi, M. Olivucci, and M. Garavelli, *J. Am. Chem Soc.* **126**, 16018 (2004).
31. C.S. Sloane and W.L. Hase, *J. Chem. Phys.* **66**, 1523 (1977).
32. L. Verlet, *Phys. Rev.* **159**, 98 (1967).
33. O. Weingart, I. Schapiro, and V. Buss, *J. Phys. Chem.* **B 111**, 3782 (2007).
34. K. Andersson, P.-A. Malmqvist, and B.O. Ross, *J. Chem. Phys.* **96**, 1218 (1992).
35. N. Försberg and P.-Å. Malmqvist, Chem. Phys. Lett. 274, 196 (1997).
36. A. Warshel, *Nature* **260**, 679 (1976).
37. A. Warshel and N. Barboy, *J. Am. Chem. Soc.* **104**, 1469 (1982).
38. T. Okada *et al.*, *J. Mol. Biol.* **342**, 571 (2004).
39. R.W. Schoenlein, L.A. Peteanu, Q. Wang, R.A. Mathies, C.V. Shank, *J. Phys. Chem.* **97**, 12087 (1993).

ULTRAFAST INVESTIGATION OF ENERGY AND CHARGE TRANSFER IN A PROTOTYPICAL PHOTOVOLTAIC BLEND

GUGLIELMO LANZANI[†] and
AJAY RAM SRIMATH KANDADA

*Center for Nano Science and Technology @ Polimi, Istituto Italiano Di Tecnologia,
Via Pascoli 70/3 20133 Milano, Italy.*

DANIELE FAZZI

*Theoretical Chemistry, Max-Planck-Institut für Kohlenforschung
(MPI-KOFO), Kaiser-Wilhelm-Platz 1
D-45470 Mülheim an der Ruhr, Germany.*

In this chapter we introduce the pump-probe technique and discuss its application to the study of energy and charge transfer at donor(D)/acceptor(A) interfaces. The pump-probe technique combines spectroscopy (transition energy, line shape and cross section) with time resolution (population lifetime and decay path), providing information on both ground and excited states. Deactivation of excited states may involve internal conversion, intersystem crossing or recombination in molecular systems, as well as inter band scattering or carrier-carrier scattering in crystals. Usually each deactivation process is quantified by a rate, in s^{-1}. The most widely used quantum mechanical expression for the rate of a transition induced by a perturbation is the golden rule [1] that depends on the square matrix element of the coupling term between initial and final state and the density of states. Indirectly, a time resolved experiment can provide a measurement of such quantities. Practically it is difficult to disentangle the two contributions. Both energy and charge transfer rate can be casted into the Fermi golden rule, which is however valid for weak perturbations that do not mix the states. When this happens the transition rate has to be worked out between the hybrid states. A typical example is inter system crossing from singlet to triplet, where the transition matrix element of the nuclear position gradient (kinetic energy) is worked out onto the mixed singlet-triplet states resulting from spin-orbit coupling [2]. A more subtle and yet not fully solved situation regards energy and charge transfer, where Coulomb coupling could be large enough to induce coherent states delocalized on several monomers. Here we do not address this problem, but focus our attention to a molecularly dispersed morphology wherein molecular interaction seems suitable for describing the interacting units. We find that energy transfer is as important or even dominating the D/A interface [3]. While looking at the electronic structure and the coupling terms this appears as very realistic, as a matter of fact the occurrence of energy transfer at D/A photovoltaic interface is rarely considered. Energy transfer is

---

[†]email: Guglielmo.Lanzani@iit.it

typically confined to special situations where wavefunction overlap is clearly negligible (chemical dyads, or in presence of inter-layers). According to our results the correct description of the photovoltaic interaction should consider both energy and charge transfer, perhaps in a comprehensive theory of Coulomb coupling.

1. The Pump-Probe Technique

There is no birthday of the pump-probe technique, as it is the most basic and general approach to time resolution. Probably it has been employed in many different ways even much before laser were invented and applied to spectroscopy. The underlying concept is very simple: A stimulus causes a non equilibrium state, a probe measures its evolution in time. Here we are dealing with the optical version, where the stimulus and the probe are light pulses. A short, resonant pulse, excites the samples and a weaker, delayed pulse is used for probing pump-induced changes. Those could regard transmission, reflection, scattering. Changes, let's say in transmission to stay with the most common setting, are measured upon changing pump-probe delay to get the time dependent characterization[4].

Note that the pump-probe delay is fixed during each data point acquisition. The detector is time-integrating the transmitted energy of the probe pulse, on a very slow time scale, order of magnitude longer then the pulse width. This is the great advantage of pump-probe. Time resolution occurs by time correlation between pulses, not by direct time resolved detection. The limiting factor in time is the pulse time duration. Information in the frequency domain can be retraced on a very narrow spectral range, limited by the observation time that is typically in the micro second time range.

The polarization in the material subject to pump and probe pulses can be described by a perturbative expansion $P(t) = P^{(1)}(t) + P^{(3)}(t) + ...$ that can be safely truncated at the third order term $P^{(3)}(t)$ for most common experiments. In some cases, when pulses have time duration comparable or even shorter than the electronic dephasing time in the sample, the perturbative approach cannot be used and the full quantum mechanical formalism is required. Here we limit ourselves to the pulse separated regime that can be accounted for by the effective linear approximation.

According to the Lambert-Beer relation, transmission T is given by

$$T = e^{-\alpha L} \tag{1}$$

where α is the absorption coefficient and L the sample thickness. Absorbance A is defined as

$$A = -\log_{10} T \qquad (2)$$

The frequency dependent absorption coefficient is given by a sum of all possible transitions between pairs of state, $i.j$, with population N_i and N_j according to:

$$\alpha = \sum_{i,j} \sigma_{ij}(\omega)(N_i - N_j) \qquad (3)$$

where σ_{ij} is the cross-section for the transition. At time $t=0$, the pump pulse causes the transmission change from T to T*. Accordingly

$$\Delta A = -\log\left(\frac{T^*}{T}\right) = -\log\left(1 + \frac{\Delta T}{T}\right) \cong -\frac{\Delta T}{T\,2.3} \qquad (4)$$

Where $\Delta T = T^* - T$ and the last approximation holds for small signal.

Now let's assume the population of the k-state is changing, due to pump excitation according to:

$$N_k^* = N_k + \Delta N_k(t) \qquad (5)$$

This reflects into a change in absorption $\Delta\alpha = \sum_{i,j} \sigma_{ij}(\omega)(\Delta N_i - \Delta N_j)$.

The probe pulse interrogates the non equilibrium state at time τ. The measured change in absorbance is:

$$\Delta A(\tau) = -\log\left\{ \frac{\int dt I_{pr}(\omega, t-\tau)\exp[-\Delta\alpha(t)L]}{\int dt I_{pr}(\omega, t)} \right\} \qquad (6)$$

Which, for small signal, can be approximated to:

$$\frac{\Delta T}{T} = -\frac{\int dt I_{pr}(t-\tau)\Delta\alpha(t)L}{\int dt I_{pr}(t)} \qquad (7)$$

If we assume that the probe pulse is much shorter then the characteristic time scale of evolution of ΔN and thus $\Delta\alpha$, we can use a delta like pulse in the integral expression:

$$\frac{\Delta T}{T}(\omega,\tau) \cong -\Delta\alpha\,L = -\sum_{i,j} \sigma_{ij}(\omega)\left[\Delta N_i(I_{Pu},\tau) - \Delta N_j(I_{Pu},\tau)\right]L \qquad (8)$$

This equation is at the heart of pump-probe: The resonant pump causes population redistribution among some of the electronic states of the sample, creating a non-equilibrium state. The probe measure new transitions associated to such population changes, and track in time the recovery of the system to equilibrium.

When the probe pulse duration cannot be disregarded, eq. 8 should include the correlation with the pulse temporal profile (in intensity) f_P:

$$\frac{\Delta T}{T}(\omega,\tau) \cong -L\sum_{i,j}\sigma_{ij}(\omega)\left[\Delta N_i(I_{Pu},\tau) - \Delta N_j(I_{Pu},\tau)\right]\otimes f_p \qquad (9)$$

Note that the time dependent population term includes the pump pulse behavior, which is in many cases represented by a convolution between the material response and the pump pulse profile.

Remember that using pulses of duration t_P you can reasonably study processes occurring on time scales $\Delta t > t_P$. With noise-free measurements (but they do not exist!) you could in principle track down processes occurring on shorter time scales, by carefully de-convoluting the time traces once both pump and probe temporal profiles are known with very high accuracy. All this never happen in reality, and the suggestion is as long as possible to study dynamics occurring on time scales longer than your pulses.

Let's now discuss how we can describe the material evolution under photoexcitation. The pump pulse acts on the sample by changing the stationary level occupation, $N_i \rightarrow N_i$ (t) where t is the time and N_i the equilibrium population (which is usually zero, except ground state). With appropriate initial conditions, and neglecting the coherent regime that would require Bloch equations, the time dependent population can be obtained from rate equations such as:

$$\frac{dN_i(t)}{dt} = G_i(t) - R_i(t) \qquad (10)$$

where $G(t)$ and $R(t)$ are the generation and deactivation rate, respectively. For instance, if the state "i" is directly populated by the pump pulse via one photon transition from the ground state the generation term for a thin sample is:

$$G_i(t) = \sigma_{0i}N_0(t)\frac{F(t)}{t_P} \qquad (11)$$

where σ_{0i} [cm^2] is the cross section for the one photon transition $0 \rightarrow i$, $N_0(t)$ [cm^{-3}] the time-dependent ground state population, which gets depleted, $F(t)$ [photon·cm^{-2}] the pump photon flux and t_p [s] the pump pulse duration.

Solving one rate equation for each involved state "i" allows to reconstruct the signal according to eq. 8 where $\Delta N_i(t) = N_i(t) - N_i$.

The spectrum associated to the state (or species) "j" is $A_j(\omega) = \sum_i \sigma_{ij}(\omega)d$ comprising all possible transitions from this state, and eq. 8 can be also written as $\frac{\Delta T}{T} = -L\sum_j A_j(\omega)N_j(t)$; an expression often used in global fitting analysis. In this analysis one get 1) population kinetics from rate equations, then 2) associate to each population a spectrum, and finally 3) re-construct the expected signal. This formula shows that the typical $\Delta T/T$ spectrum is a superposition of individual spectra ($A_j(\omega)$) from several photoexcited states, and time-dependent data must be taken at various probe wavelengths in order to single out the various contributions. Note that global fitting maybe be carried out by "blind" routine which finds mathematically the best combination of spectra. Such kind of results lacks most of the time of physical meaning and should be considered with great cautions. The "guess" on the excited species in the experiment should be assessed by general knowledge on the system, cross-checked with other experiments or varying the experimental condition.

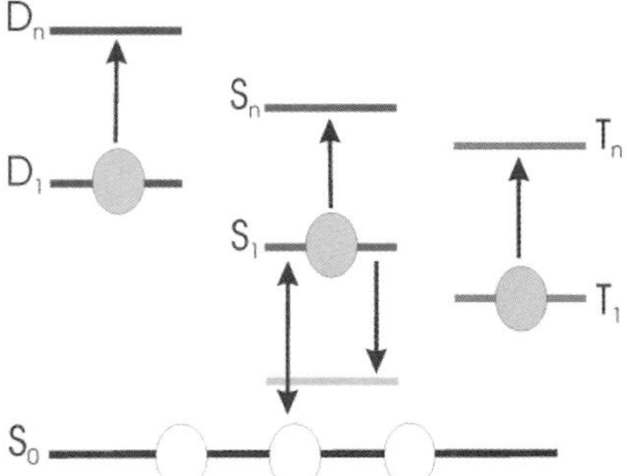

Figure 1: Probe transition following pump photoexcitation. The double arrow represents stimulated emission (downward) and bleaching (upward) of the 0-0 transition.

For $j=0$, $\Delta N_0(t)$ described ground state depletion and corresponding enhancement in probe transmission, named photobleaching (PB). PB corresponds to positive ΔT and has the spectral shape of ground state absorption when thermalization, which is usually very fast, is over. For all the other levels ($j > 0$) the probe pulse can stimulate both upwards and downwards transitions. In particular, transitions to higher lying states give rise to photoinduced absorption (PA), transitions to lower lying states give rise to stimulated emission (SE), see **Figure 1**. After thermalization, usually very fast, few lower lying states are occupied. When the lowest singlet excitation (S_1) is dipole-coupled to S_0, i.e. the system is luminescent, SE is taking place. In this case the relation between the SE spectrum and the PL spectrum can be derived from A and B Einstein coefficients as $SE(v) = \dfrac{c^3}{8\pi h}\dfrac{1}{v^3} PL(v)$. In general the existence of SE does not mean that positive ΔT is observed in the region of emission, for it depends on the spectral overlap with other absorbing transitions (often there) and on the relative cross sections involved. As a result SE may not appear at-all even in light emitting materials. Another important observation is that SE is clearly distinguished when the absorption edge is sharp and the transition occurs from the excited state to the vibrational replica of the ground state, which is red shifted. When SE occurs to the vibrationless ground state, this fully overlaps with absorption and "contribute" to photobleaching. It becomes tricky to distinguish if photoexcitation causes enhanced transmission, or light amplification. Both are positive ΔT signals. Note that light amplification (gain) requires population inversion, so the question is if the material behaves like a three level system. To check this one can measure the linear ground state absorption, α, and compare it to the nonlinear pump induced change in

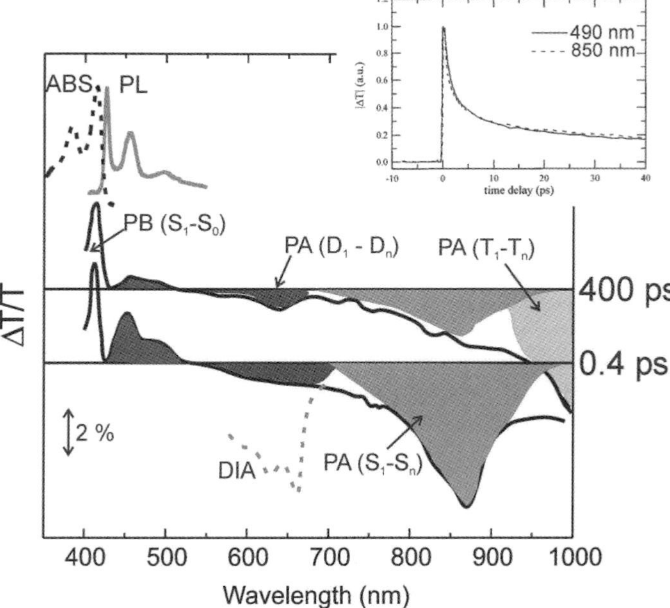

Figure 2: Pump-probe spectra of MLPPP (see chemical structure in the inset) thin film after excitation at 390 nm. Pump-probe delay is specified on the right. Ground state absorption, Photoluminescence (PL) and Doping Induced Absorption (DIA) are also reported

absorption $\Delta\alpha = -\dfrac{1}{L}\ln\left(1+\dfrac{\Delta T}{T}\right)$. If $-\dfrac{\Delta\alpha}{\alpha} > 1$ there is light amplification and gain.

As an example, we show in **Figure 2** two pump-probe spectra as measured in a conjugated polymer film[5], after resonant excitation at 390 nm. Singlet transitions associated to S_1 are SE (S_1-S_0) and PA(S_1-S_n). The kinetics at the peak wavelengths are reported in the inset. The coincidence corroborates the assignment to the same state. Photobleaching, in black, occurs below the absorption edge, matching ground state absorption (ABS, as reported in the plot). Doping induced absorption (DIA) allows identified charge signature in the transient spectrum. Finally the long wavelength peak appear at 400 ps pump-probe delay is assigned to triplet transition. It is clear that this band is however present from the very beginning, suggesting singlet fission as formation mechanism.

The pump probe technique has an extremely broad scope and it can be applied to a variety of different samples and condition. One weak point is however sensitivity. Being a differential technique with a background, it cannot compete with photoluminescence or other type of back ground free scattering. Laser stability, high repetition rates of pulse generation and a careful signal collection and analysis, can push the experiment to best performance. An alternative approach, recently implemented, is that of the multi pass transmission[3].

Transient Absorption by Multiple Pass (TrAMP) technique enables us to perform measurements on samples with optical densities below 10^{-2}. This technique draws its inspiration from absorption spectroscopy of gases where high sensitivities are achieved by increasing the optical path length by multiple reflections of the beam in the gas chamber[6]. The same principle is extended to

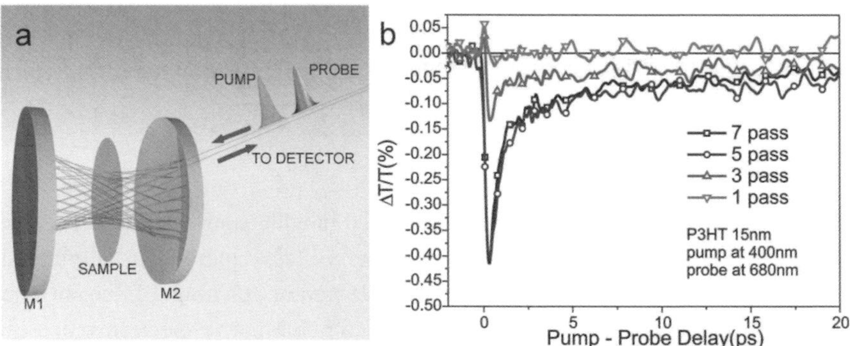

Figure 3: a) Schematic of TrAMP set up. M1 and M2 represent two curved mirrors with the sample kept at the focus. b) TrAMP kinetics collected by increasing the number of passes. The sample is a 15 nm P3HT film on a fused silica substrate (OD= 0.1).

transient absorption spectroscopy by passing the pump and probe pulses multiple times through the sample, keeping the time delay between them constant. In this way, the pump-probe signals can be amplified by a factor proportional to the number of passes through the sample. It can be shown that TrAMP improves the S/N ratio: A p-time multi-pass data would have a S/N ratio p-times better than that of a standard pump-probe repeated p-times.

Figure 3 shows the schematic lay-out of the setup that we used for realizing pump probe experiments on thin samples by adopting the Transient Absorption by Multiple Pass (TrAMP) technique. **Figure 3b** demonstrates the concept and validity of our approach. It shows the TRAMP kinetics collected at various numbers of passes in a 15nm spin coated film of P3HT on a 150μm fused silica substrate with an OD of 0.05, excited in resonance at 400 nm. Standard pump probe (single pass) cannot detect transient transmission changes in such a thin film. However, by multi pass the TrAMP technique provides an efficient amplification of the Pump-Probe signal, that is shown growing at each pass, while keeping the kinetic itself intact. This also demonstrates that possible effects on the time resolution, such as group velocity delay between pump and probe and pulse broadening, are negligible in our experiment, even in the worst condition of large spectral separation between the pump and the probe wavelength. After certain number of passes, signal amplification reaches a saturation level that can be attributed to the almost complete absorption of the pump. The threshold number of passes depends on the optical density of the sample. We use this powerful tool here to study ultrathin bilayers composed of prototypical donor and acceptor molecules used organic photovoltaics. The motivation here is to understand interface specific dynamics and in particular decipher the competition between energy transfer and charge transfer from the donor to the acceptor. Before presenting our experimental results, in the next section we briefly review the theories of energy and charge transfer and then using simple equations we calculate the respective rates for our prototypical system.

2. Charge and Energy Transfer

Transient absorption has been widely employed in understanding the mechanism of charge photogeneration in a variety of photo-voltaic materials, usually blends of donor and acceptor molecules. By the temporal tracking of the spectral signatures of charges (PA bands of charges in a pump-probe spectrum), one can directly measure the Marcus rate of electron transfer from the donor to the acceptor, the primary event for charge generation. Marcus[7] described electron transfer as an activated process with an energy activation barrier ΔG^*, which is a function of Gibbs free energy ΔG^0 and the reorganization energy λ:

$$\Delta G^* = \frac{\lambda}{4}\left(1 + \frac{\Delta G^0}{\lambda}\right)^2 \tag{12}$$

The re-organization energy refers to the energy required to bring the reactant state to the geometry of the product state without an electron transfer. Marcus used an Arrhenius relationship to extract the rate of electron transfer as a thermal activated hop. Such a classical theory was later extended to quantum systems using the golden rule formulation for the probability of transition between two states[8]. In the high temperature limit when all the vibrations can be treated classically, the semi-classical rate of electron transfer is:

$$k_{ct} = \frac{2\pi}{\hbar}V_{ct}^2\left(\frac{1}{4\pi\lambda kT}\right)^{\frac{1}{2}} exp\left(-\frac{(\lambda+\Delta G^0)^2}{4\lambda kT}\right) \tag{13}$$

V_{ct} is the Coulomb coupling between the donor and acceptor units which can be modelled as $V_{ct} = A\ exp(-\beta \cdot (r-r_0))$[9] where β is a distance scale factor and r_0 is the minimum distance dictated by the steric hindrance between the two interacting units.

Marcus theory is for a chemical reaction and does not consider the density distribution of the electronic energy levels of the donor and the acceptor systems. While considering injection from organic molecules into metal-oxide semiconductors (as in dye sensitized solar cells), Gerisher's extention of Marcus theory is to be considered [10]. Marcus electron transfer can occur from any occupied electronic energy state of the organic molecule that is matched in energy with an unoccupied receiving state in the acceptor. The number of such unoccupied states can be written as:

$$N_{unoccupied}(E) = [1 - f(E)]\rho(E) \tag{14}$$

Marcus-Gerisher rate can be calculated by integrating over all the available states at all energies and is given as:

$$k_{et} = A \int V^2[1 - f(E)]\rho(E)\ exp\left(-\frac{(\lambda+\Delta G^0)^2}{4\lambda kT}\right) dE \tag{15}$$

Apart from charge transfer, close vicinity of molecules may also result in energy transfer, where in the donor molecule is de-excited with a simultaneous excitation of the acceptor molecule. Strong intermolecular coupling can lead to a ultrafast coherent energy transfer, that has been observed in light harvesting complexes[11]. However, in most conjugated organic molecules, energy transfer occurs in the incoherent limit of weak coupling due to localized phonon interactions. In such a incoherent regime, energy transfer can be explained either by Coulomb (Förster)[12] or exchange (Dexter)[13] mechanisms. The rate of energy transfer can be expressed as sum of Förster and Dexter rates:

$$k_{EET} = k_{Förster} + k_{Dexter} = \frac{2\pi}{\hbar} \, (V_C^2 + V_D^2) J \qquad (16)$$

V_C and V_D are the electronic coupling matrix elements for Förster and Dexter mechanism and J is the spectral overlap between the emission of the donor and the absorption of the acceptor. In the limit of the point-dipole approximation, the Förster coupling can be considered as $V_C = (1/4\pi\varepsilon_0\varepsilon_r) \cdot (k \cdot \mu_D \, \mu_A / R^3)$[14] with: k the orientation factor and μ_A and μ_D the electronic transition dipole moments for the acceptor and the donor respectively. Since Dexter coupling takes into account the wavefunction overlap much similar to the coupling considered for charge transfer, $V_D = = A \exp(-\beta \cdot (r-r_0))$.

2.1. Expected charge and energy transfer rates for the prototype organic photovoltaic D/A interface

One of the widely studied organic photovoltaic blend is composed of poly (3-hexylthiophene) (P3HT), as the donor component, mixed with phenyl-C61-butyric acid methyl ester (PCBM), as the acceptor. This blend provides solar cells with 85% Internal Quantum Efficiency and power conversion efficiency of around 5%[15,16]. Such high quantum yield of charge photogeneration with respect to pure organic semiconductors results from blending the two components into a so called bulk hetero junction (BHJ). The D/A components are deposited from common solution and tend to segregate over tens of nanometers forming PCBM-rich clusters and crystalline domains of P3HT. The two phases however are not separated by sharp interfaces, because *substantial intermixing* takes place in the border region defined by the extent of diffusion of PCBM molecules into P3HT(see the illustration in **Figure 6a**)[15-20]. In this intermixed region, the two components are finely dispersed at the molecular level, as it has been clearly observed in bi-layer structures[21].

Processes at such an interface can be simplified as those involving a photo-excited P3HT molecule and a neighbouring PCBM molecule. We employ equations reported above to calculate the transfer rates of charge and energy from P3HT to PCBM as a function of intermolecular distance.

First we estimate CT rates from Equation 13. A and β parameters involved in the calculation of V_{ct} are obtained by fitting the computed semiempirical ZINDO/S electronic couplings amongst LUMO orbitals (i.e. <LUMO$_{P3HT}$| H |LUMO$_{PCBM}$>) by considering different intermolecular distances between P3HT and PCBM namely: 3.6 Å, 4.0 Å, 4.5 Å, 5.0 Å, 5.5 Å, 6.5 Å, 7.5 Å and 10.0 Å. From the fitting procedure the parameters result to be: A = 1.176·10^{-16} J and β = 20·10^9 m^{-1}. r_0 has been considered equal to 3 Å.

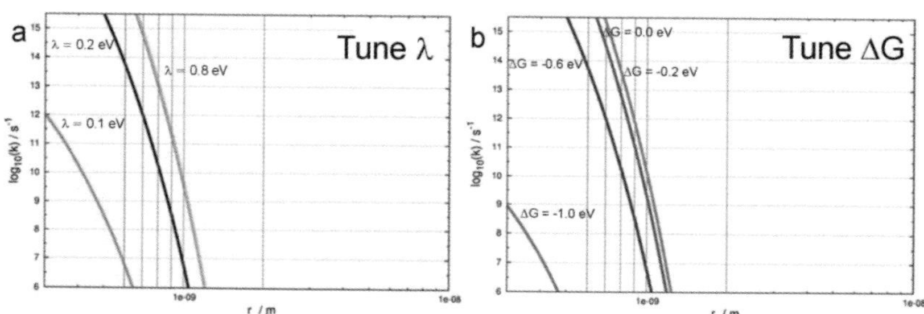

Figure 4: kct trends by varying a) reorganization energy and b) the free energy term

Reorganization energy λ and ΔG in the equation are taken from a wide set of data from the literature to be 0.2 eV and -0.6 eV respectively. However since an accurate determination of these two parameters is not trivial, we present the trend of the charge transfer rate by tuning the reorganization energy value (from 0.0 to 0.8 eV) considering ΔG = -0.6 eV, A = 1.176·10^{-16} J and β = 20·10^9 m^{-1} in **Figure 4a**, and the variation of k_{CT} transfer rate by tuning the free energy term ΔG (from 0.0 to -1.0 eV) considering λ = 0.2 eV, A = 1.176·10^{-16} J and β = 20·10^9 in **Figure 4b**. For all the different parameters considered, ultrafast electron transfer (within 1 picosecond) can be observed only when the intermolecular distance is ≤ 6 Å. This is due to the nature of the considered coupling between the interacting molecules which scales down exponentially.

Second we estimate the excitation energy transfer (EET) using Equation 16. For the Dexter term we assume that the electronic coupling V_D has the same distance-decay dependence as the charge transport matrix element V_{CT}. For this reason we consider V_D equal to V_{CT}, being both dependent basically on the two-electron integral term as the overlap matrix S_{ij}. For the Förster term, k is set to 1 for simplicity. The electronic transition dipole moments μ_{P3HT} and μ_{PCBM}, for P3HT and PCBM respectively are computed at the TD-DFT level to be equal to 1.8492·10^{-28} C·m and 3.2920·10^{-31} C·m. Spectral overlap J is estimated experimentally, by calculating the common area between the normalized emission (of P3HT) and absorption (of PCBM) spectra. **Figure 5** shows the EET rate a function of intermolecular distance. Also plotted for reference is the trend of CT rate calculated using the Marcus expression with $\lambda = 0.2$ eV and $\Delta G = -0.6$ eV.

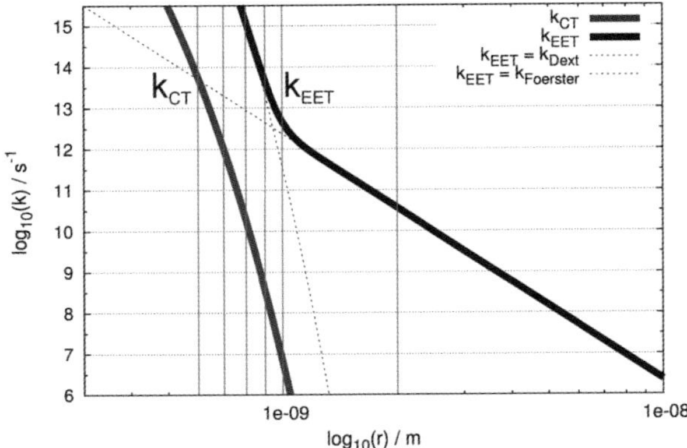

Figure 5: EET and CT rates as a function of donor-acceptor distance. Dotted lines represent Förster like and Dexter like contributions to the EET rate.

In the considered spatial range within 1 nm, EET is always faster than CT mechanism. The EET is in the ultrafast time regime (from few fs to hundreds of fs, **Figure 5**) and the CT, or the charge separation process, in the picosecond time scales for the optimized geometry of P3HT and PCBM molecules. The short range EET is due to Dexter coupling that out-competes both Förster and CT.

By and large, charge transfer is regarded as the fundamental event occurring upon donor photo-excitation in the BHJ[22,23] while energy transfer is generally ignored, at least at early times. However the proximity of D and A molecules in the blend may in principle also lead to a much faster electronic excitation energy transfer as revealed by our calculations. Indeed this has been reported for a few systems prepared *ad hoc,* such as D/A bilayers separated by an interlayer[24-26] that hampers charge separation or in chemically bound dyads[27], wherein the D and A are in a fixed positions with respect to each other. Both these cases are thus modifications of the standard BHJ, where essentially the wavefunction overlap is largely reduced. On the contrary, there are no experimental evidences of electronic excitation energy transfer as *primary event* in standard BHJ systems. Why electronic excitation energy transfer appears only in special situations and why charge transfer overwhelms it in BHJ, remains an open question.

3. Experimental Investigation of the Intermixed D/A Boundary Phase between P3HT and PCBM

In a conventional BHJ geometry, the photoexcitation dynamics is the superposition of dynamics in the pristine domains and the intermixed region at the interface and hence not reminiscent of the dynamics of the latter. We study here using the TrAMP technique, the photo-excitation dynamics of the intermixed P3HT/PCBM region obtained from a nominal 5nm/5nm bilayer. We first deposit a 5 nm P3HT layer on fused-silica substrate and then we spin coat the PCBM on top from an orthogonal solvent [21] with a nominal thickness of 5 nm. The absorption spectrum of the sample is a linear combination of the absorption spectrum of the two components with an OD of 0.02 at the absorption peak (**Figure 6b**). Because the P3HT film thickness is smaller than the PCBM molecule diffusion length, PCBM almost fully penetrate the polymer and we obtain a layer of P3HT *doped* with PCBM [21] (see cartoon in **Figure 6a**).

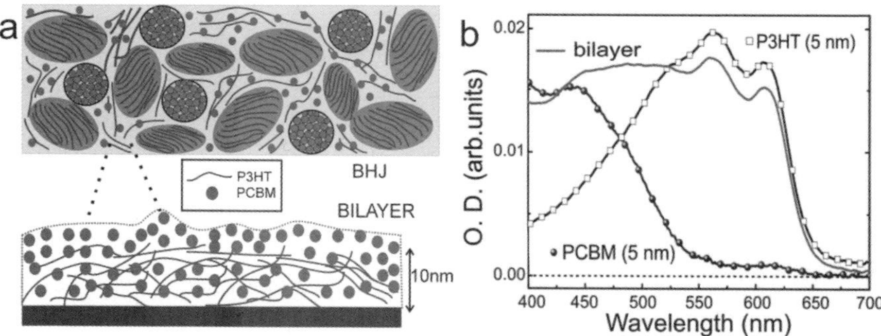

Figure 6: a) Illustration of morphologies of BHJ and bilayer architectures of P3HT/PCBM blends. b) Absorption Spectrum of 5 nm films of P3HT and PCBM and the bilayer

Figure 7(top panel) shows the transient absorption ($\Delta T/T$) spectra of 5 nm thin P3HT film as a control experiment. At wavelengths shorter than 620 nm it exhibits a positive signal due to ground state bleaching (GSB). At probe wavelengths greater than 620 nm it shows a negative photo-induced absorption (PA) signal. A comparison with dynamics, as reported in **Figure 8a**, and a rather extended set of experimental and computational data reported in literature helps in disentangling the contributions beyond spectral overlap [28,29]. Two species can be identified at early time: singlet excitons and polaron pairs (PP) as assigned by Korovyanko et al.[28] The fast decaying PA band appearing at 650 nm is assigned to geminate recombination of PP. The longer lived signal is due to singlet states, with some contribution from polarons escaping the geminate recombination as reported in the earlier TA studies on P3HT.

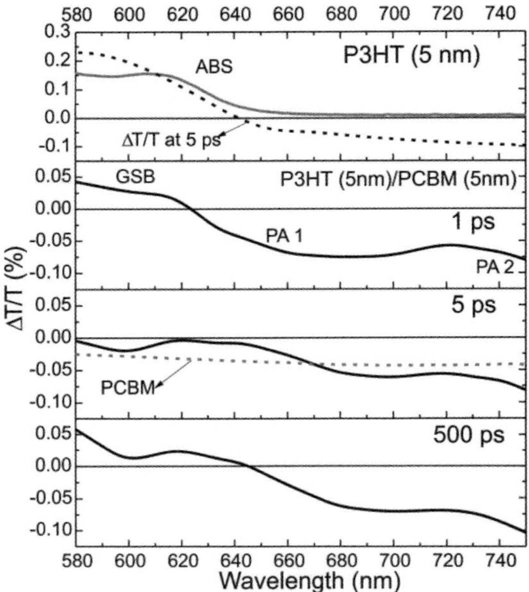

Figure 7: Transient absorption spectra of 5nm film of P3HT (top panel) and TA spectra of P3HT/PCBM bilayer at 1ps, 5 ps and 500 ps (bottom panels). Also plotted in the top panel is the absorption spectrum of P3HT. Plotted along with the 5 ps TA spectrum of the bilayer is the TA spectrum of PCBM when pumped resonantly at 400 nm.

The ΔT/T spectrum of the P3HT/PCBM intermixed layer is shown in **Figure 7**, and selected time traces are reported in **Figure 8b**. At a first glance, early spectrum resembles the pure P3HT spectrum, though with a remarkable reduction in the amplitude of GSB with respect to the PA. Within 2 ps, both the P3HT GSB and the PA band centered at 650 nm decay to zero with similar kinetics, giving rise to *a negative signal extending on the whole experimental range*. Notably the 580-620 nm spectral region that corresponds to P3HT ground state absorption shows null or negative signal. This indicates that P3HT is weakly or not excited anymore. The broad PA should correspond to an excited state population that is not perceived by the ground state of the polymer. The charge transfer process cannot explain such an observation: upon CT and hole-polaron formation, the polymer GSB stays constant due to the persistent depletion of the ground state population.

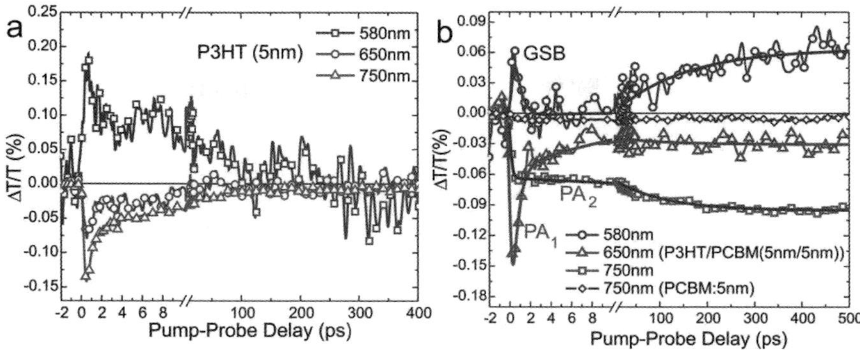

Figure 8: a) TA kinetics at specific wavelengths of 5 nm P3HT film. b) TA kinetics of the bilayer. Also plotted is the kinetic at a probe wavelength of 750, of a 5nm PCBM film pumped at 520 nm.

Our observation can however be explained by *electronic excitation energy transfer* from P3HT to PCBM. Accordingly, P3HT GSB would decay and a new PA due to ($S_n \leftarrow S_1$) in PCBM would appear. As a reference, ΔT/T spectrum of pristine PCBM, measured upon resonant excitation (400 nm), is reported in *Figure 3b*. It is a broad photoinduced absorption band, extending over the whole experimental range. This fits reasonably well with our observation, supporting our explanation. It is fair to note a discrepancy between the lineshape of the pure PCBM PA band and our data, in the 580-620 nm spectral region. This can be ascribed to a residual GSB of the polymer, indicating that following polymer photoexcitation there is a branching between electronic energy transfer and charge transfer towards PCBM, here in favor of the former.

Concominant excitation of PCBM, when we excite the P3HT, should also be evaluated as a possible explanation for the reported experimental results. At this purpose we studied a 5 nm thick PCBM film, with a similar optical density of the intermixed sample and pumping at 520 nm. The plot in **Figure 8b** contains the time trace at 750 nm, assigned to PCBM $S_n \leftarrow S_1$ PA peak. The signal amplitude, without any normalization, clearly shows that the role of direct PCBM excitation is negligible. We infer that direct photo-excitation of PCBM contributes less than 10% of the signal measured in the P3HT/PCBM sample.

The electronic energy transfer occurs on a time scale faster than our resolution ($\approx$ 200 fs). The broad PA band assigned to PCBM is formed during pump excitation and it is present from the very beginning in our transient spectra. This explains the different ratio between GSB and PA in the P3HT/PCBM sample with respect to the one in pristine P3HT.

On the longer timescale, we observe a build-up of the GSB with a time constant of 130 ps along with a correlated increase in the PA band centered at 750 nm. The return of GSB in transient spectra is a rather unusual phenomenon that is associated to back charge transfer following the initial energy transfer in D-A dyads of different type [27,31]. Here it can be understood as the generation of polaron states in P3HT via hole-transfer from PCBM. Though the absorption bands of P3HT$^+$ and PCBM singlet overlap, due to higher oscillator strength of the polaron, its contribution dominates the dynamics[3]. The overlap of the bands can also explain the apparent slower growth rate in the polaron PA with respect to the GSB. This is also confirmed by the slow PA growth at 900 nm (see **Figure 9a**) after the initial PP recombination, and this will be discussed in detail later. In neat P3HT the dominant contribution at 900 nm is the $(S_n \leftarrow S_1)$ transition (**Figure 9b**), but in the ultrathin diffuse layer such a contribution is absent, consistent with the conjectured electronic energy transfer.

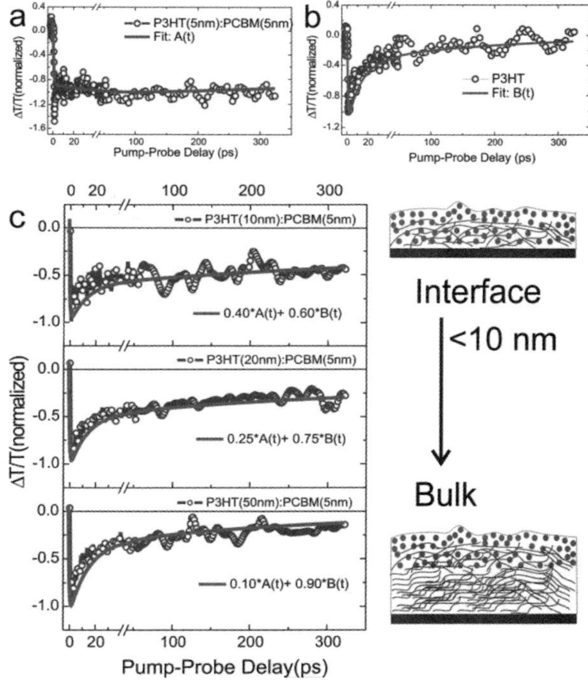

Figure 9: a) TrAMP kinetics at 900 nm probe for 5nm/5nm bilayer of P3HT/PCBM; a multi-exponential fit A(t) of the data is also shown. b) Similar kinetic and fit B(t) for a 5 nm P3HT film. c) TrAMP kinetics for bilayers with 10 nm, 20 nm and 50 nm thicknesses of P3HT with a constant 5 nm PCBM on the top. Also shown are the fits obtained by the linear superposition of A(t) and B(t) in each of the three cases.

Hence, in the considered molecularly-mixed morphology, the photophysical picture for charge generation can be interpreted as a two-step process. First, electronic energy transfer happens, as the dominant process, from P3HT to PCBM ($D^* + A \rightarrow D + A^*$). This is also well supported by our simple calculations of charge and energy transfer rates. Then it is followed by a hole transfer from the excited PCBM back to P3HT, ($A^* + D \rightarrow D^+ + A^-$).

In the standard BHJ the intermixed doped region is in between the donor and acceptor clusters [32]. In order to include exciton diffusion into our picture, we perform TrAMP measurements on P3HT/PCBM bilayers by increasing the thickness of the P3HT layer. Four samples are considered where-in 5 nm, 10 nm, 20 nm and 50 nm thick P3HT films are covered with 5 nm of PCBM. Each of these samples simulates the spatial extent for exciton diffusion in P3HT rich domains in a BHJ. The probe wavelength in these experiments is tuned to 900 nm, to have a signature of singlet exciton absorption in P3HT. As discussed earlier however, other species may also contribute at this wavelength.

The 5/5 nm bilayer shows an instantaneous PA that reduces by only a few percent within 2 ps, it stays flat and then it increases in 100 ps time scale (see **Figure 9a**). According to our model this signal contains contributions from PP recombination at initial time, and from P3HT polarons at longer delays. In **Figure 9b** we show the signal obtained from a pure 5 nm P3HT film. The samples in **Figures 9a, b** may represent the two distinct morphologies present in a BHJ: the well intermixed P3HT/PCBM interface and the pristine P3HT domain. The other three samples (10/5, 20/5 and 50/5 nm) show intermediate behavior, that can be simulated as a linear combination of the traces in 4a and 4b plots. This is shown by first fitting multi-exponential functions, A(t) and B(t), to the time traces in **Figures 9a,b** respectively. Linear combination of A(t) and B(t) reproduces quite well all the other data, enabling us to infer a relative weight of the two components in each of the samples. The 10 nm sample (**Figure 9c**) shows almost equal contribution from the pristine phase and the intermixed region. The 20 nm (**Figure 9d**) shows 25% contribution from the intermixed region and the 50 nm sample (**Figure 9e**) shows a meager 10% contribution. The *interface specific* dynamics are perceivable only when the thickness of the P3HT layer is less than 10 nm. This length scale stands for the balance between the electronic energy transfer and the exciton diffusion limited processes. We speculate that in a standard BHJ blend such dynamics are averaged out in the pump probe measurements and due to the spectral overlap and the consequent congestion of signals, charge transfer seems the dominant event.

4. Conclusions

In conclusion, employing the novel TrAMP technique, we measured the photo-excitation dynamics occurring in the finely intermixed phase of P3HT and PCBM. We show that the *expected* charge generation process is mediated by an initial ultrafast electronic energy transfer from the polymer to the acceptor molecule. The initial energy transfer ($D^* + A \rightarrow D + A^*$) takes place in less than 200 fs. Charge separation is a secondary process occurring between the excited acceptor and the polymer domain on the 100 ps time scale. Such intermixed domains are the actual border regions between D and A clusters in the photovoltaic active layer based on BHJ and hence the observed ultrafast EET should be an important event in the primarily step for charge generation, usually underestimated in the description of the photovoltaic event. This brings to some considerations on the role and actual values assigned so far to the exciton diffusion length. The critical interface length scale of 10 nm we found here is actually the convolution among the exciton diffusion, the molecular inter-diffusion and the optimum inter-molecular distance for efficient EET at the interface. We also note that Ward et al. [33] have recently shown a possible energy transfer pathway from copolymers such as PCDTDBT to PCBM. Thus, the process we highlight here is not necessarily specific to P3HT/PCBM, but it may be considered as an unavoidable paradigm for the concept of organic photovoltaic.

References

1. A. Ito and T. J. Meyer, Phys. Chem. Chem. Phys. 14, 13731-13745 (2012).
2. J. J. Sakurai, "Modern Quantum Mechanics", Adisson-Wesley Publishing Company, Revised Edition (1994).
3. A. R. Srimath Kandada et al, Sci. Rep. 3, 2073 (2013).
4. G. Lanzani, "The Photophysics behind Photovoltaics and Photonics", Wiley-VCH Verlag GmbH (2012).
5. G. Cerullo, S. Stagira, M. Nisoli, S. De Silvestri, G. Lanzani, G. Kranzelbinder, W. Graupner, G. Leising, Phys. Rev. B. 57, 12806(1998).
6. D. Herriot, H. Kogelnik and R. Kompfner, Applied Optics 3, 523-526 (1964).
7. R. A. Marcus, J. Chem. Phys. 24, 966-978 (1956).
8. M. R. Wasielewski, Chem. Rev. 92, 435-461 (1992).
9. J. L. Bredas et al. Proc. Natl. Acad. Sci. 99, 5804-5809 (2002).
10. H. Gerischer, J. Phys. Chem. 95, 1356-1359 (1991).
11. G. D. Scholes, Ann. Rev. Phys. Chem. 54, 57-87 (2003).
12. T. Förster, Annalen der Physik 437, 55-75 (1948).

13. D. Dexter, J. Chem. Phys. 21, 836 (1953).

14. A. Olaya-Castro and G. D. Scholes, Int. Rev. Phys. Chem. 30, 49-77 (2011).

15. I. A. Howard, R. Mauer, M. Meister, and F. Laquai, J. Am. Chem. Soc. 132, 14866–14876(2010).

16. W. Ma et al., Adv. Funct. Mater. 15, 1617 -1622(2005).

17. T. Erb et al., Adv. Funct. Mater. 15, 1193 - 1196(2005).

18. X. Yang. et al., Nano Lett. 5, 579 – 583(2005).

19. G. Dennler, M. C. Scharber, and C. J. Brabec, Adv. Mater. 21, 1323–1338(2009).

20. T. Savenije, J. E. Kroeze, X. Yang, and J. Loos, Adv. Funct. Mater. 15, 1260 - 1266(2005).

21. C. W. Rochester, S. C. Mauger, and A. J. Moulé, J. Phys. Chem. C. 116, 7287–7292(2012).

22. H. Ohkita et al., J. Am. Chem. Soc. 130, 3030–3042(2008).

23. J. Guo, H. Ohkita, H. Benten, and S. Ito, J. Am. Chem. Soc. 132, 6154-6164(2010).

24. D. C. Coffey, A. J. Ferguson, N. Kopidakis and G. Rumbles, ACS Nano. 4, 5437-5445(2010).

25. X. Y. Liu, M. A. Summers, S. R. Scully, and M. D. McGehee, J. Appl. Phys. 99, 093521(2006).

26. S. R. Scully, P. B. Armstrong, C. Edder, J. M. J. Fréchet and M. D. McGehee, Adv. Mater. 19, 2961-2966(2007).

27. P. A. van Hal et al., Phy. Rev. B. 64, 075206(2001).

28. O. J. Korovyanko et al., Phys. Rev. B. 64, 235122 (2001).

29. J. Cabanillas-Gonzalez, G. Grancini and G. Lanzani, Adv Mater. 23, 5468-5485(2011).

30. A. Bakulin et al., Adv. Func. Mater. 20, 1653-1660 (2010).

31. S. Pillai et al., Phys. Chem. Chem. Phys. 15, 4775-4784 (2013).

32. G. Grancini et al., J. Phys. Chem. Lett. 2, 1099-1105 (2011).

33. A. J. Ward et al., J. Phys. Chem. C. 116, 23931-23937 (2012).

VACANCY-DOPED PLASMONIC COPPER CHALCOGENIDE NANOCRYSTALS WITH TUNABLE OPTICAL PROPERTIES

ILKA KRIEGEL,[1⊥*] JESSICA RODRÍGUEZ-FERNÁNDEZ,[1]

[1] *Photonics and Optoelectronics Group, Department of Physics and CeNS, Ludwig-Maximilians-Universität München, Munich, Germany, Nanosystems Initiative Munich (NIM), Munich, Germany*

⊥*Current address: Dipartimento di Fisica, Politecnico di Milano, Piazza Leonardo da Vinci 32, 20133 Milano, Italy*

CHENGYANG JIANG[2]

[2]*Department of Chemistry, University of Chicago, Chicago, Illinois 60637, United States*

RICHARD SCHALLER[3,4]

[3]*Argonne National Laboratory, Center for Nanoscale Materials, Argonne, Illinois 60439, United States*

[4]*Department of Chemistry, Northwestern University, Evanston, Illinois 60208, United States*

ENRICO DA COMO[5]

[5]*Department of Physics, University of Bath, Claverton Down, BA2 7AY Bath, United Kingdom*

DMITRI V. TALAPIN[2,3]

[2] *Department of Chemistry, University of Chicago, Chicago, Illinois 60637, United States*

[3]*Argonne National Laboratory, Center for Nanoscale Materials, Argonne, Illinois 60439, United States*

JOCHEN FELDMANN[1]

[1] *Photonics and Optoelectronics Group, Department of Physics and CeNS, Ludwig-Maximilians-Universität München, Munich, Germany*

1. Introduction

Nowadays, nanoparticles can be synthesized with well-defined shapes and sizes. At such nanoscale dimensions their optical properties are strongly altered, depending on the physical properties of the bulk material [1-17]. In semiconductor NCs, the confinement of the electronic motion to nanometer ranges results in the quantization of the energy levels, strongly affecting their optical response [3]. The optical properties of noble metal nanoparticles on the other hand are governed by a completely different process. Here the high carrier density plays a dominating role. When excited by an electromagnetic wave the carriers are stimulated to collective oscillations, resulting in intense resonances [5, 13], the so called localized surface plasmon resonances (LSPR) [5, 13]. So far LSPRs have been mostly investigated in noble metal nanostructures, where the high carrier density of around 1022–1023 cm−3 leads to LSPRs in the visible spectral range. These resonances can be partially influenced by the size and geometry of those nanostructures. This provides to a certain degree tunability of the optical properties [18], limited however to the stage of synthesis. Fundamentally, the high carrier density that is responsible for the plasmonic response of the system is not restricted to noble metal nanoparticles only. Actually, doped semiconductor NCs appear to be an attractive alternative [19, 20], although doping through the insertion of impurity atoms is far from being trivial in NCs [21]. Relatively low carrier densities are achieved in this way and plasmon resonances in the mid to far-infrared (IR) are observed [22-24]. Another appealing option is vacancy doping, where vacancies in the structure are responsible for an increased carrier density of around $1021 cm^{-3}$ [19, 23, 25]. This very high number of carriers confined to a small volume gives rise to LSPRs in the near-infrared (NIR). Thus, vacancy doped semiconductor NCs bridge the gap between noble metal nanoparticles and impurity doped semiconductor NCs. A highly attractive feature of vacancy doped semiconductor NCs is the opportunity to tailor the carrier density by controlling the level of doping in the structure [19, 20]. This delivers a tool to actively tune the LSPR over a wide range of frequencies, including the transition from a metallic to a purely semiconducting behavior, depending on their level of doping [20-29]. This highlights this type of plasmonic material over conventional metallic nanostructures [19, 23, 25, 26, 28, 30].

In this chapter we introduce copper chalcogenides NCs, namely copper sulfide ($Cu_{2-x}S$), copper selenide ($Cu_{2-x}Se$), and copper telluride ($Cu_{2-x}Te$) as a class of vacancy doped semiconductor NCs. The high level of copper vacancies in the structure, depicted by *(2-x)* in the chemical formula are responsible for the increased carrier density [19, 20, 26, 28, 31-40]. Indeed, the spectra of copper chalcogenide NCs are dominated by strong resonances in the NIR. In the first part of this chapter we give evidence on the plasmonic nature of those NIR resonances. This is achieved by probing typical plasmonic characteristics, such as refractive index sensitivity, size dependence, carrier dynamics, or interparticle plasmon coupling. In the second part we introduce the key advantage of vacancy doped semiconductor NCs. This results in the possibility to tune their LSPR over a wide range of spectral frequencies in the NIR. Indeed, the purely semiconducting properties, such as photoluminescence and excitonic features become apparent and dominating upon a complete suppression of the plasmon resonance, and are strongly affected by its presence.

2. Plasmonic Properties of Vacancy-Doped Copper Chalcogenide NCs

Vacancy-doping is a key feature of the family of copper chalcogenide nanocrystals (NCs), namely copper sulfide ($Cu_{2-x}S$), copper selenide ($Cu_{2-x}Se$), and copper telluride ($Cu_{2-x}Te$). (2-x) in the chemical formula accounts for a copper deficiency in the NC lattice, which in turn is responsible for an increased charge carrier density of $\sim 10^{21}/cm^3$ [19, 41, 42]. For small spherical metallic nanoparticles in the dipole approximation the plasmon resonance condition is achieved when

$$\varepsilon'(\omega) = -2\varepsilon_m \qquad (I)$$

is satisfied [43]. $\varepsilon'(\omega)$ is the real part of the dielectric function and ε_m the dielectric constant of the medium. The resonance frequency ω_{sp} at which the resonance condition (I) is satisfied is given by:

$$\omega_{sp} = \sqrt{\frac{\omega_p^2}{1 + 2\varepsilon_m} - \gamma^2} \qquad (II)$$

and denotes the frequency at which the LSPR is found. γ depicts the damping constant, which for the assumption of free carriers is equal to the plasmon linewidth. The bulk plasmon resonance ω_p is given as:

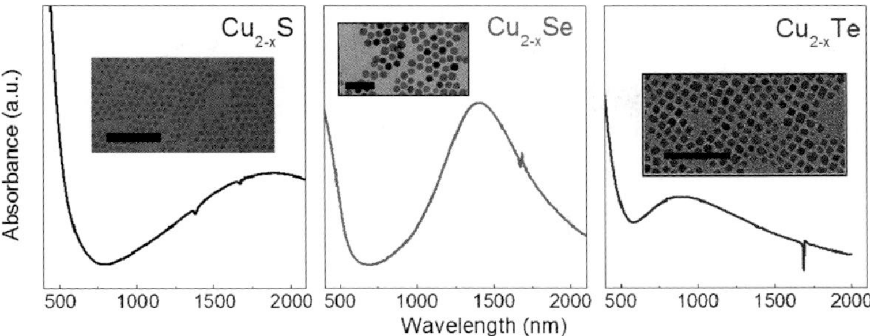

Figure 1 Absorption spectra for (a) Cu2-xS NCs (d = 5 nm), (b) Cu2-xSe NCs (d = 12 nm), and (c) Cu2-xTe NCs (d = 5 nm). The insets show representative TEM images (scalebar 50 nm). Reprinted with permission from [28]. Copyright 2012. American Chemical Society.

$$\omega_p^2 = \frac{ne^2}{\varepsilon_0 m_e} \tag{III}$$

with n being the carrier density, m_e the electron mass and e the electron charge [43]. From this equation it is clear the dependence of ω_{sp} on the carrier density n and the medium refractive index ε_m. With carrier densities of around $\sim 10^{21}/cm^3$ in copper chalcogenide NCs, ω_{sp} is expected to be located in the NIR. In Figure 1 typical absorption spectra of $Cu_{2-x}S$ (d = 5 nm, where d is the NCs' diameter), $Cu_{2-x}Se$ (d = 12 nm), and $Cu_{2-x}Te$ NCs (d = 5nm) are given [28]. The spectra are dominated by a steep rise in the blue wavelength region, attributed to the interband absorption, and a dominant resonance in the NIR. With ω_{sp} (maximum of the NIR resonance) and γ the linewidth, we can calculate the charge carrier density n according equation (II). Indeed, for $Cu_{2-x}S$, $Cu_{2-x}Se$, and $Cu_{2-x}Te$ NCs values of around $10^{21}\,cm^{-3}$ are determined, agreeing with previous findings for bulk copper chalcogenides [19, 44]. In the following we provide further evidence on the plasmonic nature of the NIR plasmon resonances in vacancy-doped copper chalcogenide NCs.

Formula (II) clearly shows the dependence of the resonance frequency of the LSPR ω_{sp}, on the dielectric constant of the surrounding medium (ε_m) [45]. The sensitivity of the NIR resonance to the medium refractive index is attested by comparing the absorption spectra of $Cu_{2-x}S$ and $Cu_{2-x}Se$ NCs in solvents with different refractive indexes, namely hexane (1.38), toluene (1.50) and CS_2 (1.63) (Figure 2a). For both materials, $Cu_{2-x}S$ and $Cu_{2-x}Se$ NCs, a shift of the absorption maximum, *i.e.*, LSPR with ε_m is observed. In Figure 2 the maximum peak position of the LSPR of $Cu_{2-x}S$ and $Cu_{2-x}Se$ NCs is plotted vs. the refractive

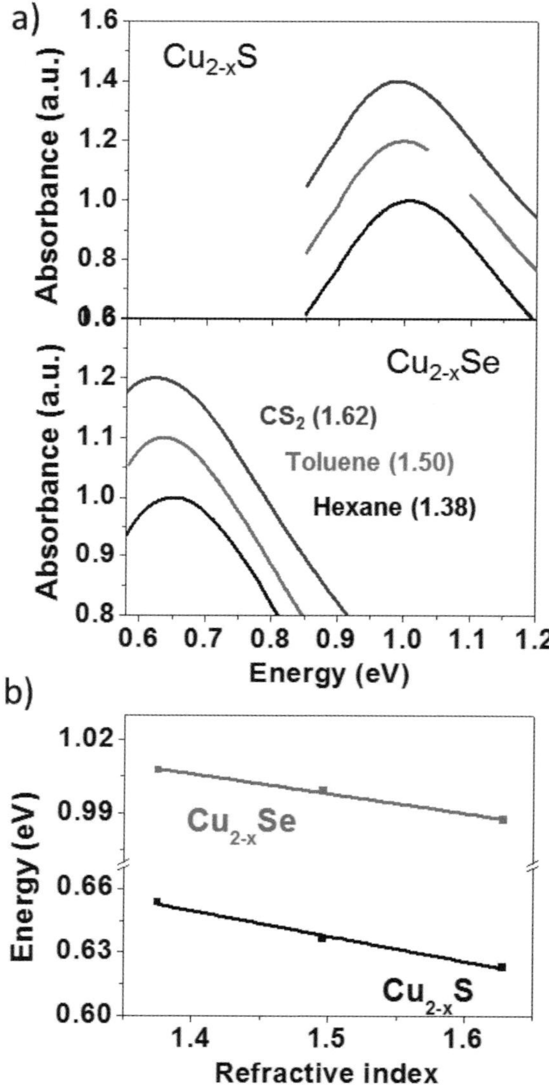

Figure 2 Plasmon resonances of (a) Cu$_{2-x}$S and Cu$_{2-x}$Se NCs in solvents of different refractive indices, CS$_2$ (blue curve), toluene (red curve) and hexane (black curve). (b) Evolution of the LSPR maximum (in energy units) as a function of the solvent refractive index for Cu$_{2-x}$S (black line) and Cu$_{2-x}$Se (red line) [28]. Reprinted with permission from [28]. Copyright 2012. American Chemical Society.

index of the solvent and reveals a red-shift with increasing solvent refractive index, a behavior indicative for plasmon resonances [45-47]. This dependence has so far mainly been demonstrated for noble metal nanoparticles, and clearly demonstrates the sensitivity of the NIR resonance in vacancy-doped copper chalcogenide NCs to the medium refractive index [28]. This highlights their plasmonic nature and opens the route for sensing applications with copper chalcogenide NCs.

On a different front, it is known that below a certain diameter, the so called 'intrinsic size effect' of plasmonic resonances becomes critical [47, 48]. This is particularly important when the nanoparticle diameter falls below the mean free path of the carriers. When the mean free path of the carriers exceeds the nanoparticle diameter additional scattering with the surface leads to a faster loss in coherency and a broadening and red-shift of the plasmon resonance with decreasing nanoparticle size [49, 50]. Figure 3 shows the size dependent optical spectra of $Cu_{2-x}S$ NCs of various sizes, namely 4.5 nm (red curve, Figure 3), 3 nm (brown curve, Figure 3), and 2.5 nm (black curve, Figure 3) in diameter (see TEM images as insets in Figure 3). A broadening and progressive red-shift of the LSPR with decreasing NC size is observed. For NCs with diameters < 2.5 nm the plasmon resonance is completely damped (not shown). Taken altogether this result demonstrates the 'intrinsic size effect' in the optical response of $Cu_{2-x}S$ NCs [19, 28]. This effect is attributed to the effect of surface scattering of free carriers [49-51]. It is particularly obvious for NCs below 2.5 nm, where a complete damping of the LSPR is observed.

Plasmonic coupling is a phenomenon of great interest especially for biolabeling or surface enhanced Raman spectroscopy [52, 53]. It occurs when two or more plasmonic nanoparticles are brought into close vicinity. So far, plasmon coupling has been typically investigated on plasmonic nanoparticles of noble metals. On the other hand, under experimentally favorable conditions monodisperse semiconductor NCs tend to self-organization into ordered superstructure arrays of densely packed nanoparticles [6]. Provoked by slow evaporation of the solvent, natural forces such as entropy, electrostatics, and van der Waals interactions the NCs are triggered to order into close packed structures [6, 54]. A major advantage here is that the NCs may couple with each other creating a multifunctional response, while retaining the exceptional, physical properties of the corresponding building blocks [6]. Self-assembly of nanoparticles appears attractive for the study of plasmonic coupling phenomena [6, 54-60], though it typically requires a high level of monodispersity. Copper chalcogenide NCs can be nowadays produced with a good level of monodispersity and therefore, under suitable experimental conditions,

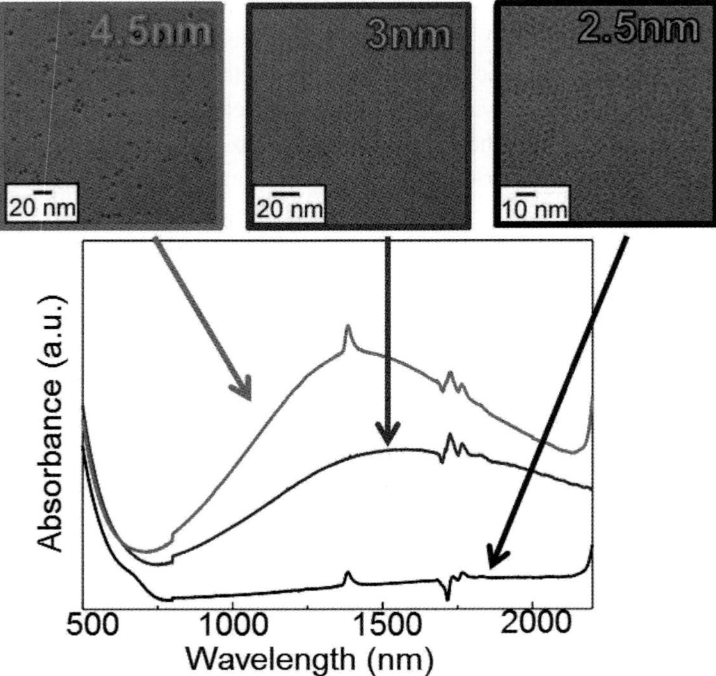

Figure 3 Absorption spectra of $Cu_{2-x}S$ NCs of varying sizes ranging from 4.5 nm (red curve), 3 nm (brown curve) to 2.5 nm (black curve) in diameter with respective TEM images. Below 2.5 nm in diameter the plasmon resonance is completely damped [28]. Reprinted with permission from [28]. Copyright 2012. American Chemical Society.

self-assembly can occur, thus giving rise to plasmon coupling, as we discuss next. In Figure 4 a TEM micrograph of a superlattice arrangement of $Cu_{2-x}S$ NCs with hexagonal ordering is shown (red frame). Due to the high monodispersity of the 12 nm sized $Cu_{2-x}S$ NCs the particles tend to self-assemble into superstructures ranging between 500 and 1000 nm. Notably, self-assembly of such NCs occurs in solution upon synthesis completion [39]. To monitor the behavior of the plasmon resonance of the NCs upon reducing interparticle plasmon coupling, the careful disassembly of the NCs can be triggered through a post-synthetic addition of ligands in excess (oleylamine, OAm). The change in the optical properties given in Figure 4 is monitored upon the ligand-induced disassembly of the NCs (red curve to blue curve). The initial sample of close packed NCs (red curve) displays a plasmon band maximum at ~1540 nm. Upon disassembly, triggered through the addition of ligands in excess, the plasmon

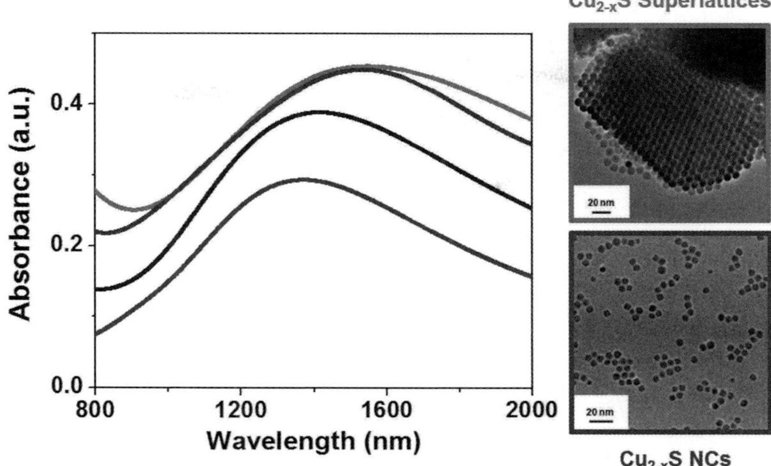

Figure 4 Changes of the absorbance spectra of Cu$_{2-x}$S NCs (12 nm in diameter) tracked with time after the addition of an excess of ligands, together with representative TEM images of the superstructures (red frame) and dispersed NCs (blue frame). The extinction spectrum of close packed NCs (red curve) shows a red-shifted plasmon resonance with respect to the dispersed NCs (blue curve) [39]. Reprinted with permission from [39]. Copyright 2012. American Chemical Society.

band gradually shifts until a value of 1370 nm (blue curve), corresponding to a blue-shift of 170 nm [39, 61]. Interparticle plasmon coupling occurs when the plasmonic nanoparticles are close enough to interact and leads to a red-shift of the plasmonic resonance with respect to the non-interacting particles [52]. However, reports on this phenomenon for plasmonic semiconductor NCs have remained fairly unexplored. To the best of our knowledge, plasmon coupling in vacancy-doped copper chalcogenide NCs is reported for the first time by us in ref. [39]. The given results underline the plasmonic character of Cu$_{2-x}$S NCs and present an efficient route to create plasmonically coupled structures via NC self-assembly [39].

Plasmon resonances in metallic nanoparticles show a very characteristic dynamical behavior in the region of the LSPR, which can be probed by transient absorption spectroscopy [62]. In Figure 5a transient absorption spectra of Cu$_{2-x}$Se NCs (d = 12 nm) at different delay times after excitation with a femtosecond laser pulse in the interband absorption regime at 400 nm are given [28]. An optical non-linearity in the spectral range of the LSPR is observed. The corresponding time evolution of the maximum non-linearity is also shown

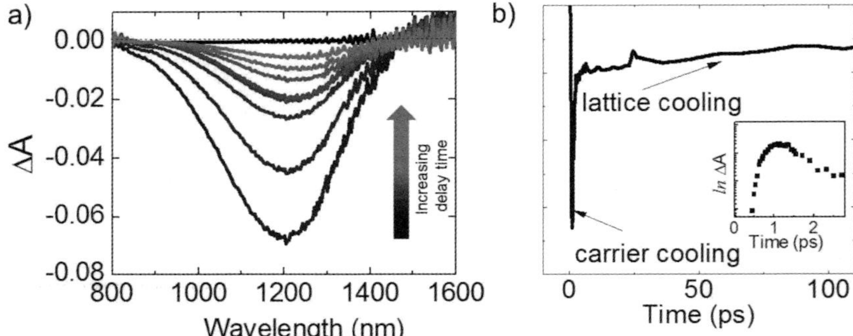

Figure 5 a) Transient absorption spectra of $Cu_{2-x}Se$ NCs (x > 0, d = 12 nm) at different time delays after excitation at 400 nm with a femtosecond pulsed laser together with b) the time decay of the transient absorption signal at the absorption maximum (1200 nm). Inset displays the initial rise plotted in logarithmic scale [28]. Reprinted with permission from [28]. Copyright 2012. American Chemical Society.

(Figure 5b) and shows a characteristic two step decay. The initial fast decay occurs within a few picoseconds, while the second longer decay takes place on a hundred picoseconds time frame [28]. The observed dynamics are very similar to those of noble metal nanoparticles in the spectral region of the LSPR [63], described by a two-step decay as well [13]. The non-linearity is commonly ascribed to a pump-pulse induced heating of the carrier gas. The initial fast decay results from the carrier relaxation via carrier-phonon scattering, leaving the lattice at elevated temperatures. The second longer decay is ascribed to the lattice cooling due to phonon emission to the surrounding medium [5, 63]. Furthermore, an initial rise is observed in the transient signal on shorter time scales (inset to Figure 5b, logarithmic scale), which is attributed to the carrier thermalization [5]. This is the time the carriers take to relax resulting in a Fermi distribution of the carrier gas at elevated temperatures [64]. Taken altogether, in vacancy-doped copper chalcogenide NCs the overall observations are consistent with the plasmon dynamics observed in noble metal nanoparticles, with both material types showing similar plasmon dynamics [35-37]. These results are in agreement with other works on vacancy-doped copper chalcogenide NCs [36, 37].

3. Tunable Optical Properties of Vacancy-Doped Plasmonic Copper Chalcogenide NCs

A key feature of vacancy doped copper chalcogenide NCs is their ability to hold *localized surface plasmon resonances (LSPRs) in the NIR* resulting from a high level of vacancy doping, as discussed above [19, 20, 28, 30, 31, 33]. Assuming that with each Cu-ion an electron is removed from the top of the valence band (leaving a hole behind) the carrier density is expected to be proportional to the copper vacancy density in the structure [19]. By taking into account the stoichiometry factor x, the particle volume V, the molar mass M and the density ρ, the copper vacancy density in copper chalcogenide NCs can be estimated to be in the range of $\sim 10^{21}$ cm^{-3}, clearly correlating with the carrier density. Thus, the stoichiometry factor x, can be considered as a measure of the vacancy, *i.e.* carrier density of the system. Taking into account the above discussion the opportunity to chemically tailor the vacancy density in copper chalcogenide NCs appears as an appealing characteristic of vacancy doped semiconductor NCs. More precisely, by controlling the copper vacancy density in the system, the carrier density and with this the LSPR can be tuned over a wide range of frequencies [20-29].

This is demonstrated in Figure 6a for $Cu_{2-x}Se$ NCs. Experimentally, $Cu_{2-x}Se$ NCs have been synthesized with a stoichiometry factor of $x = 0$. Thereafter the system is exposed to oxygen, which drives copper vacancy formation and with this an increase in carrier density upon air exposure [28]. The increased carrier density is detected by an increasing absorption in the NIR attributed to the formation of the LSPR. A gradual blue-shift and increase in intensity of the NIR resonance agrees with a progressive increase of carrier density [28]. Structuraly the oxidation process and the crystallographic changes occurring upon copper vacancy formation can be tracked via XRD. The initial, non-treated sample displays an XRD pattern ascribed to the stoichiometric tetragonal phase $Cu_{2-x}Se$ NCs ($x = 0$) (black curve in Figure 6b). Upon exposure to oxygen, i.e. upon copper vacancy formation the crystal structure progressively transforms into the non-stoichiometric, cubic phase $Cu_{2-x}Se$ ($x = 0.2$), in agreement with previous reports (Figure 6) [65]. The given results demonstrate the correlation between the evolution of the LSPR, as a measure for the carrier density, and the copper vacancy formation given by the solid state transformation from stoichiometric $Cu_{2-x}Se$ ($x = 0$) to the copper deficient $Cu_{2-x}Se$ ($x > 0$) upon oxidation via air exposure [28].

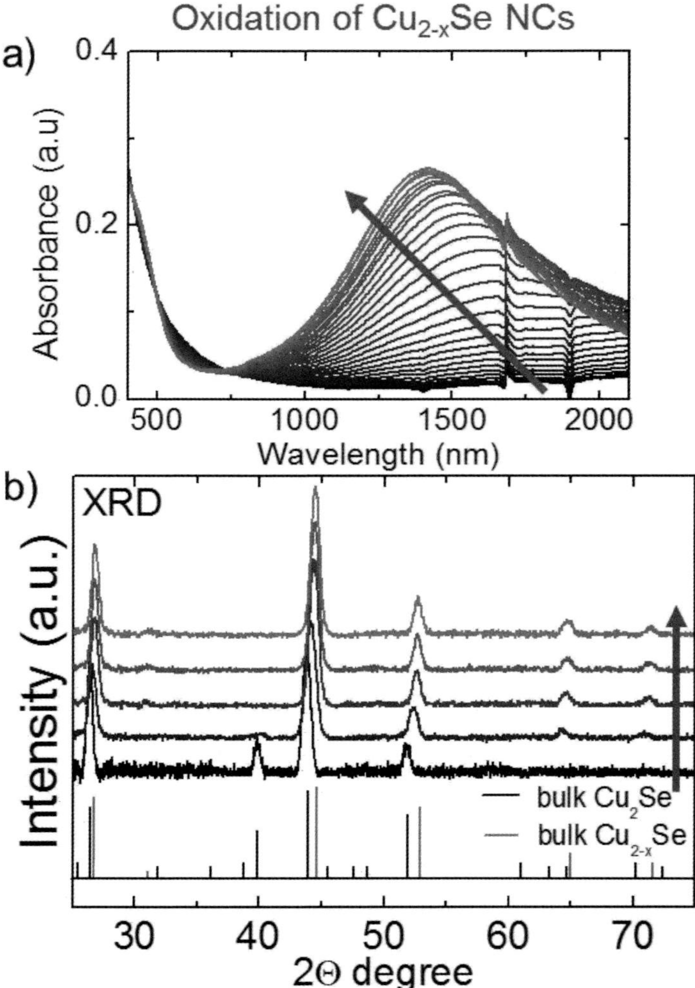

Figure 6 a) Time evolution of the absorption spectra of stoichiometric Cu$_{2-x}$Se (x =0, black curve) NCs during oxidation. The plasmon resonance gradually increases in intensity and blue-shifts with increasing exposure time, indicated from black to red. b) Time evolution of the XRD patterns of the Cu$_{2-x}$Se NCs during oxidation, from the black to the red curve upon air exposure. The vertical bars are the corresponding, color-coded, bulk reference patterns [28]. Reprinted with permission from [28]. Copyright 2012. American Chemical Society.

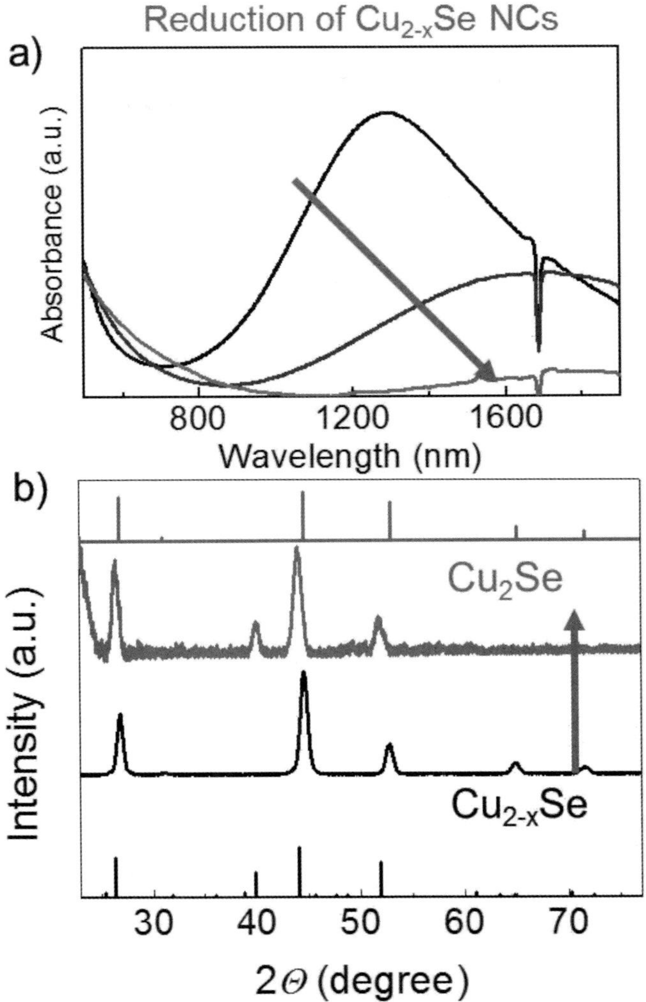

Figure 7 a) Evolution of the absorption spectra upon the addition of a strong reducing agent to a Cu$_{2-x}$Se (x > 0) NCs dispersion. The initial copper deficient (x > 0) sample shows a pronounced plasmon resonance (black curve), while after some time of reaction the plasmon resonance is suppressed, indicating a decrease in carrier density and copper deficiencies. b) Diffraction patterns of copper deficient Cu$_{2-x}$Se (x > 0) NCs (black curves), and the reduced Cu$_{2-x}$Se NCs (red curves) together with the respective reference patterns [28]. Reprinted with permission from [28]. Copyright 2012. American Chemical Society.

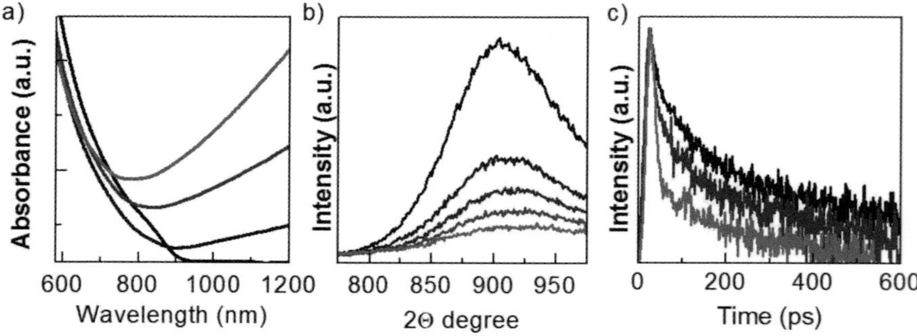

Figure 8 a) Absorption spectrum of $Cu_{2-x}S$ NCs upon oxidation (black to red curves), with the corresponding b) steady state and c) time resolved PL spectra. Upon oxidation a NIR plasmon resonance is evolving resulting in a disappearance of the excitonic transition and a blue shift of the absorption onset. Simultaneously, the PL is decreasing in intensity concomitant with a shortening of the exciton lifetime. Reprinted with permission from [28]. Copyright 2012. American Chemical Society.

Interestingly, as demonstrated in Figure 7 this process can be reversed upon the addition of a strong reducing agent, detected in both, absorption (Figure 7a) and XRD (Figure 7b). This process can be explained as follows: the strong reducing agent (namely diisobutylaluminium hydride, DIBAH) induces the reduction of Cu(II) at the NC surface to Cu(I). This likely results in vacancy filling, and thus in a decrease in charge carrier density that, in turn, is responsible for the observed plasmonic red-shift and intensity decrease upon reduction [28, 65]. The filling of copper vacancies further results in the decrease of the carrier density and an ultimate reduction of the LSPR. This is agreement with the XRD results, where the stoichiometric, non-doped tetragonal $Cu_{2-x}Se$ NCs (x = 0) phase is recovered upon complete reduction (Figure 7b) [28].

So far we have mainly discussed the plasmonic properties of vacancy doped copper chalcogenide NCs. In the following, however, we want to focus on the excitonic properties of $Cu_{2-x}S$ NCs. A closer look to the absorption spectrum of the stoichiometric $Cu_{2-x}S$ NCs with $x = 0$ (black curve in Figure 8a) and a diameter of 5 nm reveals a steep rise in the blue wavelength region and a weak excitonic resonance with an absorption onset at ca. 900 nm [28]. This absorption edge is blue shifted with respect to the bulk bandgap (1.2 eV) [41] suggesting size quantization effects and an exciton Bohr radius of at least 5 nm [31]. The corresponding photoluminescence (PL) spectrum is given in Figure 8b, black curve. The PL peaks at ca. 920 nm and shows a very low fluorescence quantum yield [28]. In the following, we will discuss the effect of LSPR evolution on the excitonic transitions of the NCs. A detailed investigation in this regard is very important, especially in relation to the application of copper chalcogenide NCs in solar applications [66, 67]. The results of such an investigation are shown in Figure 8[28].

The evolution of the plasmon resonance (Figure 8a, black to red curves) is triggered through the exposure of the sample to oxygen as described above. Noteworthy, the excitonic peak observed in the stoichiometric $Cu_{2-x}S$ NCs ($x = 0$) gradually disappears upon the formation and gradual evolution of the LSPR, indicating a loss in oscillator strength as an effect of either the LSPR or the increased carrier density on the exciton. Further exposure to air leads to a gradual blue-shift and decrease in intensity in the region of the interband transitions [28]. This is related to the Burstein-Moss effect of heavily doped semiconductors, where heavy doping leads to an increase of the effective energy required to excite an electron from the top of the valence to the bottom of the conduction band, ultimately resulting in a blue shift of the absorption onset [68]. Simultaneously, a gradual quenching of PL intensity is observed with the evolving LSPR, until a complete suppression is reached (Figure 8, black to red curves) [28]. This observation is further evaluated by means of time-resolved PL spectroscopy (Figure 8c). Herein three samples are taken into account. The stoichiometric $Cu_{2-x}S$ NCs ($x = 0$) with no NIR plasmon resonance and two samples with increasing plasmonic contribution (black to red curves). A shortening of the exciton lifetime is observed along with an increasing vacancy doping in the NC. The three traces given in Figure 8c are described by a bi-exponential decay with a fast initial component, which corresponds to the time resolution of the setup, and a second longer decay [28]. With increasing plasmonic contribution the amplitude of the fast initial component gains intensity. Such an observation is characteristic for a multiparticle interaction, such as Auger recombination [69]. Indeed, Auger recombination is likely to happen in this system, as an increased plasmonic absorption occurs along with an increased density of holes. Those might interact with the exciton, resulting in a non-radiative decay of the electron-hole pair [28]. Carrier trapping related to the number of copper vacancies created in the structure or an altered surface chemistry upon oxidation might further play an important role here. A partial influence of the PL quenching by energy transfer from the exciton to the plasmon cannot be excluded [28]. However, the observed effect upon LSPR evolution on the excitonic is an important one and, hence, it should be taken into consideration in relation to the application of these NCs in photovoltaics [66, 67]. As a final remark, the suppression of the NIR plasmon resonance of $Cu_{2-x}S$ NCs under reductive conditions results in a complete regeneration of the initial excitonic transitions and the PL intensity. These findings indicate a complete recovery of the initial interband transitions after full reduction. For further information the reader is referred to the references [28, 61].

4. Conclusions

In conclusion, vacancy doped copper chalcogenide NCs, with carrier densities of $\sim 10^{21} cm^{-3}$, are characterized by intense localized surface plasmon resonances (LSPRs) in the near infrared (NIR). These resonances represent typical plasmonic characteristics, such as refractive index sensitivity, the intrinsic size effect, interparticle plasmon coupling, and plasmon dynamics [28, 39]. These findings emphasize their potential for plasmonic applications, such as photothermal therapy [33] or surface enhanced Raman spectroscopy [70]. Furthermore, the chemical tailoring of the copper vacancies in the structure, by means of oxidation and reduction represents a way to tune the plasmon resonance over a wide range of frequencies [28], and thus highlights the potential of this type of NCs over plasmonic nanoparticles of noble metals. Moreover, the control over their structural characteristics allows for a reversible tuning between their excitonic and plasmonic properties. Indeed, the ability to hold excitons and plasmons on demand points to copper chalcogenide NCs as a new material system for the investigation of exciton-plasmon interactions [28]. Nevertheless, these excitonic characteristics are heavily disturbed upon LSPR formation, resulting in a loss of the excitonic absorption and a quenching of photoluminescence together with a shortening of fluorescence lifetime. These findings are very relevant for their prospective use in photoenergy conversion [66, 67].

Acknowledgments

The Nanosystems Initiative Munich (NIM) is kindly acknowledged. We also acknowledge the Bavarian Ministry of Science, Research and Arts for financial support through the project Solar Technologies Go Hybrid and the EU commission for funding through the Marie-Curie research training network ICARUS. We gratefully acknowledge Johann M. Szeifert and Andrey Lutich for their contribution to this work.

References

1. Alivisatos, A.P., *Semiconductor Clusters, Nanocrystals, and Quantum Dots.* Science, 1996. **271**(5251): p. 933-937.

2. Murray, C.B., C.R. Kagan, and M.G. Bawendi, *Synthesis and characterization of monodisperse nanocrystals and close-packed nanocrystal assemblies.* Annual Review of Materials Science, 2000. **30**(1): p. 545-610.

3. Efros, A.L. and M. Rosen, *The electronic structure of semiconductor nanocrystals.* Annual Review of Materials Science, 2000. **30**(1): p. 475-521.

4. Nozik, A.J., *Quantum dot solar cells.* Physica E-Low-Dimensional Systems & Nanostructures, 2002. **14**(1-2): p. 115-120.

5. Link, S. and M.A. El-Sayed, *Optical properties and ultrafast dynamics of metallic nanocrystals.* Annual Review of Physical Chemistry, 2003. **54**(1): p. 331-366.

6. Talapin, D.V., *LEGO Materials.* ACS Nano, 2008. **2**(6): p. 1097-1100.

7. Norris, D.J., A.L. Efros, and S.C. Erwin, *Doped Nanocrystals.* Science, 2008. **319**(5871): p. 1776-1779.

8. Rogach, A.L., *Semiconductor Nanocrystal Quantum Dots.* 2008: Springer-Verlag/Wien.

9. Hillhouse, H.W. and M.C. Beard, *Solar cells from colloidal nanocrystals: Fundamentals, materials, devices, and economics.* Current Opinion in Colloid & Interface Science, 2009. **14**(4): p. 245-259.

10. Rogach, A.L., et al., *Energy transfer with semiconductor nanocrystals.* Journal of Materials Chemistry, 2009. **19**(9).

11. Zhao, Y. and C. Burda, *Development of plasmonic semiconductor nanomaterials with copper chalcogenides for a future with sustainable energy materials.* Energy & Environmental Science, 2012. **5**(2): p. 5564-5576.

12. Rivest, J.B. and P.K. Jain, *Cation Exchange on the Nanoscale: an Emerging Technique for New Material Synthesis, Device Fabrication, and Chemical Sensing.* Chemical Society Reviews, 2013. **42**(1): p. 89-96.

13. Burda, C., et al., *Chemistry and Properties of Nanocrystals of Different Shapes.* Chemical Reviews, 2005. **105**(4): p. 1025-1102.

14. Peng, X., et al., *Shape control of CdSe nanocrystals.* Nature, 2000. **404**(6773): p. 59-61.

15. Manna, L., et al., *Controlled Growth of Tetrapod-Branched Inorganic Nanocrystals.* Nature Materials, 2003. **2**(6): p. 382-385.

16. Deka, S., et al., *Octapod-Shaped Colloidal Nanocrystals of Cadmium Chalcogenides via "One-Pot" Cation Exchange and Seeded Growth.* Nano Letters, 2010. **10**(9): p. 3770-3776.

17. Pelaz, B., et al., *The State of Nanoparticle-Based Nanoscience and Biotechnology: Progress, Promises, and Challenges.* ACS Nano, 2012. **6**(10): p. 8468-8483.

18. Myroshnychenko, V., et al., *Modelling the optical response of gold nanoparticles.* Chemical Society Reviews, 2008. **37**(9): p. 1792-1805.

19. Luther, J.M., et al., *Localized Surface Plasmon Resonances Arising from Free Carriers in Doped Quantum Dots.* Nature Materials, 2011. **10**(5): p. 361-366.

20. Routzahn, A.L., et al., *Plasmonics with Doped Quantum Dots.* Israel Journal of Chemistry, 2012. **52**(11-12): p. 983-991.

21. Rowe, D.J., et al., *Phosphorus-Doped Silicon Nanocrystals Exhibiting Mid-Infrared Localized Surface Plasmon Resonance.* Nano Letters, 2013. **13**(3): p. 1317-1322.

22. Li, S.Q., et al., *Infrared Plasmonics with Indium–Tin-Oxide Nanorod Arrays.* ACS Nano, 2011. **5**(11): p. 9161-9170.

23. Garcia, G., et al., *Dynamically Modulating the Surface Plasmon Resonance of Doped Semiconductor Nanocrystals.* Nano Letters, 2011. **11**(10): p. 4415-4420.

24. Buonsanti, R., et al., *Tunable Infrared Absorption and Visible Transparency of Colloidal Aluminum-Doped Zinc Oxide Nanocrystals.* Nano Letters, 2011. **11**(11): p. 4706-4710.

25. Manthiram, K. and A.P. Alivisatos, *Tunable Localized Surface Plasmon Resonances in Tungsten Oxide Nanocrystals.* Journal of the American Chemical Society, 2012. **134**(9): p. 3995-3998.

26. Dorfs, D., et al., *Reversible Tunability of the Near-Infrared Valence Band Plasmon Resonance in $Cu_{2-x}Se$ Nanocrystals.* Journal of the American Chemical Society, 2011. **133**(29): p. 11175-11180.

27. Hsu, S.-W., K. On, and A.R. Tao, *Localized Surface Plasmon Resonances of Anisotropic Semiconductor Nanocrystals.* Journal of the American Chemical Society, 2011. **133**(47): p. 19072-19075.

28. Kriegel, I., et al., *Tuning the Excitonic and Plasmonic Properties of Copper Chalcogenide Nanocrystals.* Journal of the American Chemical Society, 2012. **134**(3): p. 1583-1590.

29. Hsu, S.-W., W. Bryks, and A.R. Tao, *Effects of Carrier Density and Shape on the Localized Surface Plasmon Resonances of $Cu_{2-x}S$ Nanodisks.* Chemistry of Materials, 2012. **24**(19): p. 3765-3771.

30. Xie, Y., et al., *Copper Sulfide Nanocrystals with Tunable Composition by Reduction of Covellite Nanocrystals with Cu+ Ions.* Journal of the American Chemical Society, 2013.

31. Zhao, Y., et al., *Plasmonic $Cu_{2-x}S$ Nanocrystals: Optical and Structural Properties of Copper-Deficient Copper(I) Sulfides.* Journal of the American Chemical Society, 2009. **131**(12): p. 4253-4261.

32. Li, W., et al., *Morphology Evolution of $Cu_{2-x}S$ Nanoparticles: From Spheres to Dodecahedrons.* Chemical Communications, 2011. **47**(37): p. 10332-10334.

33. Hessel, C.M., et al., *Copper Selenide Nanocrystals for Photothermal Therapy.* Nano Letters, 2011. **11**(6): p. 2560-2566.

34. Shaviv, E., et al., *Absorption Properties of Metal–Semiconductor Hybrid Nanoparticles.* ACS Nano, 2011. **5**(6): p. 4712-4719.
35. Scotognella, F., et al., *Plasmonics in heavily-doped semiconductor nanocrystals.* The European Physical Journal B, 2013. **86**(4): p. 1-13.
36. Della Valle, G., et al., *Ultrafast Optical Mapping of Nonlinear Plasmon Dynamics in Cu2–xSe Nanoparticles.* The Journal of Physical Chemistry Letters, 2013: p. 3337-3344.
37. Scotognella, F., et al., *Plasmon Dynamics in Colloidal $Cu_{2-x}Se$ Nanocrystals.* Nano Letters, 2011. **11**(11): p. 4711-4717.
38. Kriegel, I., et al., *Shedding Light on Vacancy-Doped Copper Chalcogenides: Shape-Controlled Synthesis, Optical Properties, and Modeling of Copper Telluride Nanocrystals with Near-Infrared Plasmon Resonances.* ACS Nano, 2013. **7**(5): p. 4367-4377.
39. Kriegel, I., et al., *Tuning the Light Absorption of $Cu_{1.97}S$ Nanocrystals in Supercrystal Structures.* Chemistry of Materials, 2011. **23**(7): p. 1830-1834.
40. Kriegel, I., et al., *Cation exchange synthesis and optoelectronic properties of type II CdTe-Cu2-xTe nano-heterostructures.* Journal of Materials Chemistry C, 2013.
41. Mulder, B.J., *Optical properties of crystals of cuprous sulphides (chalcosite, djurleite, $Cu_{1.9}S$, and digenite).* Physica Status Solidi (a), 1972. **13**(1): p. 79-88.
42. Gorbachev, V.V. and I.M. Putilin, *Some parameters of band structure in copper selenide and telluride.* Physica Status Solidi (a), 1973. **16**(2): p. 553-559.
43. Kreibig, U. and M. Vollmer, *Optical Properties of Metal Clusters.* Springer Series in Materials Science. Vol. 25. 1995.
44. Mansour, B.A., S.E. Demian, and H.A. Zayed, *Determination of the effective mass for highly degenerate copper selenide from reflectivity measurements.* Journal of Materials Science: Materials in Electronics, 1992. **3**(4): p. 249-252.
45. Underwood, S. and P. Mulvaney, *Effect of the Solution Refractive Index on the Color of Gold Colloids.* Langmuir, 1994. **10**(10): p. 3427-3430.
46. Chen, H., et al., *Shape- and Size-Dependent Refractive Index Sensitivity of Gold Nanoparticles.* Langmuir, 2008. **24**(10): p. 5233-5237.
47. Kelly, K.L., et al., *The Optical Properties of Metal Nanoparticles: The Influence of Size, Shape, and Dielectric Environment.* The Journal of Physical Chemistry B, 2003. **107**(3): p. 668-677.
48. Link, S. and M.A. El-Sayed, *Size and Temperature Dependence of the Plasmon Absorption of Colloidal Gold Nanoparticles.* The Journal of Physical Chemistry B, 1999. **103**(21): p. 4212-4217.
49. Link, S. and M.A. El-Sayed, *Spectral Properties and Relaxation Dynamics of Surface Plasmon Electronic Oscillations in Gold and Silver Nanodots*

and Nanorods. The Journal of Physical Chemistry B, 1999. **103**(40): p. 8410-8426.

50. Kreibig, U. and C.v. Fragstein, *The limitation of electron mean free path in small silver particles.* Zeitschrift für Physik, 1969. **224**(4): p. 307-323.

51. Berciaud, S., et al., *Observation of Intrinsic Size Effects in the Optical Response of Individual Gold Nanoparticles.* Nano Letters, 2005. **5**(3): p. 515-518.

52. Jain, P.K. and M.A. El-Sayed, *Plasmonic coupling in noble metal nanostructures.* Chemical Physics Letters, 2010. **487**(4–6): p. 153-164.

53. de la Rica, R. and M.M. Stevens, *Plasmonic ELISA for the ultrasensitive detection of disease biomarkers with the naked eye.* Nat Nano, 2012. **7**(12): p. 821-824.

54. Rogach, A.L., et al., *Organization of Matter on Different Size Scales: Monodisperse Nanocrystals and Their Superstructures.* Advanced Functional Materials, 2002. **12**(10): p. 653-664.

55. Shevchenko, E.V., et al., *Structural Characterization of Self-Assembled Multifunctional Binary Nanoparticle Superlattices.* Journal of the American Chemical Society, 2006. **128**(11): p. 3620-3637.

56. Shevchenko, E.V., et al., *Structural diversity in binary nanoparticle superlattices.* Nature, 2006. **439**(7072): p. 55-59.

57. Shevchenko, E.V., et al., *Quasi-ternary Nanoparticle Superlattices Through Nanoparticle Design.* Advanced Materials, 2007. **19**(23): p. 4183-4188.

58. Shevchenko, E.V., et al., *Self-Assembled Binary Superlattices of CdSe and Au Nanocrystals and Their Fluorescence Properties.* Journal of the American Chemical Society, 2008. **130**(11): p. 3274-3275.

59. Kovalenko, M.V., M.I. Bodnarchuk, and D.V. Talapin, *Nanocrystal Superlattices with Thermally Degradable Hybrid Inorganic-Organic Capping Ligands.* Journal of the American Chemical Society, 2010. **132**(43): p. 15124-15126.

60. Murray, C.B., C.R. Kagan, and M.G. Bawendi, *Self-Organization of CdSe Nanocrystallites into Three-Dimensional Quantum Dot Superlattices.* Science, 1995. **270**(5240): p. 1335-1338.

61. Kriegel, I., *Near-infrared plasmonics with vacancy doped semiconductor nanocrystals* Ph. D. thesis, Munich, 2013.

62. Cerullo, G., et al., *Time-resolved methods in biophysics. 4. Broadband pump-probe spectroscopy system with sub-20 fs temporal resolution for the study of energy transfer processes in photosynthesis.* Photochemical & Photobiological Sciences, 2007. **6**(2): p. 135-144.

63. Perner, M., et al., *Optically Induced Damping of the Surface Plasmon Resonance in Gold Colloids.* Physical Review Letters, 1997. **78**(11): p. 2192-2195.

64. Link, S. and M.A. El-Sayed, *Shape and size dependence of radiative, non-radiative and photothermal properties of gold nanocrystals.* International Reviews in Physical Chemistry, 2000. **19**(3): p. 409-453.

65. Riha, S.C., D.C. Johnson, and A.L. Prieto, *Cu_2Se Nanoparticles with Tunable Electronic Properties Due to a Controlled Solid-State Phase Transition Driven by Copper Oxidation and Cationic Conduction.* Journal of the American Chemical Society, 2011. **133**(5): p. 1383-1390.

66. Teranishi, T. and M. Sakamoto, *Charge Separation in Type-II Semiconductor Heterodimers.* The Journal of Physical Chemistry Letters, 2013. **4**(17): p. 2867-2873.

67. Rivest, J.B., et al., *Assembled Monolayer Nanorod Heterojunctions.* ACS Nano, 2011. **5**(5): p. 3811-3816.

68. Lukashev, P., et al., *Electronic and crystal structure of $Cu_{2-x}S$: Full-potential electronic structure calculations.* Physical Review B, 2007. **76**(19): p. 195202-14.

69. Schaller, R.D., et al., *High-Efficiency Carrier Multiplication and Ultrafast Charge Separation in Semiconductor Nanocrystals Studied via Time-Resolved Photoluminescence.* The Journal of Physical Chemistry B, 2006. **110**(50): p. 25332-25338.

70. Li, W., et al., *CuTe Nanocrystals: Shape and Size Control, Plasmonic Properties, and Use as SERS Probes and Photothermal Agents.* Journal of the American Chemical Society, 2013. **135**(19): p. 7098-7101.